T0298344

Understanding CATIA

Science, Technology and Management Series

Series Editor: J. Paulo Davim, Professor, Department of Mechanical Engineering, University of Aveiro, Portugal

This book series focuses on special volumes from conferences, workshops and symposiums, as well as volumes on topics of current interest in all aspects of science, technology and management. The series will discuss topics such as mathematics, chemistry, physics, materials science, nano sciences, sustainability science, computational sciences, mechanical engineering, industrial engineering, manufacturing engineering, mechatronics engineering, electrical engineering, systems engineering, biomedical engineering, management sciences, economical science, human resource management, social sciences and engineering education. The books will present principles, model techniques, methodologies and applications of science, technology and management.

Advanced Mathematical Techniques in Engineering Sciences
Edited by Mangey Ram and J. Paulo Davim

Soft Computing Techniques for Engineering Optimization
Edited by Kaushik Kumar, Supriyo Roy and J. Paulo Davim

Handbook of IOT and Big Data
Edited by Vijender Kumar Solanki, Vicente García Díaz and J. Paulo Davim

Digital Manufacturing and Assembly Systems in Industry 4.0
Edited by Kaushik Kumar, Divya Zindani and J. Paulo Davim

Optimization Using Evolutionary Algorithms and Metaheuristics
Edited by Kaushik Kumar and J. Paulo Davim

Integration of Process Planning and Scheduling
Approaches and Algorithms
Edited by Rakesh Kumar Phanden, Ajai Jain and J. Paulo Davim

Understanding CATIA
A Tutorial Approach
Edited by Kaushik Kumar, Chikesh Ranjan, and J. Paulo Davim

For more information about this series, please visit https://www.routledge.com/Science-Technology-and-Management/book-series/CRCSCITECMAN

Understanding CATIA

A Tutorial Approach

Edited by

Kaushik Kumar, Chikesh Ranjan,
and J. Paulo Davim

CRC Press is an imprint of the
Taylor & Francis Group, an **informa** business

First edition published 2021
by CRC Press
6000 Broken Sound Parkway NW, Suite 300, Boca Raton, FL 33487-2742
and by CRC Press
2 Park Square, Milton Park, Abingdon, Oxon, OX14 4RN

Library of Congress Cataloging-in-Publication Data

Library of Congress Cataloging-in-Publication Data
Names: Kumar, K. (Kaushik), 1968- editor. | Ranjan, Chikesh, editor. |
Davim, J. Paulo, editor.
Title: Understanding CATIA : a tutorial approach / edited by Kaushik Kumar,
Chikesh Ranjan, J. Paulo Davim.
Description: First edition. | Boca Raton : CRC Press, 2021. | Series:
Science, technology, and management | Includes bibliographical
references and index.
Identifiers: LCCN 2020055463 (print) | LCCN 2020055464 (ebook) | ISBN
9780367487942 (hbk) | ISBN 9781003121657 (ebk)
Subjects: LCSH: Computer-aided design—Computer programs. | CATIA (Computer file)
Classification: LCC TA345.5.C38 U53 2021 (print) | LCC TA345.5.C38
(ebook) | DDC 620/.00420285536—dc23
LC record available at https://lccn.loc.gov/2020055463
LC ebook record available at https://lccn.loc.gov/2020055464

ISBN: 978-0-367-48794-2 (hbk)
ISBN: 978-1-003-12165-7 (ebk)

Typeset in Times LT Std
by KnowledgeWorks Global Ltd.

Contents

SECTION I Introduction

SECTION II Sketcher

SECTION III Part Design

SECTION IV Assembly

SECTION V Drafting

SECTION VI Case Study

Preface

The authors are pleased to present the book *Understanding CATIA: A Tutorial Approach* as a part of the *Science, Technology and Management Series*. The title of this book was chosen keeping in mind the current importance of three-dimensional (3D) modeling as well as popularity of one of the most sought after software used in the industrial and manufacturing world.

The pictorial representation of an object is one of the simplest and most common modes of communication adopted by mankind. Moreover, it was the only common mode of communication because of its simplicity and universality, irrespective of the differences in the topography, language and knowledge level of people. The power of pictorial communication has been understood by engineers and further improvised to model engineering objects. 3D objects were presented as two-dimensional (2D) drawings, known as engineering drawings. Since the main purpose of engineering drawing is to create or manufacture the object for human use, these drawings are to be realistic with the true dimensions, unlike the other pictorial drawings created by the artist whose focus is only the appreciation and not the realism. The engineering drawings can be further tuned to make the shapes esthetically pleasing and ergonomically viable. Since at the universal level the facts conveyed by means of such drawings are to be the same, the drawings are drawn using the universal standardization procedures and specifications. Some of the best-known engineering drawings are that of Leonardo da Vinci, for his designs of military machines. Industrial drawings were prepared using pencil, T-square, set squares and scales, whereas permanent drawings and tracings were made by ink. A major advancement on the drafting device led to the development of Universal Drafting Machine. This device was basically developed to combine the T-square, set squares, scales and protractors, which allowed different engineering disciplines to develop their own approaches to design and drafting.

Physical objects are of 3D nature but are required to be represented on a 2D drawing sheet. This led to the development of various views (full, half sectional and full sectional) for complete picture. Hence, it became a subject of expertise and "reading" an engineering drawing was not a common man's cup of tea. Organizations had to recruit and retain expert draftsmen for translation of an engineering drawing for the manufacturing and production units.

With the invention of computers and development of different types of software, engineering practice also changed. As the manual drawings required large workspace and hardware, consumed more time and labor, needed strenuous effort in editing or modifying drawings and posed a lot of storage or transfer problems, the computer-based graphic system came as a solution to produce fast, simple, accurate and repeatable engineering drawings. The Computer-Aided Drafting (CAD) software were developed under two categories: coordinate-based system and parametric-based system. Coordinate-based software (AutoCAD, STADD etc.) created the object with x-, y- and z-coordinates, securely placing the it at a specific point with respect to the origin, whereas parametric-based software (CATIA, CREO, SOLIDWORKS etc.)

used dimensional parameters, like length, radius etc., for object creation. For 2D modeling, coordinate-based software are preferred and hence architectural drawing extensively uses them. But while working with 3D modeling, coordinate-based software becomes quite difficult to handle especially when editing is required. The parametric-based software are preferred as they do not specify coordinate hence models get updated just by changing the dimensions. This feature has increased their acceptability to engineering and industrial community. Further to modeling ease, manufacturing processes like computer-aided manufacturing (CAM), rapid prototyping (RP) also require output from parametric-based software.

Due to the increasing acceptability, nowadays the software market is flooded with 3D modeling software, namely, CATIA, CREO, MECHANICAL INVENTOR, SOLIDWORKS and UNIGRAPHICS. Although the basic features offered by all of these are more or less same, yet CATIA has become the most popular as it is easy to operate and more user-friendly. Hence, most of the industries and academic institutions, both globally and locally, prefer CATIA as their working tool.

It is true to say that in many instances the best way to learn complex behavior is by means of imitation. For instance, most of us learned to walk, talk, run, etc. solely by imitating the actions and speech of those around us. The same approach can be adopted to learn using CATIA software by imitation, i.e., using the examples provided in this book. This is the essence of the philosophy and the innovative approach used in this book. The authors have attempted to provide readers with a comprehensive cross-section of various models in a variety of engineering areas in order to provide a broad choice of examples to be imitated in one's own work. In developing these examples, the authors' intent has been to explain most of the features and templates provided by the developer. By displaying these features in an assortment of disciplines and modeling, the authors hope to give readers the confidence to employ these program enhancements in their own applications. The primary aim of this book is to assist in learning the use of CATIA software through examples taken from various areas of engineering. The content and treatment of the subject matter are considered to be most appropriate for university students studying engineering and practicing engineers who wish to learn the use of CATIA. This book is exclusively structured around CATIA, and no other solid modeling software currently available is considered.

The book is segregated in six sections containing nine chapters out of which seven chapters describe the various features of the software, whereas the last two are dedicated to modeling two complete engineering components that are quite complicated but are very popular amongst students, researchers and industrialists and are generally included in the course curriculum of undergraduate courses of different universities and institutions. The six sections are *Section I – Introduction*, *Section II – Sketcher*, *Section III – Part Design*, *Section IV – Assembly*, *Section V – Drafting* and *Section VI – Case Study*. The sections cover the complete package. *Section I* contains Chapter 1, *Section II* comprises Chapters 2–4, *Section III* contains only Chapter 5, *Section IV* has Chapter 6 and *Section V* has Chapter 7. The last section of the book, *Section VI* provides Chapters 8 and 9. The e-version of the book also includes videos of Chapters 8 and 9.

Chapter 1 introduces the software to the readers and educates on the system requirements. The chapter also guides the users about various types of documents, shortcut keys, usage of three-button mouse and functions of keyboard. Further, the various workbenches of the software are provided and relevant toolbars are discussed in brief.

For modeling any product, initially the base sketch is required to be drawn; hence, Chapter 2 discusses in detail the *Sketcher Workbench.* After discussing the ways of accessing the workbench, *Profile Toolbar, Circle Toolbar, Spline Toolbar, Conic, Line* and *Point* have been elaborately explained. All the subparts of each of the toolbars have been discussed with the screenshots to make the user understand every step involved.

The next chapter of the section, Chapter 3, provides details of the editing tools or, in the terms of CATIA language, *Operation Toolbar.* It elaborates on options like *Corner, Chamfer, Trim, Break* and *Quick Trim* apart from operations like *Relimitations, Transformation* etc.

To properly define the sketch, Constraints are very essential requirements and Chapter 4, the last chapter of the section, deals exclusively with the same. It explains the usage and operations of the menus under the *Constraints Toolbar,* including *Geometric Constraint, Dimensional Constraints* and *Auto Constraint.* The chapter finally provides insight about *Editing Multi-Constraint* and *Dimensions Modifying* after placement.

The book now enters into Section III. Chapter 5 explains the 3D modeling part with *Part Design.* Like the chapter on Sketcher, here also the chapter discusses in detail the *Part Design Workbench.* After discussing the ways of accessing the workbench, *Sketch-Based Features Toolbar* containing *Pad, Pocket, Shaft, Groove, Hole, Rib, Slot* has been dealt with. Features like *Solid Combine, Stiffener, Multi-Section Solid* are also explained. Moreover, associated toolbars like *Dress-up Features Toolbar, Transformation Features Toolbar* like *Mirror, Pattern Toolbar* and *Scale Toolbar* have been discussed.

After creating various solids, the next step in engineering is creating assembly and the next chapter, i.e., Chapter 6, deals exclusively with this. After discussing the ways of accessing the relevant workbench, the chapter discusses about recalling the components, moving and updating them. The *Constrains Toolbar, Measure Toolbar, Space Analysis Toolbar* and *Annotation Toolbars* are then explained with the aid of screenshots.

Till this point the book deals with creation of 3D models but manufacturing of the same requires 2D drawings along with the relevant dimensions. Chapter 7 deals with the depiction of assembly of product in a 2D drawing known as *Drafting.* The chapter talks about the *Views* and *Dimensioning* toolbar.

In the last section of the book, i.e., *Section VI,* Chapters 8 and 9 pick up two products that are very common and form a part of the academic curriculum of undergraduate study. Chapter 8 creates an *Oldham's Coupling* from scratch and creates the assembly along with the manufacturing drawing of the same; Chapter 9 does the same for a *Clamping Device.*

Acknowledgements

First and foremost, we would like to thank God. It was your blessing that provided us the strength to believe in passion, hard work and pursue dreams. We thank our families for their patience as we undertook yet another challenge that decreases the amount of time we could spend with them. They were our inspiration and motivation. We would like to thank our parents and grandparents for allowing us to follow our ambitions. We would like to thank all the contributing authors as they are the pillars of this structure. We would also like to thank them for their belief in us. We would like to thank all of our colleagues and friends in different parts of the world for sharing ideas in shaping our thoughts. Our efforts will be rewarded if the students, researchers and professionals of all fields related to 3D modeling, in particular, and product development, in general, benefit from this book.

We owe a huge thanks to *Dassault Systèmes* for permitting us to use the screenshot from software CATIA developed by them, in our book. We are also indebted to Editorial Advisory Board Members, Book Development Editor and the team of CRC Press for their availability for work on this huge project. All of their efforts helped to make this book complete and we could not have done it without them.

Last, but definitely not least, we would like to thank all individuals who spared their time to help us during the process of writing this book; without their support and encouragement, we would have probably given up the project.

Kaushik Kumar

Chikesh Ranjan

J. Paulo Davim

About the Editors

Kaushik Kumar, (B.Tech. [Mechanical Engineering, REC (Now NIT), Warangal], MBA [Marketing, IGNOU] and Ph.D. [Engineering, Jadavpur University]), is presently an Associate Professor in the Department of Mechanical Engineering, Birla Institute of Technology, Mesra, India. He has 19 years of teaching and research experience and more than 11 years of industrial experience in a manufacturing unit of global repute. His areas of teaching and research interest are composites, optimization, non-conventional machining, CAD/CAM, rapid prototyping and quality management systems. He has 9 patents, more than 35 books, 30 edited books, 55 book chapters, 150 international journal publications, 22 international and 1 national conference publications to his credit. He is on the editorial board and review panel of seven international and one national journal of repute. He has been felicitated with many awards and honors (Web of Science core collection 102 publications/h-index 10+, SCOPUS/h-index 10+, Google Scholar/h-index 23+).

Chikesh Ranjan, (BE [Mechanical Engineering, Marathwada Institute of Technology, Aurangabad, India], M.E. [Design of Mechanical Equipment, Birla Institute of Technology, Mesra, India]), is presently pursuing Ph.D. at Birla Institute of Technology, Mesra. He is an Assistant Professor in the Department of Mechanical Engineering, RTC Institute of Technology, Ranchi, India and has over 7 years of teaching and research experience. His areas of interest are product and process design, CAD/CAM/CAE, rapid prototyping and composites. He has 11 books, 15 international journal and 4 international conference publications to his credit.

J. Paulo Davim is a Full Professor at the University of Aveiro, Portugal. He is also distinguished as honorary professor in several universities/colleges in China, India and Spain. He received his Ph.D. degree in Mechanical Engineering in 1997, M.Sc. degree in Mechanical Engineering (materials and manufacturing processes) in 1991, Mechanical Engineering degree (5 years) in 1986, from the University of Porto (FEUP), the Aggregate title (Full Habilitation) from the University of Coimbra in 2005 and the D.Sc. (Higher Doctorate) from London Metropolitan University in 2013. He is Senior Chartered Engineer by the Portuguese Institution of Engineers with an MBA and Specialist titles in Engineering and Industrial Management as well as in Metrology. He is also Eur Ing by FEANI-Brussels and Fellow (FIET) of IET-London. He has more than 30 years of teaching and research experience in manufacturing, materials, mechanical and industrial engineering, with special emphasis in machining & tribology. He has also interest in management, engineering education and higher education for sustainability. Dr Davim has guided large numbers of postdoctotal, Ph.D. and master's students as well as coordinated and participated in several financed research projects. He has received several scientific awards and honors. He has worked as evaluator of projects for European Research Council and other international research agencies as well as examiner of Ph.D. thesis for many universities in different countries. He is the editor in chief of several international journals, guest editor of journals, books editor, book series editor, and scientific advisory for many international journals and conferences. At present, he is an editorial board member of 30 international journals and acts as reviewer for more than 100 prestigious Web of Science journals. In addition, he has also published as editor (and co-editor) more than 150 books and as author (and co-author) more than 15 books, 100 book chapters and 500 articles in journals and conferences (more than 280 articles in journals indexed in Web of Science core collection/h-index 57+/10,500+ citations, SCOPUS/h-index 62+/13,000+ citations, Google Scholar/h-index 80+/21,500+ citations).

Section I

Introduction

1 Introducing CATIA Basics

1.1 INTRODUCTION TO CATIA

CADD Software CATIA "*Computer Aided Three-Dimensional Interactive Application*" is prepared by one of the best leading companies, Dassault Systèmes. This software has a large number of users due to its flexibility. It also provides flexibility for product design feature-based and parametric design. Due to this feature, any change made in any part of the product design is updated quickly in all workbenches.

1.2 SYSTEM REQUIREMENTS

For smooth running of CATIA software, the following system requirements are as discussed:

1. OPERATING SYSTEM: Windows 7 or Windows 10(64 or 32 bit)
2. PROCESSOR: Multicore processor (Intel i5, i7 or Xenon)
3. RAM: 4–8 GB or more.
4. HARD DISK SPACE: 10GB
5. MONITOR SIZE:17 inches or more

1.3 TYPES OF DOCUMENTS

While working with CATIA's different workbenches, you will work with different types of documents. Here is a short description of the most common ones:

- *CAT Part*: This is the most common file, representing a part. This file contains all the geometrical information of an object. This is the file obtained while using the Part Design workbench (as well as the Sketcher).
- *CAT Product*: This file is created while using the Assembly Design workbench. This file does not contain any geometrical elements but is rather referencing CAT Part files together, thus creating the assembly.
- *CAT Drawing*: This document is used with the Generative Drawing and Interactive Drafting workbenches. It is essentially a 2D drawing.
- *CAT Analysis*: This document is used while analyzing a part or an assembly (i.e., stress analysis and kinematic analysis).It will contain all the parameters of the analysis done on the object.
- *CAT Material*: This document is a material library that can contain customized material not existing in CATIA standard material library.
- *Catalog*: This document can contain standard parts (like nuts, bolts, etc.), which can thereafter be used while creating an assembly. It is used to regroup elements of the same family.

1.4 SHORTCUT KEYS

Shortcut keys are very much used for quick product design. You can use these action keys for performing action in the CATIA software. The action and their action keys are as shown in Figure 1.1.

1.5 FUNCTION OF MOUSE BUTTON

To work in CATIA V5 design workbenches, it is very much essential to know the operation of mouse button function. This will reduce the product design time. The different combinations are discussed next.

ACTION	ACTION KEYS
Abort Current Process	Esc
CATIA V5 Assistance	F1
Specification Tree Toggle	F3
Hide/Show	F9
Swap Variable Space	F10
Rotate to Left	Shift + Left
Rotate to Right	Shift + Right
Rotate Upward	Shift + Up
Rotate Downward	Shift + Down
Start Macros	Alt + F8
Visual Basic	Alt + F11
Zoom In	Ctrl + Page Up
Zoom Out	Ctrl + Page Down
Pan Left	Ctrl + Left
Pan Right	Ctrl + Right
Pan Up	Ctrl + Up
Pan Down	Ctrl + Down
Rotate About Z – Axis Counter-clockwise	Ctrl + Shift + Left
Rotate About Z – Axis Clockwise	Ctrl + Shift + Right
Swap Windows	Ctrl + Tab
New Document	Ctrl + N
Open Document	Ctrl + O
Document Save	Ctrl + S
Document Print	Ctrl + P
Search	Ctrl + F
Update	Ctrl + U
Cut	Ctrl + X
Copy	Ctrl + C
Paste	Ctrl + V
Redo	Ctrl + Y
Undo	Ctrl + Z

FIGURE 1.1 List of shortcut keys.

Center the display	Select and release the middle mouse button on the location that you want to be centered and it will move to the center of the display
Pan	Select and hold the middle mouse button and you can move your display around by moving the mouse
Rotate	Select and hold the middle mouse button and then select and hold either the first or third mouse buttons and you can rotate the display around by moving the mouse. You should see a rotational ball appear for reference. Both buttons will be held down simultaneously.
Zoom	Select and hold the middle mouse button and then select and release either the first or third mouse buttons and you can zoom in or out by moving the mouse up or down. Only the middle mouse button will be held down.
Rotate and Zoom	While on a geometrical entity you can press and hold the Shift key and then press the middle mouse button to perform a rotation and zoom using a viewpoint control.

FIGURE 1.2 Three-button mouse shortcut keys.

1.5.1 Three-Button Mouse

The three-button mouse action and their action keys are as shown in Figure 1.2.

1.6 FUNCTION OF KEYBOARD

The functions of keyboard keys are shown in Figure 1.3.

Pan	Press and hold the Ctrl key and select the arrows to pan up, down, right or left
Rotate around the vertical	Press and hold the Shift key and select the left or right arrow
Rotate around the horizontal	Press and hold the Shift key and select the up or down arrow
Rotate around the normal	Press and hold the Ctrl and Shift keys and select the left or right arrow
Zoom In	Press and hold the Ctrl key and select the Page Up key
Zoom Out	Press and hold the Ctrl key and select the Page Down key

FIGURE 1.3 Functions of keyboard keys.

1.7 WORKBENCHES IN CATIA

CATIA software provides different workbenches for designing different types of products.

The workbench is an environment having different types of toolbar consisting of different types of tools. These tools are used for designing different types of product. The workbenches present in CATIA V5 are as follows:

- *"Sketcher workbench"*
- *"Part Design workbench"*
- *"Wireframe and Surface Design workbench"*
- *"Assembly Design workbench"*
- *"Drafting workbench"*
- *"Generative Sheet metal Design workbench"*

1.7.1 Sketcher Workbench

The Sketcher workbench is used for creating 2D modeling. This workbench is used by other workbenches for designing parts.

1.7.2 Part Design Workbench

The Part Design workbench is used for designing solid model. The basic needs for designing solid model are to draw a sketch, for that you must use Sketcher workbench. The different types of tools can be used for sketching the sketch in Sketcher workbench. After drawing the sketch, exit the Sketcher workbench and convert the sketch into solid model. The tools present in the Part Design workbench are used to design the sketch in to a solid model.

1.7.3 Wireframe and Surface Design Workbench

This workbench is used for designing surface model. The different types of tools are present in workbench for designing surface model. The working with tools is like Part Design workbench.

1.7.4 Assembly Design Workbench

This workbench is used to assemble the components deign in Part Design workbench using different types of constraints available in this workbench. Followings are the types of assembly design as discussed:

- *Bottom-up*
- *Top-down*

In bottom-up assembly, products are created in Part Design workbenches and assembly of these parts in assembly workbench.

In top-down assembly, products are created and assembly of these parts in assembly workbench.

1.7.5 Drafting Workbench

This workbench is used for producing front, top, side and isometric view of the parts or assemblies. Followings are the types of drafting as discussed:

- Generative Drafting
- Interactive Drafting

In Generative Drafting, drawings of parts and assemblies are generated automatically. The bill of material can also be generated.

In Interactive Drafting, drawings and dimensions of parts and assemblies are generated manually.

1.7.6 Generative Sheet Metal Design Workbench

This workbench is used for the designing of the Sheet metal components.

1.8 TOOLBARS

CATIA is very user-friendly in designing software. The different types of toolbar present in different workbenches are used for designing products. All Tools under different toolbar are discussed in respective chapters.

Section II

Sketcher

2 Sketcher Workbench

2.1 INTRODUCTION

The Sketcher workbench is used for sketching the product design. The Sketcher workbench has different types of tools like line, rectangle, circle, parallelogram, arc, and spline. In this chapter, we will discuss different types of sketching tools step by step using CATIA software.

2.2 ACCESSING THE SKETCHER WORKBENCH

After starting CATIA, productl, a new screen will appear (Figure 2.1).

To access the Sketcher workbench, select in sequence Start, Mechanical Design and Sketcher (Figure 2.2).

A new screen in the name of New Part will appear as displayed in Figure 2.3.

Enter part name in New Part dialog box and press on OK. The part design workbench is open, as displayed in Figure 2.4.

Select Sketch tool from Sketcher toolbar. Select the planes from tree or from the geometry area as shown in Figure 2.5.

Open the Sketcher Screen by selecting the xy plane as shown in Figure 2.6.

After selecting the Sketcher, you can select the mode as per your drawing requirement. For modify units, you must select Tools and Options from the above menu bar. Choose the units as per your requirements. In option screen, select Units tab and select the unit for length, angle, time and mass using different options as displayed in Figure 2.7.

FIGURE 2.1 Starting CATIA software.

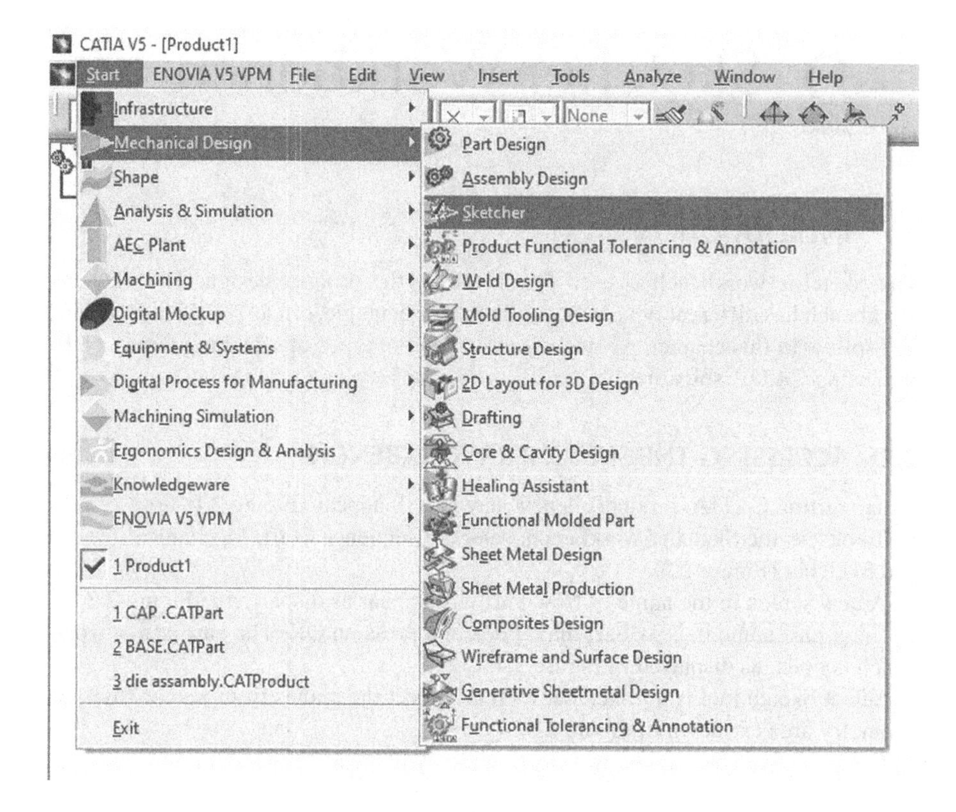

FIGURE 2.2 Access the Sketcher.

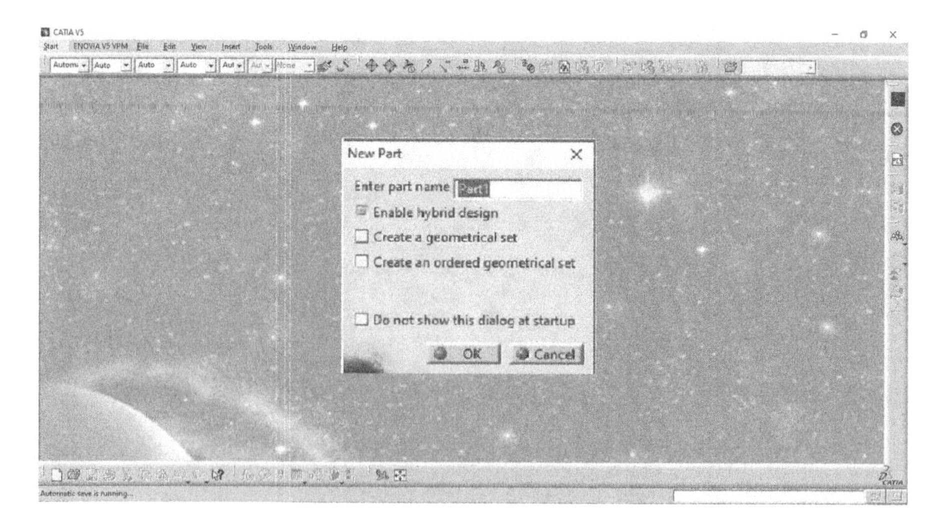

FIGURE 2.3 New Part.

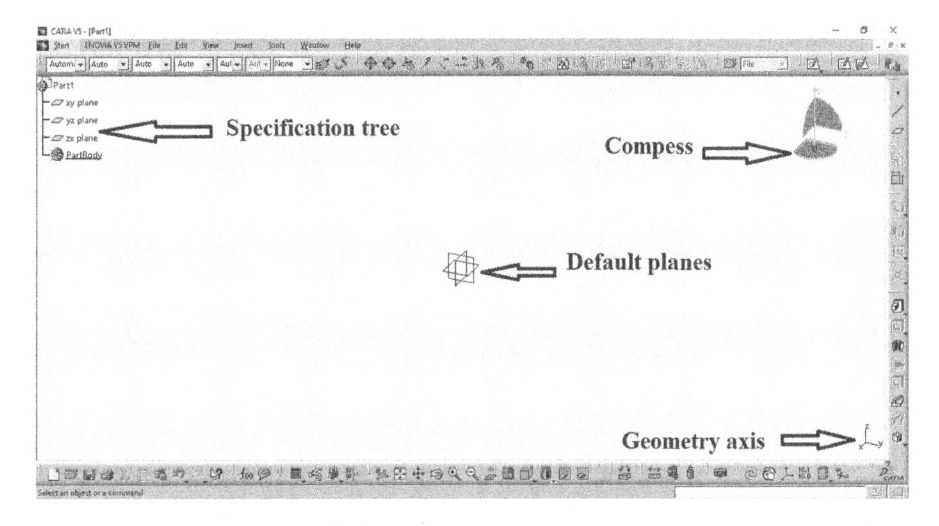

FIGURE 2.4 New Part displayed Screen.

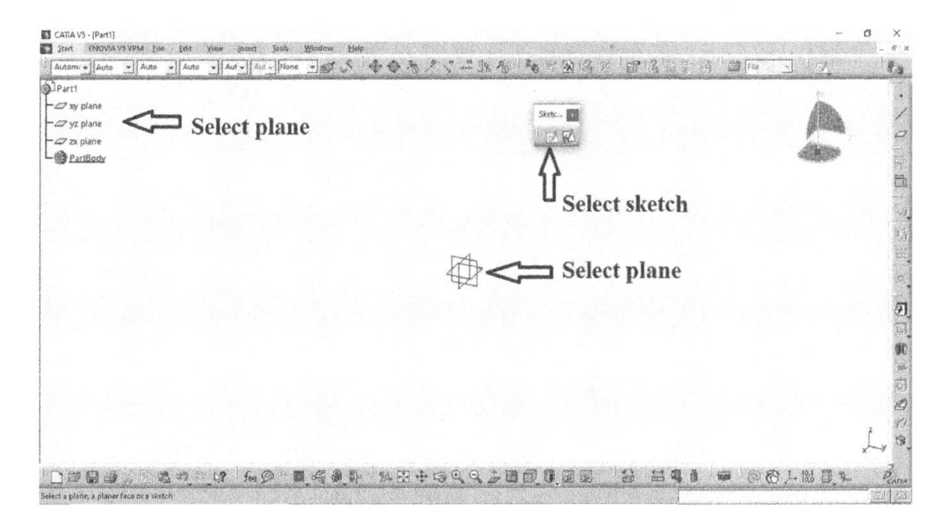

FIGURE 2.5 Sketcher workbench invoked Sketcher tool and plane.

2.3 PROFILE TOOLBAR

The Tools present in this toolbar as shown in Figure 2.8 are used to create geometry in Sketcher workbench. The Tools in this toolbar are Predefined Profile, Circle, Spline, Conic and Line.

2.3.1 Profile Toolbar

In this toolbar, Tools are used for making continuous line and arc. For drawing profile, select Profile tool from Profile toolbar as shown in Figure 2.8. After selecting the tool, select First Point, Second Point and Third Point as displayed in Figure 2.9.

FIGURE 2.6 Sketcher workbench invoked using the xy plane.

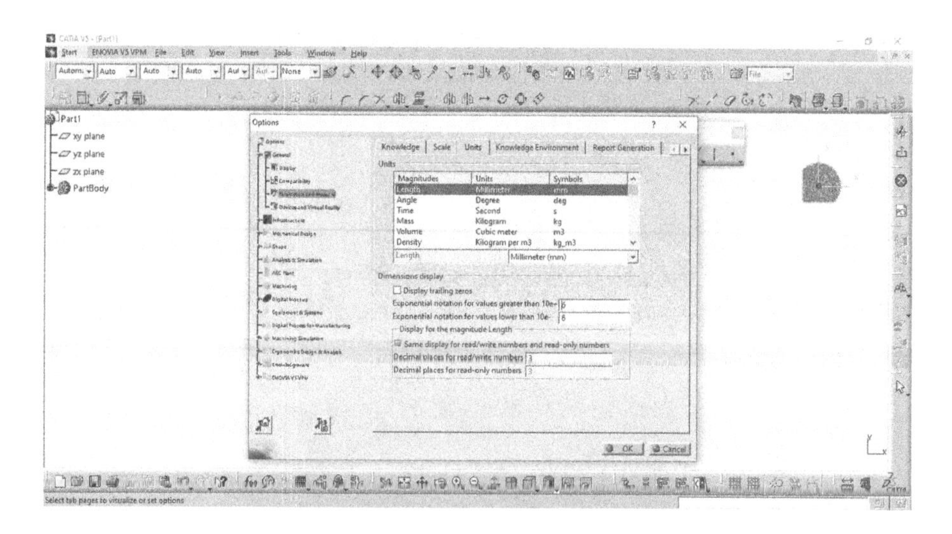

FIGURE 2.7 Displayed Options dialog box with the units.

2.3.2 Predefined Toolbar

The tools under Predefined toolbar are shown in Figure 2.10.

2.3.2.1 Rectangle

The rectangle has four sides, two are horizontal and two are vertical, and it is constructed by two points. For drawing a rectangle, select Rectangle tool from Predefine toolbar as shown in Figure 2.10. Click the geometry area to indicate the first corner of the rectangle followed by second click to indicate the diagonally opposite corner as displayed in Figure 2.11.

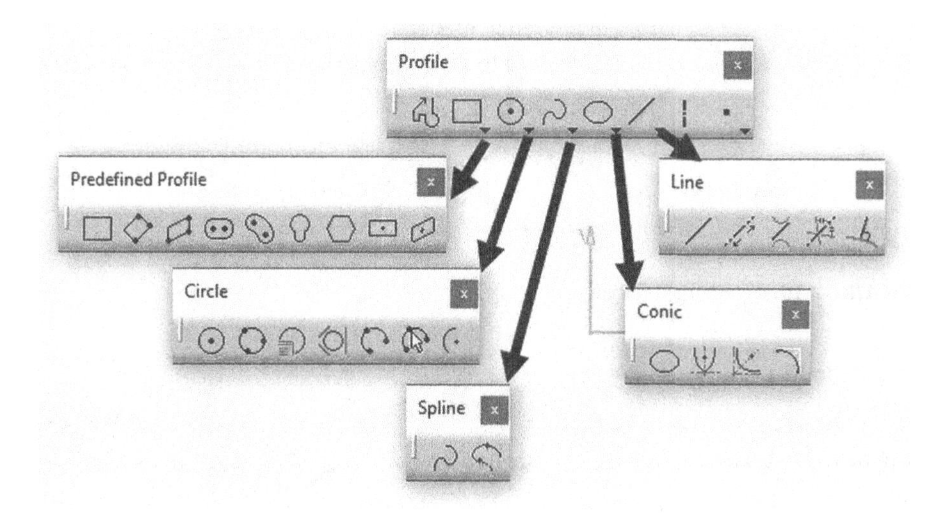

FIGURE 2.8 Profile toolbar.

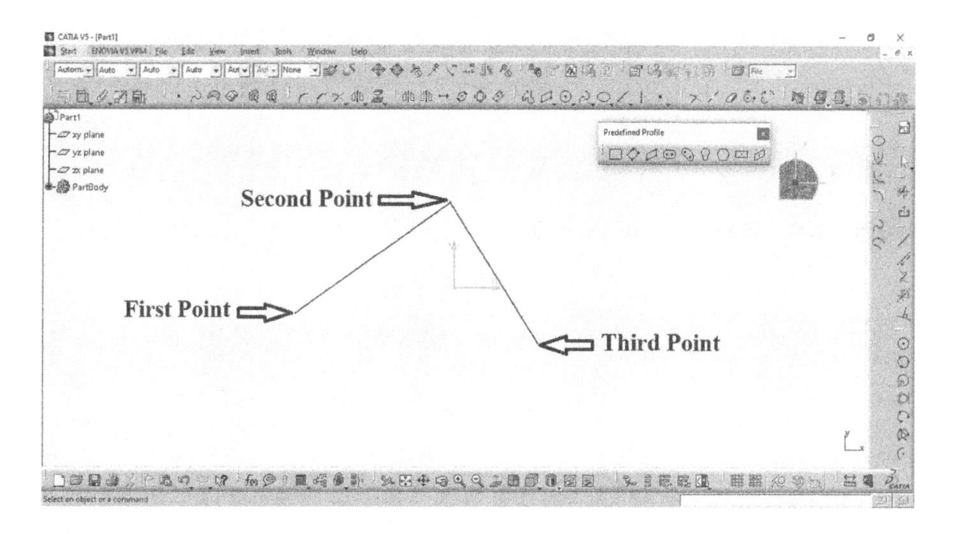

FIGURE 2.9 An open profile drawn using the Profile tool.

2.3.2.2 Oriented Rectangle

It is constructed by three corner points. For drawing an oriented rectangle, select Oriented rectangle tool from Predefine toolbar as shown in Figure 2.10. Click the geometry area to indicate the first corner of the rectangle followed a second click to complete the first side. Finally, click a location to indicate the diagonally opposite corner as displayed in Figure 2.12.

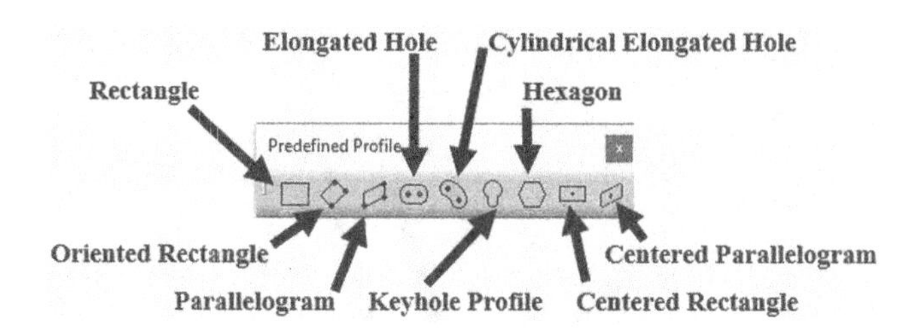

FIGURE 2.10 Predefine toolbar.

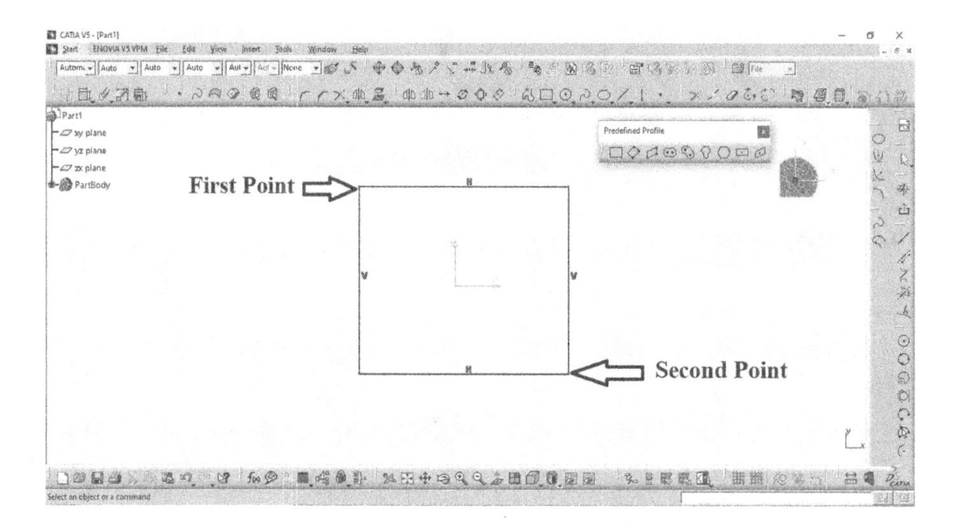

FIGURE 2.11 Constriction of rectangle.

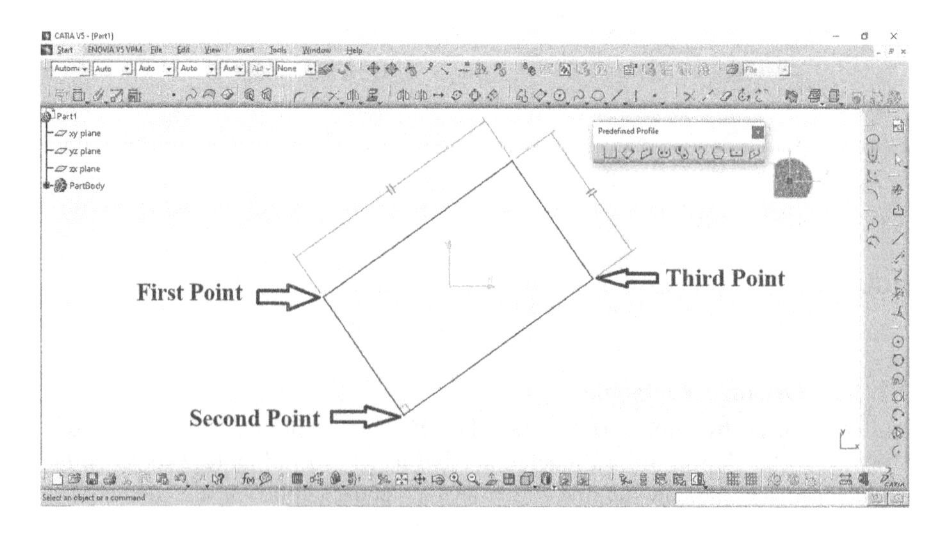

FIGURE 2.12 Constriction of oriented rectangle.

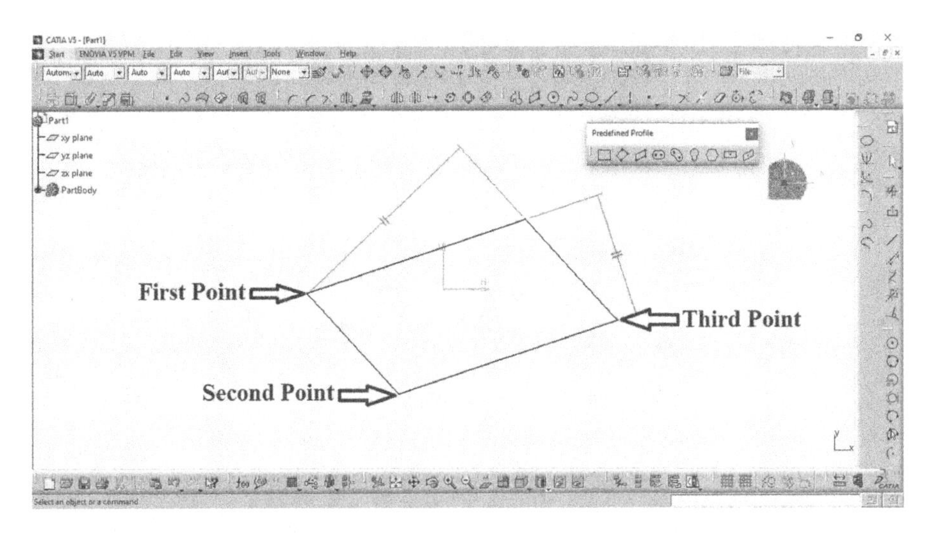

FIGURE 2.13 Constriction of parallelogram.

2.3.2.3 Parallelogram

The parallelogram is defined by three corner points and opposite sides parallel to each other. For drawing a parallelogram, select Parallelogram tool from Predefine toolbar as shown in Figure 2.10. Click the geometry area to indicate the first corner of the parallelogram followed by the second click to complete the first side. Finally, click a location to indicate the diagonally-opposite corner as shown in Figure 2.13.

2.3.2.4 Elongated Hole

It is constructed by two points and a radius. The elongated hole or slot is defined by two points and a radius. For drawing an elongated hole, select Elongated hole tool from Predefine toolbar as shown in Figure 2.10. The first two points selected using the mouse will define the position and length of the slot axis. The third selection controls the size of the slot as shown in Figure 2.14.

2.3.2.5 Cylindrical Elongated Hole or Slot

It is constructed by a cylindrical radius, two point and a hole radius. For drawing a cylindrical elongated hole, select the Cylindrical elongated hole tool from Predefine toolbar as shown in Figure 2.10. The first selection indicates the center point of the radial axis of the slot. The second and third selections define the radius and radial length of the slot. The final selection defines the size of the slot as shown in Figure 2.15.

2.3.2.6 Keyhole Profile

It is constructed by two center points and two radii. For drawing a keyhole profile, select Keyhole profile tool from Predefine toolbar as shown in Figure 2.10. The first selection indicates the center of the large radius of the profile. The second indicates the center of the small radius, the third selection defines the size of the small radius and, finally, the last selection indicates the size of the large radius as shown in Figure 2.16.

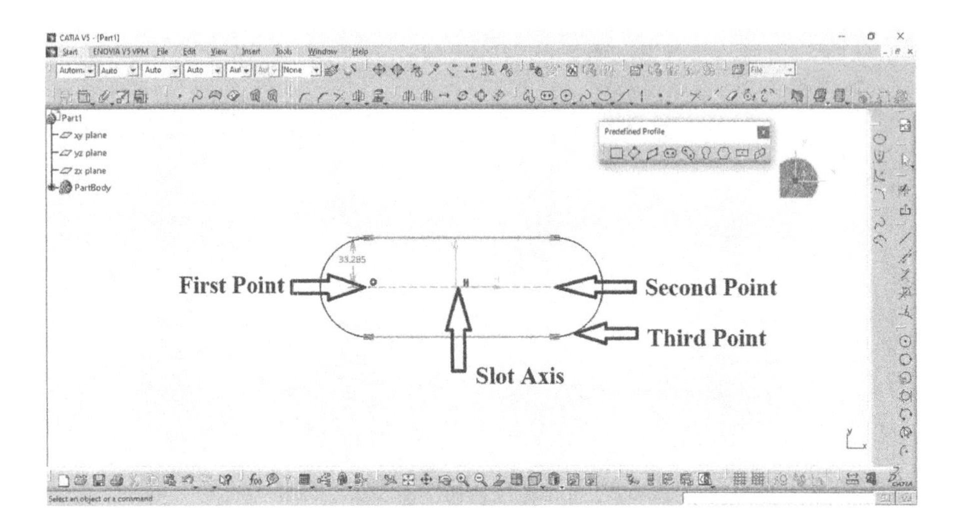

FIGURE 2.14 Constriction of elongated hole or slot.

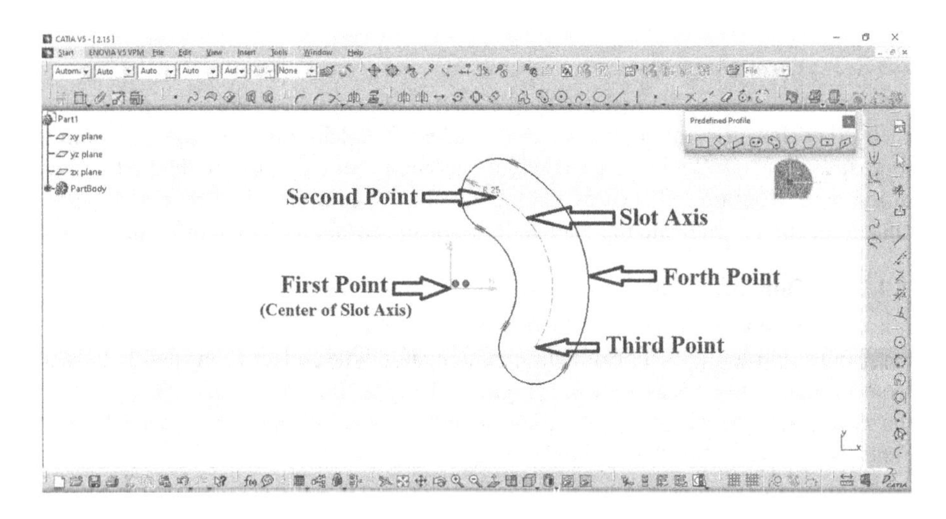

FIGURE 2.15 Constriction of cylindrical elongated hole or slot.

2.3.2.7 Hexagon

It is constructed by a center point and the radius of an inscribed circle. For drawing a hexagon, select Hexagon tool from Predefine toolbar as shown in Figure 2.10. The first selection indicates the center of the hexagon and the second define the size of the profile as shown in Figure 2.17.

2.3.2.8 Centered Rectangle

It is constructed by a center point and a corner point. For drawing a centered rectangle, select Centered rectangle tool from Predefine toolbar as shown in Figure 2.10. The first selection indicates the center of the rectangle and the second point defines the size of the rectangle profile as shown in Figure 2.18.

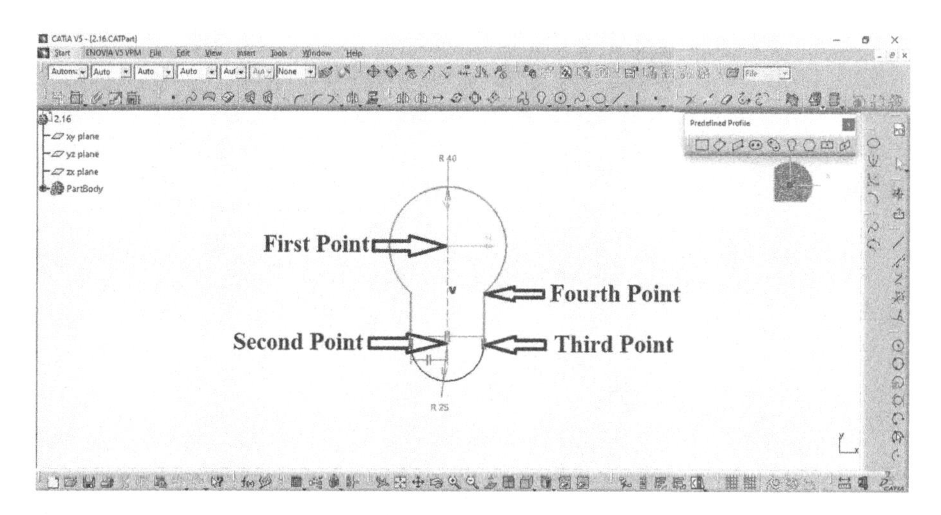

FIGURE 2.16 Constriction of keyhole profile.

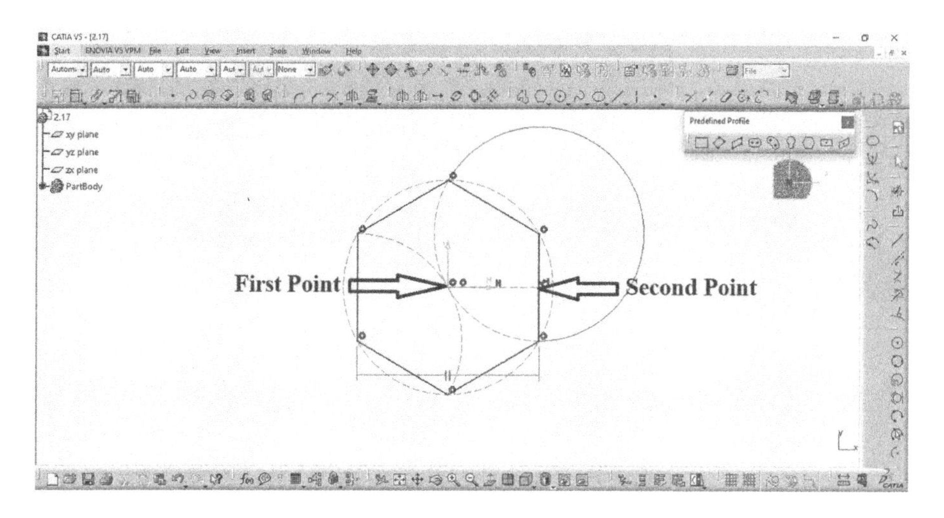

FIGURE 2.17 Constriction of hexagon.

2.3.2.9 Centered Parallelogram

It is constructed by an intersecting two-line center point and a corner point. For drawing a centered parallelogram, select Centered parallelogram tool from Predefine toolbar as shown in Figure 2.10. The first selection indicates the intersecting two line center of the rectangle and the second point defines the size of the centered parallelogram profile as shown in Figure 2.19.

2.3.3 Circle Toolbar

The tools in this toolbar are shown in Figure 2.20.

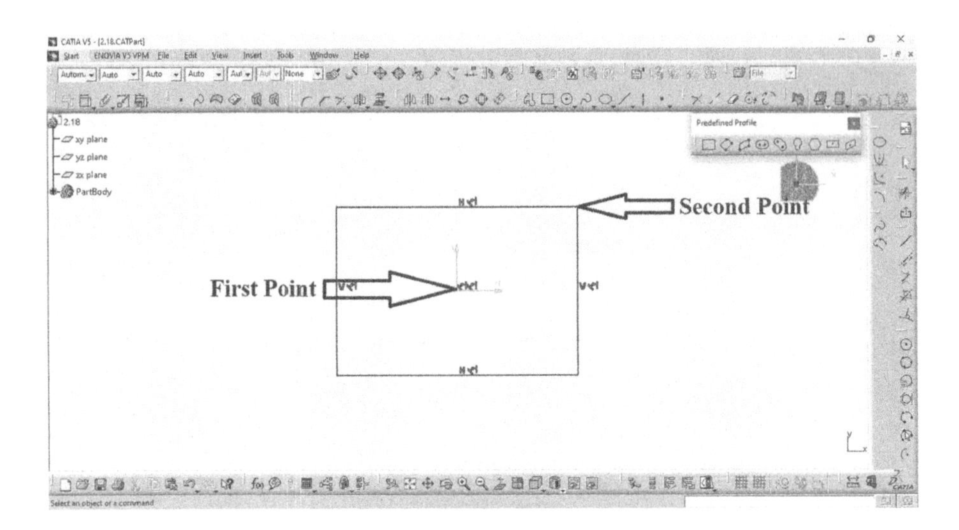

FIGURE 2.18 Constriction of centered rectangle.

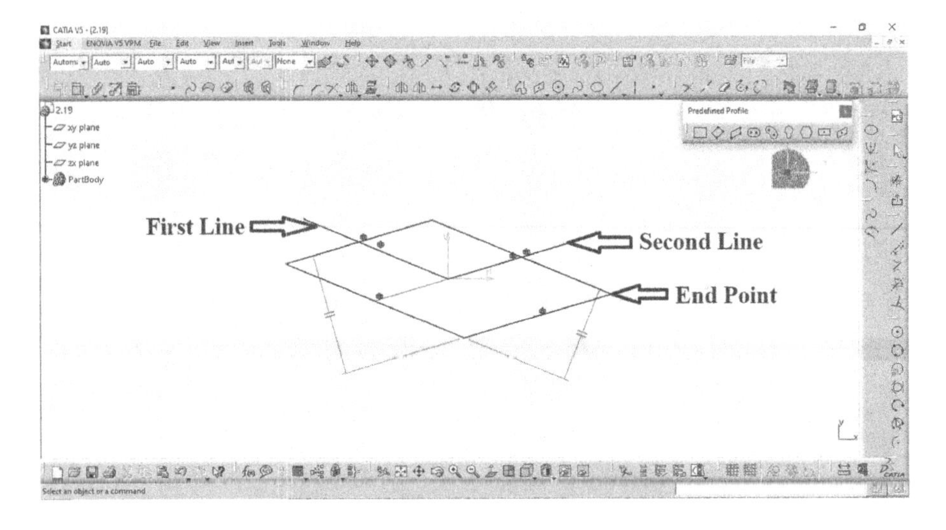

FIGURE 2.19 Constriction of centered parallelogram.

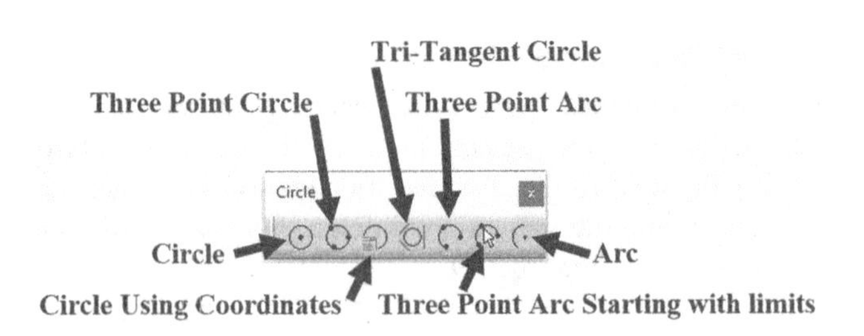

FIGURE 2.20 Circle toolbar.

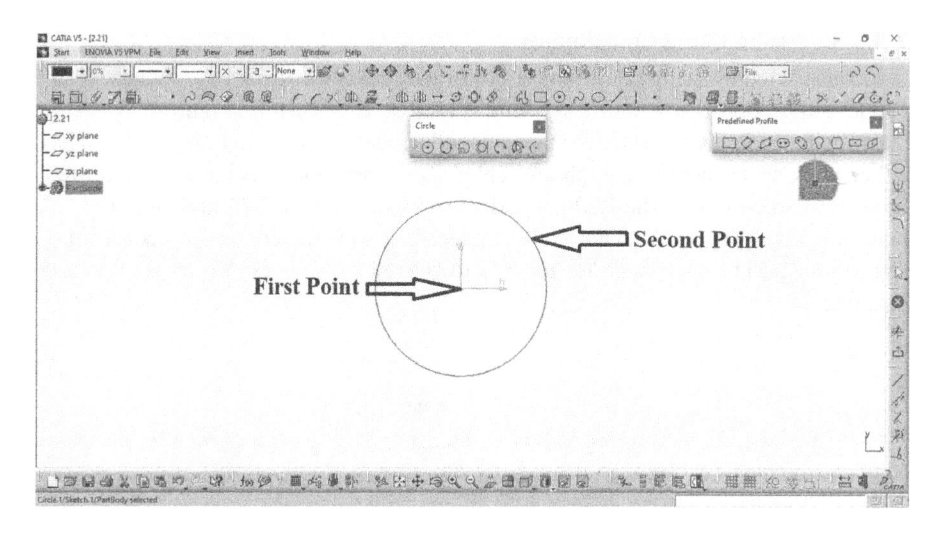

FIGURE 2.21 Constriction of circle.

2.3.3.1 Circle

It is constructed by a center point and a radius. For drawing a circle, select Circle tool from Circle toolbar as shown in Figure 2.20. The first point selection indicates the center of the circle and second point defines its size as shown in Figure 2.21.

2.3.3.2 Three-Point Circle

It is constructed by three circumferential points. To draw such a three-point circle, you need to draw three points. For drawing a three-point circle, select Three-point circle tool from Circle toolbar as shown in Figure 2.20. After selecting the icon, select first point, second point and third point as shown in Figure 2.22.

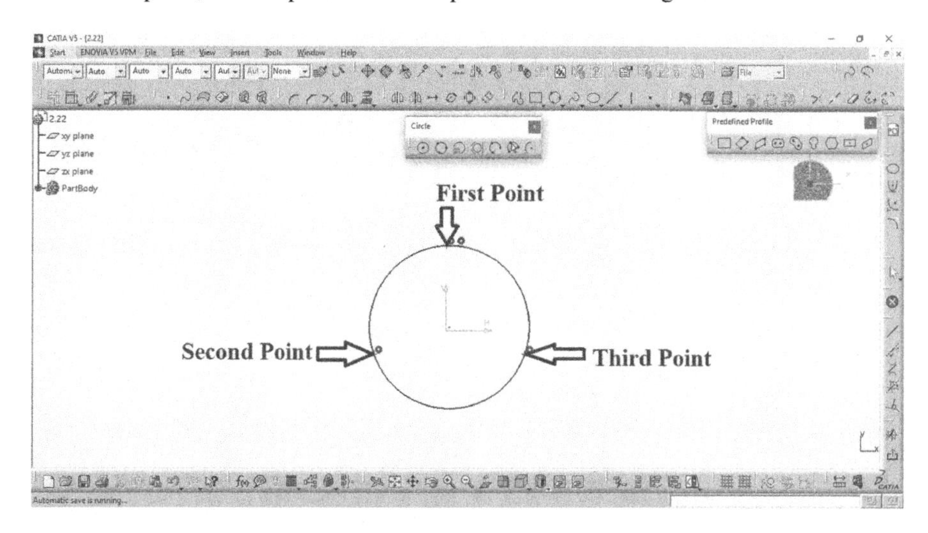

FIGURE 2.22 Constriction of three-point circle.

2.3.3.3 Circle Using Coordinates

It is constructed by using coordinates. For drawing a circle using coordinates, select Circle using coordinates tool from Circle toolbar as shown in Figure 2.20. After selecting the icon, a Circle Definition panel will appear as shown in Figure 2.23. Select either the Cartesian or Polar tab and enter the center point ordinates as required. Enter the radius size in the Radius field as shown in Figure 2.24 and click the OK button to insert the circle. The circle is generated with the controlling constraints. The center point constraints are relative to the sketch axis H and V as shown in Figure 2.25.

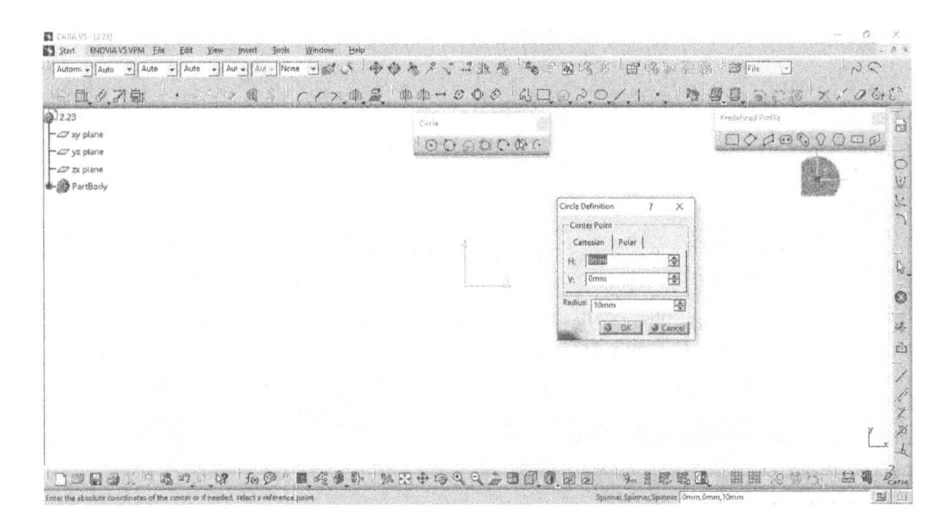

FIGURE 2.23 Circle Definition panel.

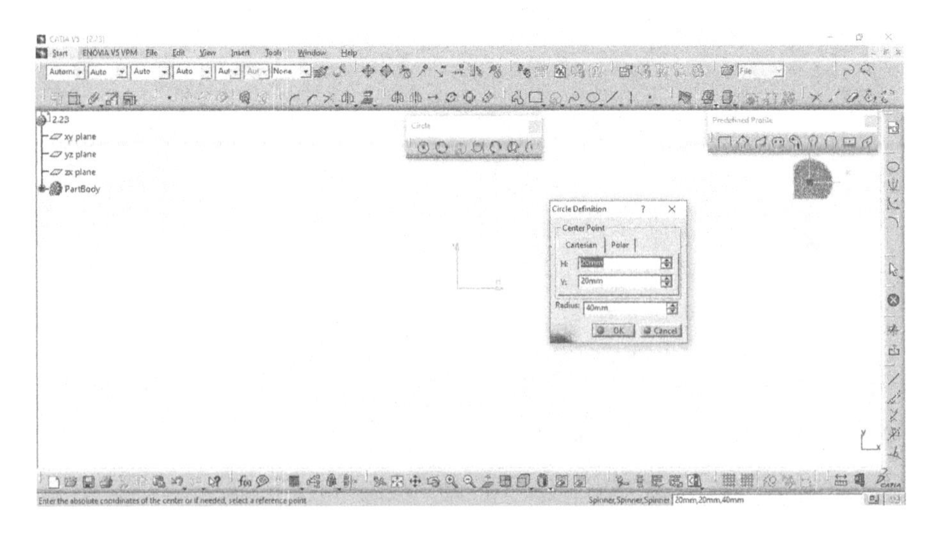

FIGURE 2.24 Circle Definition panel data filled.

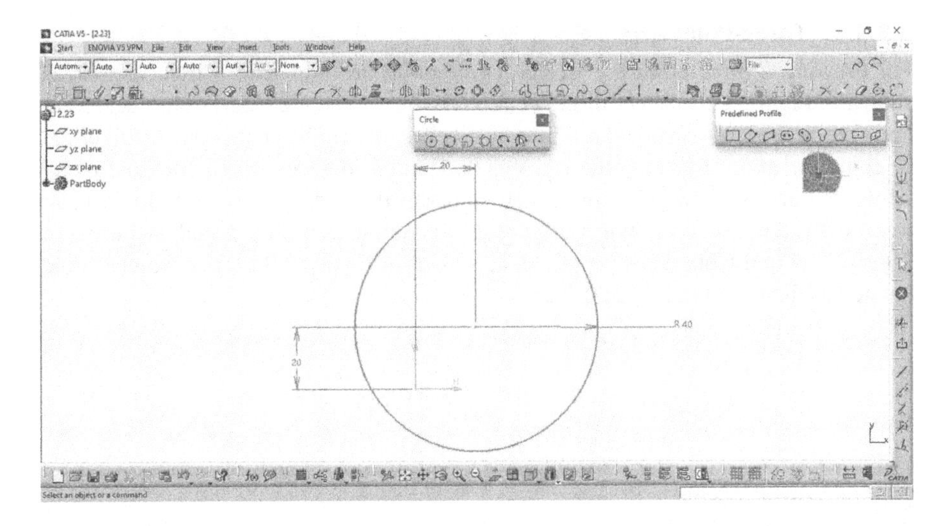

FIGURE 2.25 Constriction circle using coordinates.

2.3.3.4 Tri-Tangent Circle

For drawing a tri-tangent circle, select Tri-tangent circle tool from Circle toolbar as shown in Figure 2.20. To draw such a tri-tangent circle, you need to draw three sketches, i.e., lines, circle or another sketch. After selecting the icon, choose the three-line sketched elements as shown in Figure 2.26. The tri-tangent constructed circle is shown in Figure 2.26.

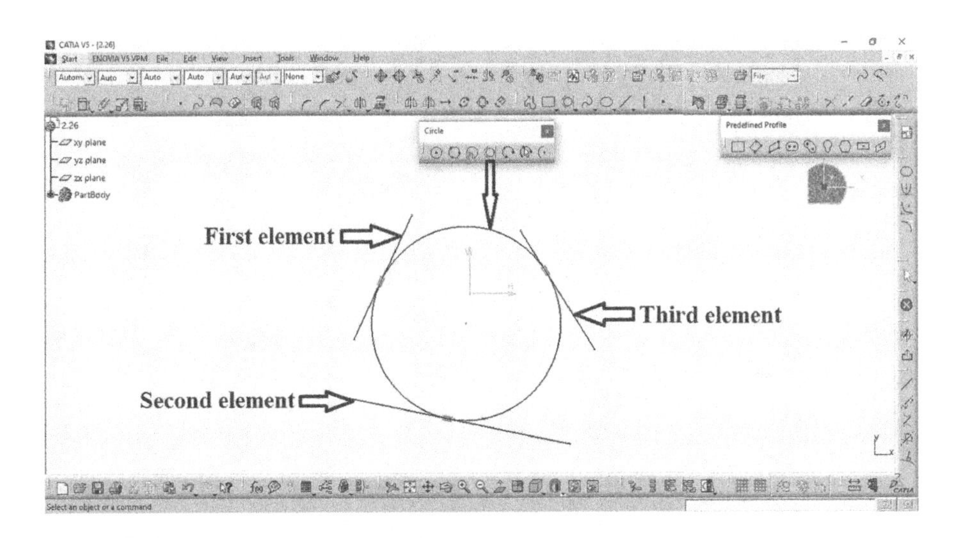

FIGURE 2.26 Constriction of tri-tangent circle.

2.3.3.5 Three-Point Arc

For drawing a three-point arc, select Three-point arc tool from Circle toolbar as shown in Figure 2.20. Select three locations to create the arc (Start point, Midpoint and then End point) as shown in Figure 2.27. The arc can be changed to a full circle or compliment arc by using the right mouse button to access the contextual menu whilst the arc is highlighted in orange, you can then select the Circle.1object tab to access the Close as shown in Figure 2.28, complete circle created as shown in Figure 2.29 and Complement options as shown in Figure 2.30, complement circle created as displayed in Figure 2.31.

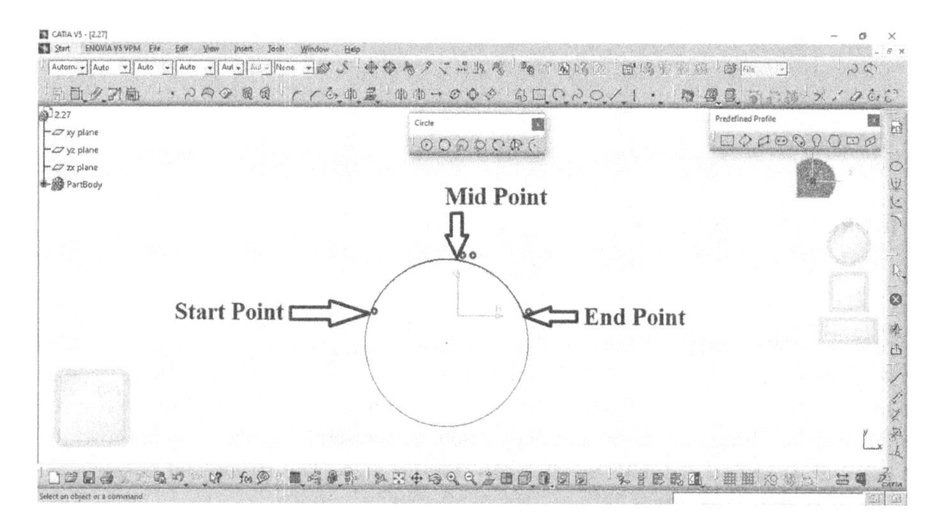

FIGURE 2.27 Constriction of three-point arc.

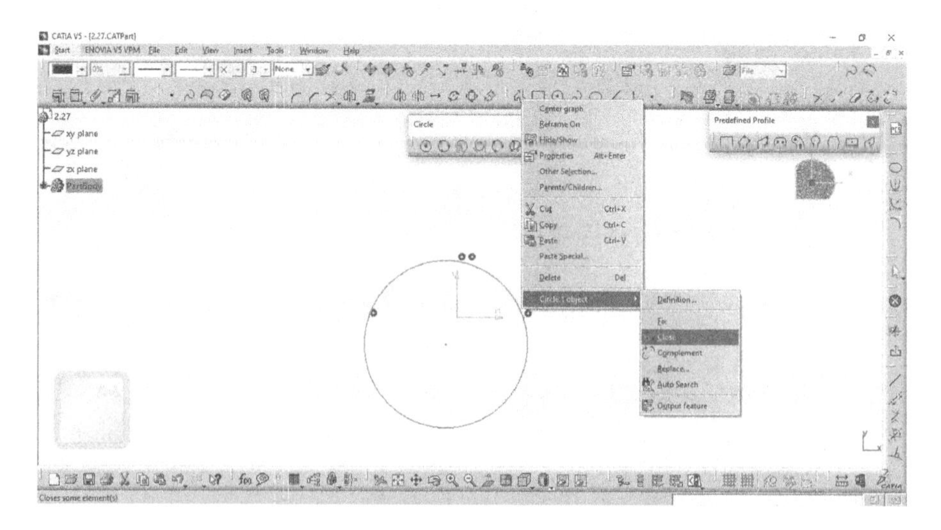

FIGURE 2.28 Constriction of three-point arc circle.

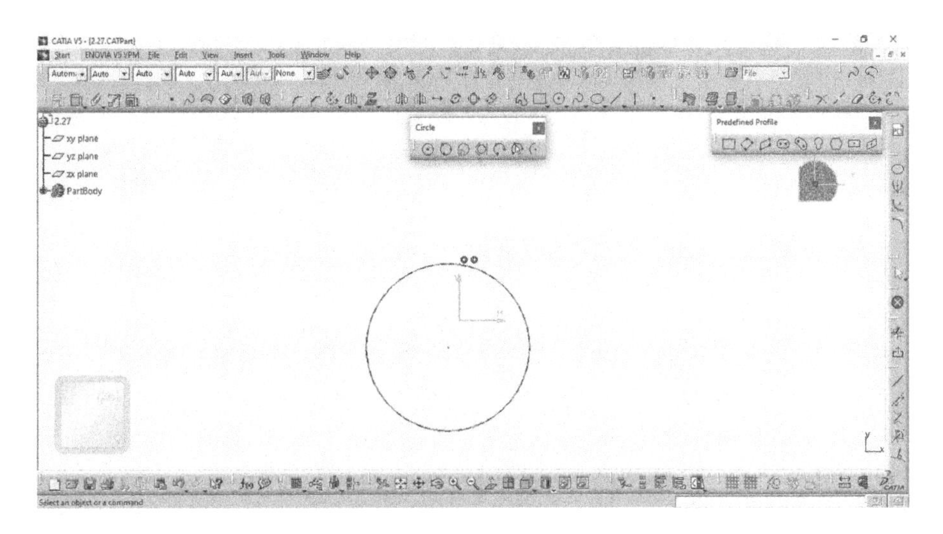

FIGURE 2.29 Constriction of three-point arc circle.

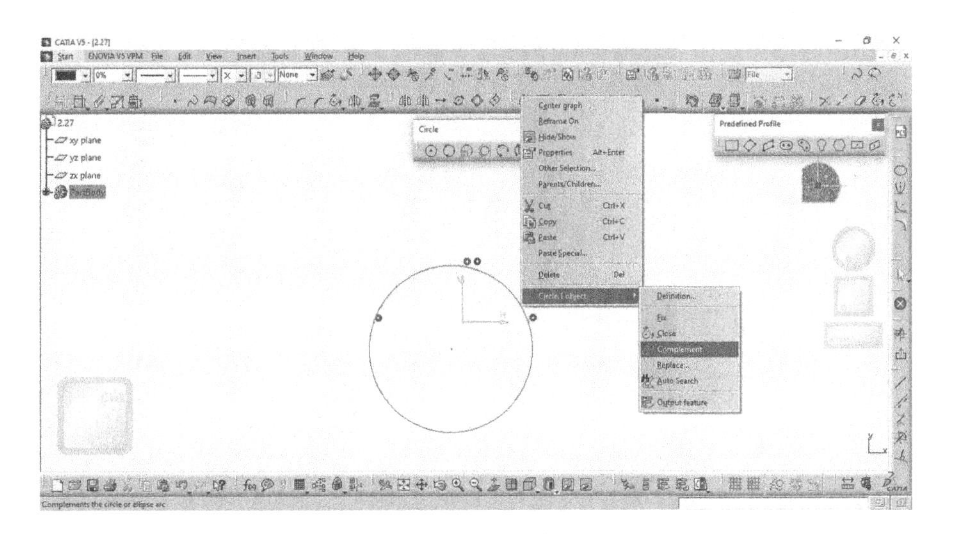

FIGURE 2.30 Constriction of three-point arc complement circle.

2.3.3.6 Three-Point Arc Starting with Limits

For drawing a three-point arc starting with limits select Three-point arc starting with limits tool from Circle toolbar as shown in Figure 2.20. Select three locations to create the arc (Start point, Midpoint and then End point). The Start Point indicates start point of arc, Midpoint indicated end of arc and End point indicates size or limit of arc as shown in Figure 2.32. The arc can be changed to a full circle or compliment arc by using the right mouse button to access the contextual menu whilst the arc is

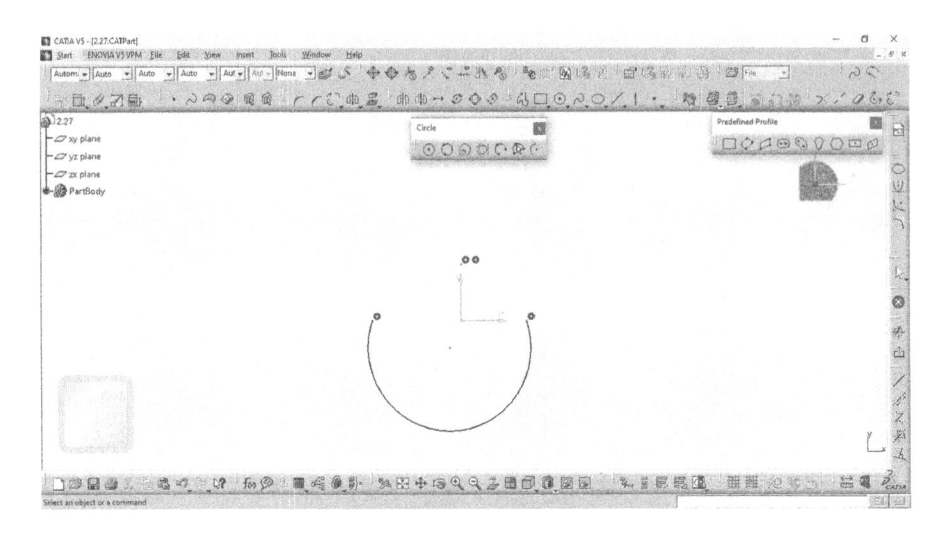

FIGURE 2.31 Constriction of three-point arc complement circle.

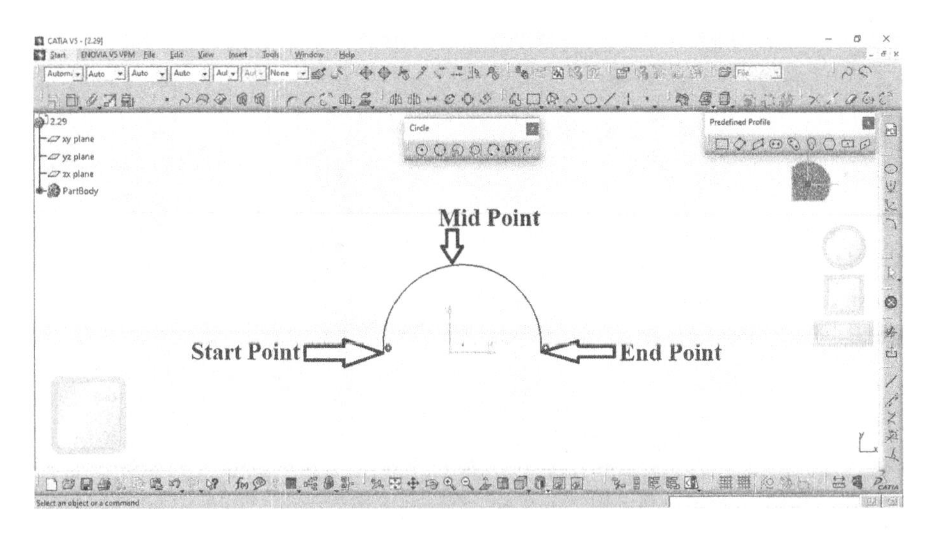

FIGURE 2.32 Constriction of three-point arc starting with limits.

highlighted in orange, you can then select the Circle.1object tab to access the Close and Complement options as shown in Figure 2.33. Again, the contextual menu can be used to close to a circle or create a complement arc as discussed above.

2.3.3.7 Arc

For drawing an arc with limits, select Arc tool from Circle toolbar as shown in Figure 2.20. After selecting tool, choose the first, second and third points as center, starting and end points of constructed arc, as shown in Figure 2.34.

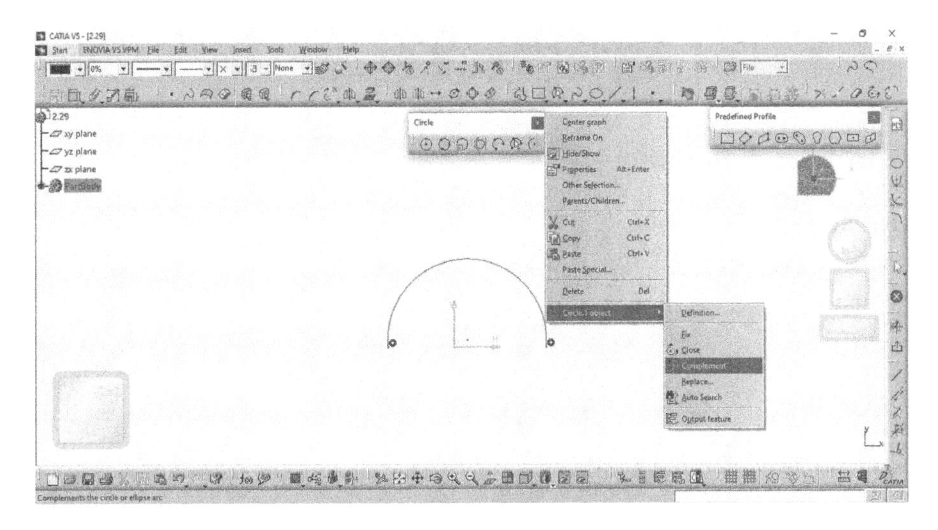

FIGURE 2.33 Constriction of three-point arc starting with limits edit.

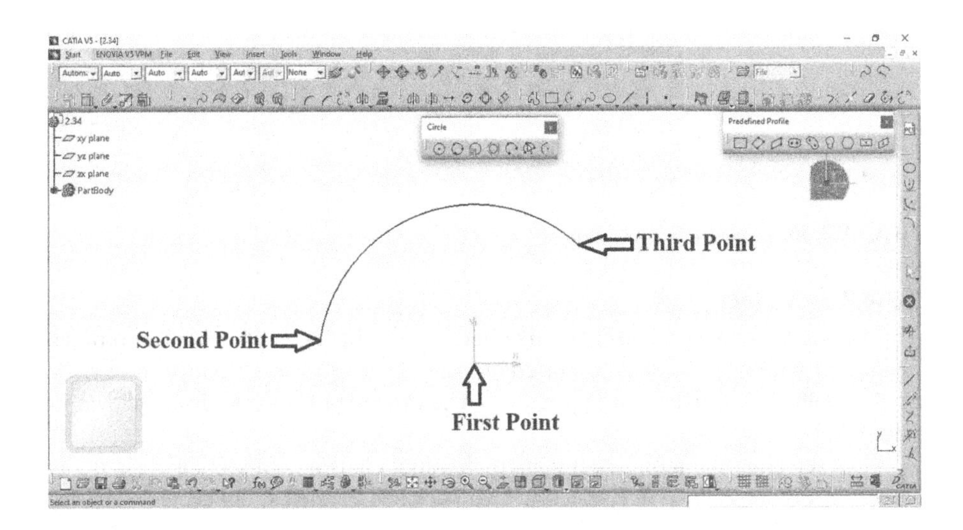

FIGURE 2.34 Constriction of arc.

2.3.4 Spline Toolbar

The tools in this toolbar are displayed in Figure 2.35.

2.3.4.1 Spline

It is constructed by three or more points. The tangency and curvature radius at each point may be specified. For drawing a spline, select Spline tool from Circle toolbar as shown in Figure 2.20. Select a series of locations known as Control Points through which a Spline is generated as you define the control point locations indicated with red arrow as displayed in Figure 2.36. Double click and select the finish the Spline.

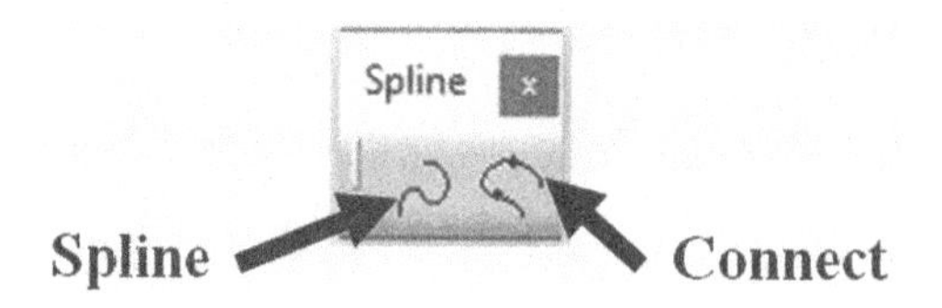

FIGURE 2.35 Spline toolbar.

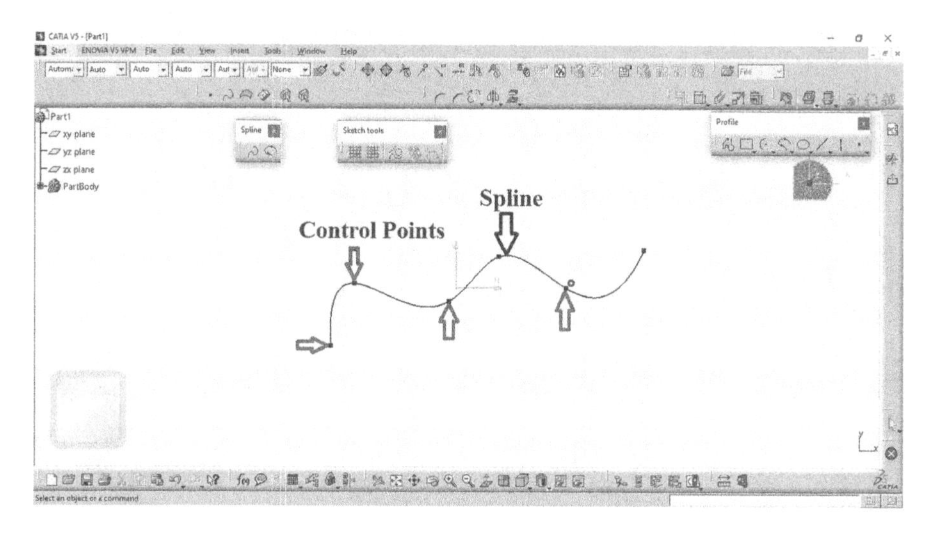

FIGURE 2.36 Constriction of spline.

2.3.4.2 Connect

This is used to connect two sketches like lines, arcs, ellipses, circle, points or profiles connected with a spline sketch. For connecting a sketch, select Connect tool from Spline toolbar as displayed in Figure 2.35. The expended Sketch tools are shown in Figure 2.37.

After selecting the tool, select constructed first element and second element, as shown in Figure 2.38. After selecting two sketched elements, connected two elements is as displayed (Figure 2.39).

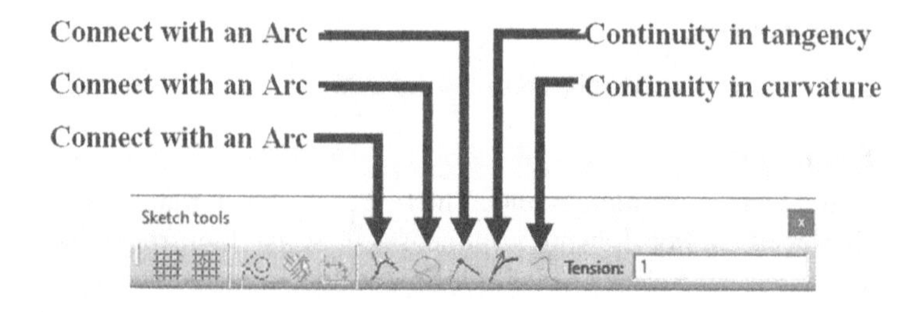

FIGURE 2.37 After selecting Connect tool Sketch tools displayed.

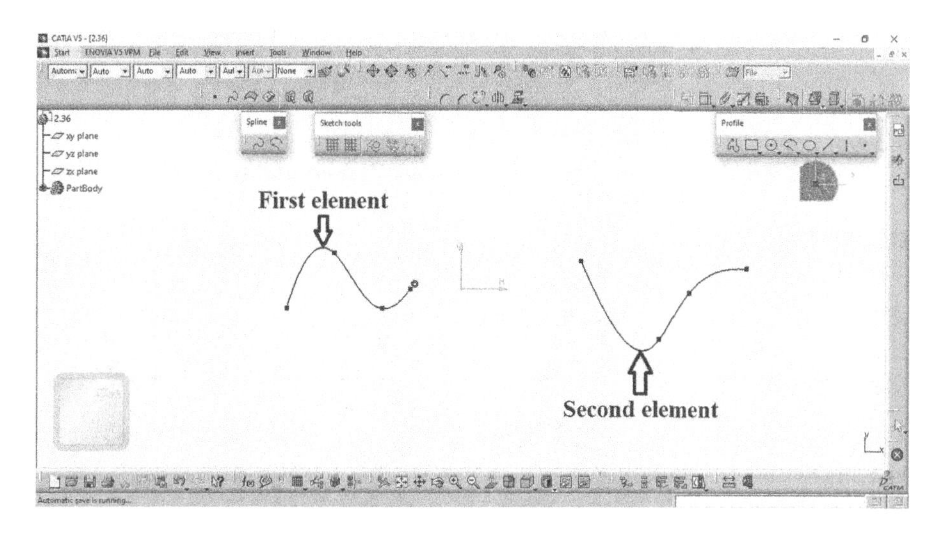

FIGURE 2.38 Selection of first element and second element.

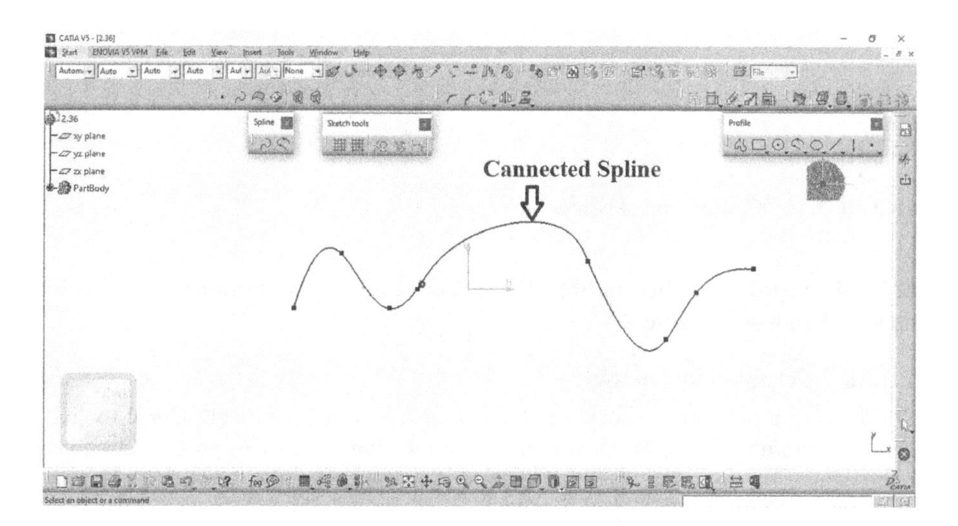

FIGURE 2.39 Constriction of Connected Spline.

2.3.5 Conic

The tools in this toolbar are displayed in Figure 2.40.

2.3.5.1 Ellipse

For drawing an ellipse, select Ellipse tool from Conic toolbar as shown in Figure 2.40. Firstly, select a location to indicate the center of the ellipse followed by a location to define the major axis radius and finally a location to define the minor axis radius, as shown in Figure 2.41. For editing geometric values of the ellipse as it is

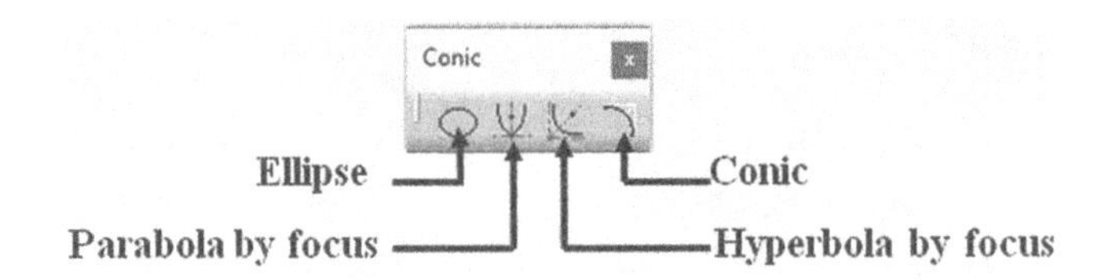

FIGURE 2.40 Conic toolbar.

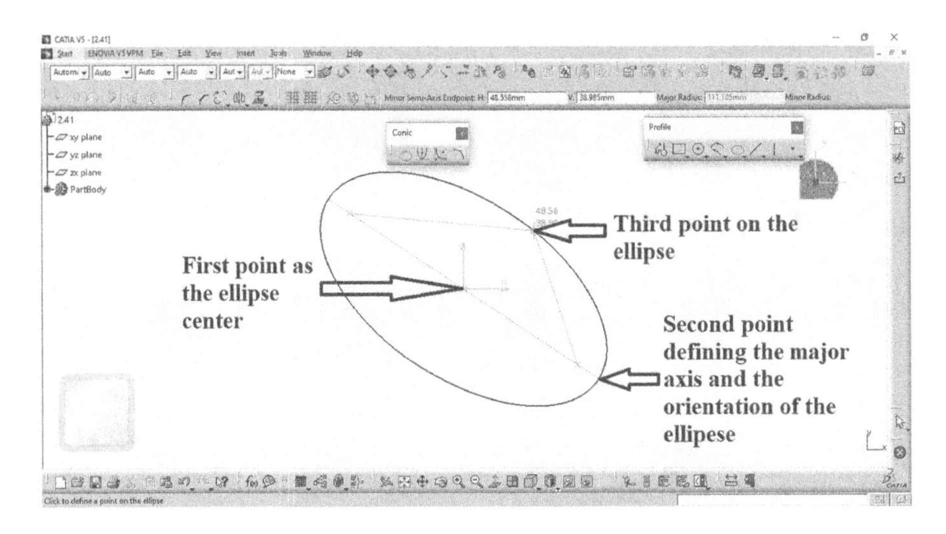

FIGURE 2.41 Constriction of ellipse.

being defined double click on the ellipse sketch; an Ellipse Definition panel will appear as shown in Figure 2.42.

2.3.5.2 Parabola by Focus

For drawing a parabola by focus, select Parabola by focus tool from Conic toolbar as shown in Figure 2.40. The first point selection defines the focus point, the second point defines the apex, and the third and fourth points define the start and end points of the Parabola as shown in Figure 2.43.

2.3.5.3 Hyperbola by Focus

For drawing a hyperbola by focus, select Hyperbola by focus tool from Conic toolbar as shown in Figure 2.40. First select the focus point, followed by the center intersect point. The third selection defines the apex, and the fourth and fifth define the start and end points of the Hyperbola as shown in Figure 2.44.

2.3.5.4 Conic

For drawing a conic, select Conic tool from Conic toolbar as shown in Figure 2.40. The first, second, third and fourth points indicate the start and end points of the conic and the fifth defines the shape of the conic as shown in Figure 2.45.

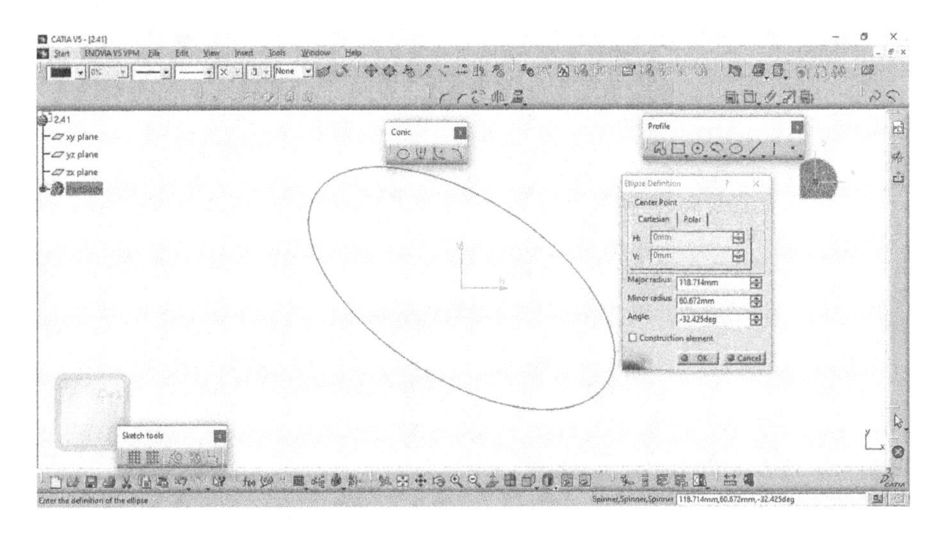

FIGURE 2.42 Edit of ellipse.

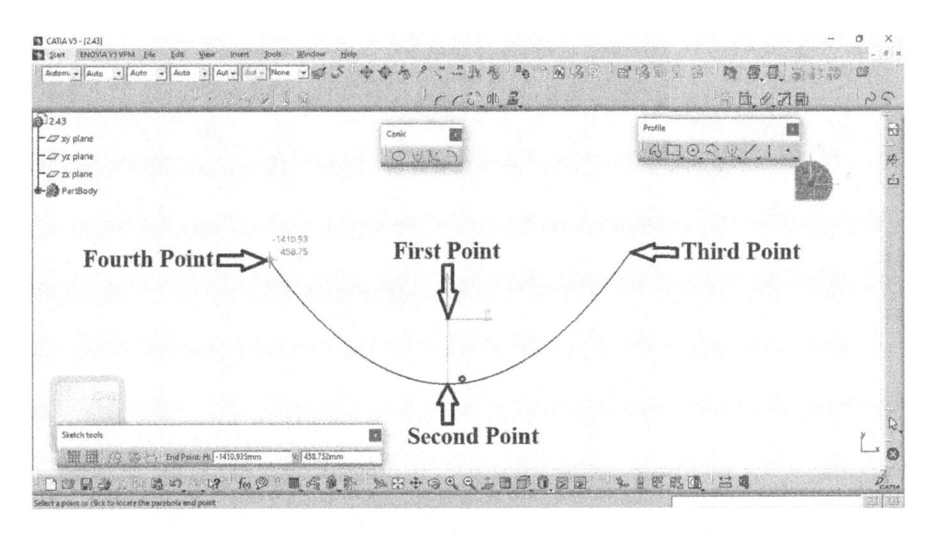

FIGURE 2.43 Constriction of parabola by focus.

2.3.6 Line

The tools in this toolbar are shown in Figure 2.46.

2.3.6.1 Line

The line is constructed by two points. For drawing a line, select Line tool from Line toolbar as shown in Figure 2.46. Create a line by defining its start and end points (default option). The first selection indicates the start point and the second defines the end point of line as shown in Figure 2.47.

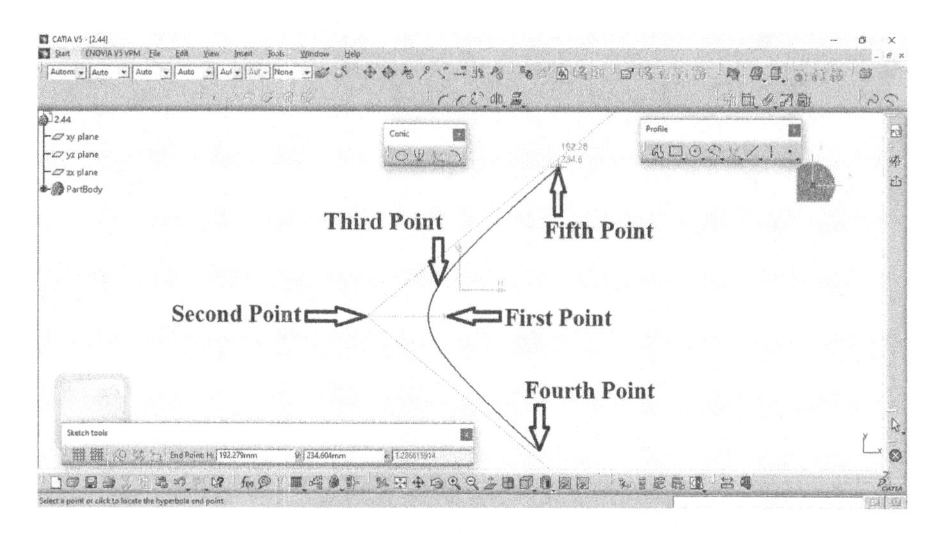

FIGURE 2.44 Constriction of hyperbola by focus.

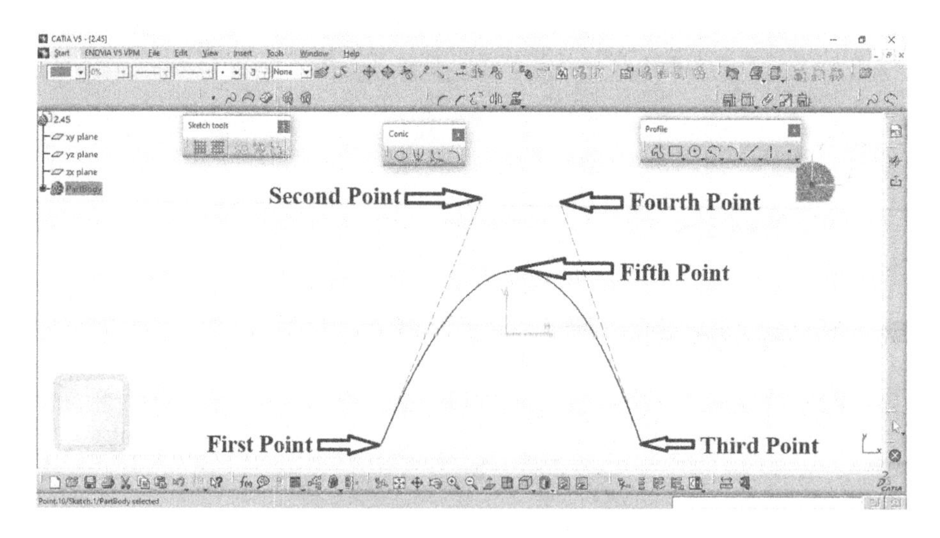

FIGURE 2.45 Constriction of conic.

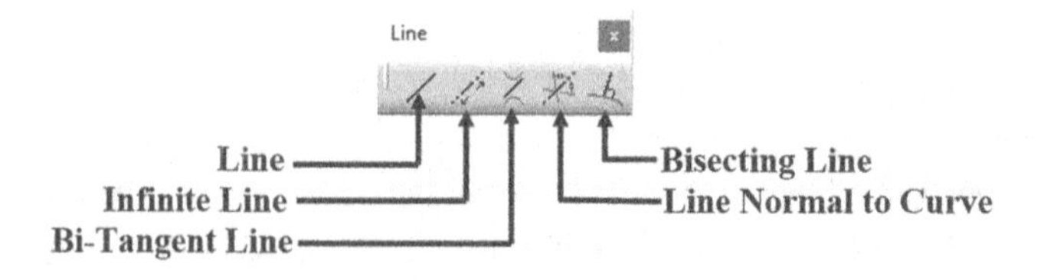

FIGURE 2.46 Line toolbar.

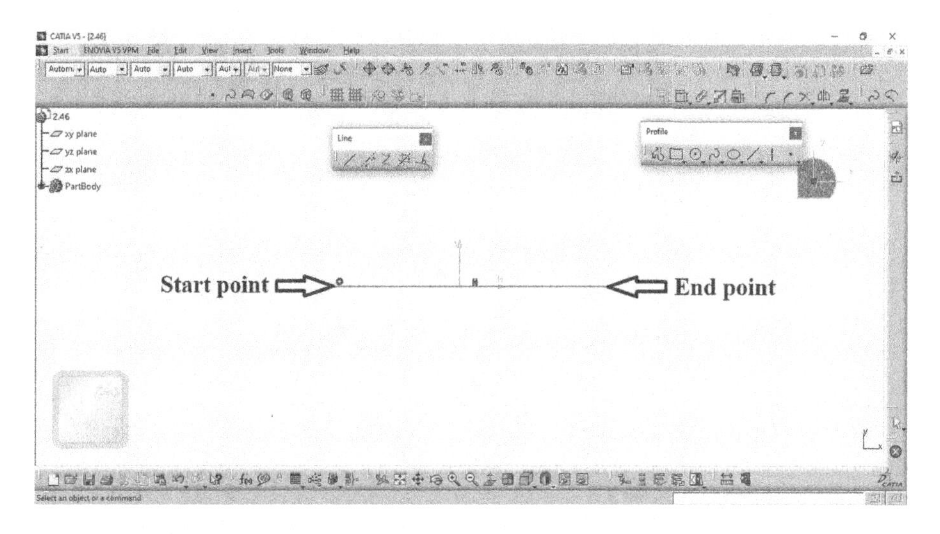

FIGURE 2.47 Constriction of line.

2.3.6.2 Infinite Line

For drawing infinite lines, select Infinite lines tool from Line toolbar as shown in Figure 2.46. To create horizontal and vertical lines, select the Insert Infinite Line icon followed by the Horizontal Line (default option) or Vertical Line icon on the Sketch tools and select a location for the line. To create an angled line after selecting the Insert Infinite Line icon, select Line Through Two Points icon on the Sketch tools toolbar and select two locations to generate the line as shown in Figure 2.48.

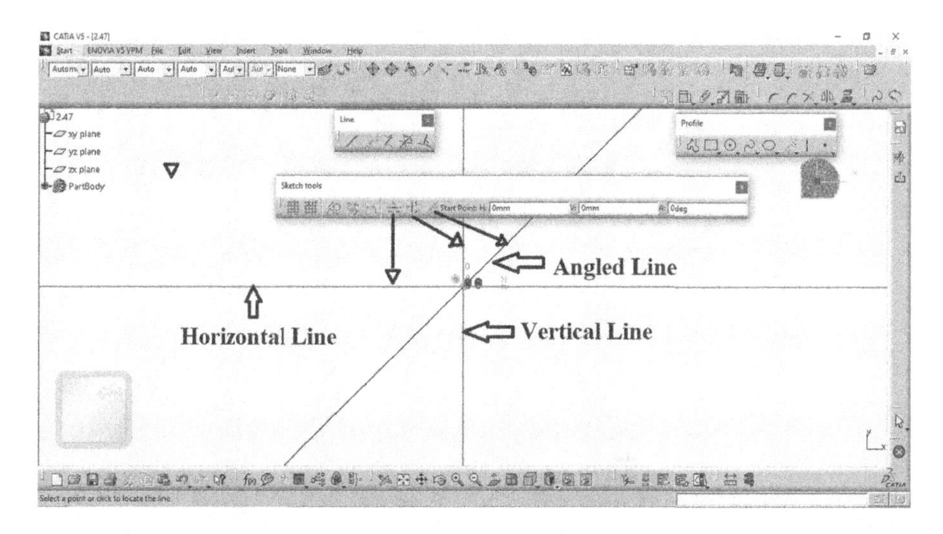

FIGURE 2.48 Constriction of infinite line.

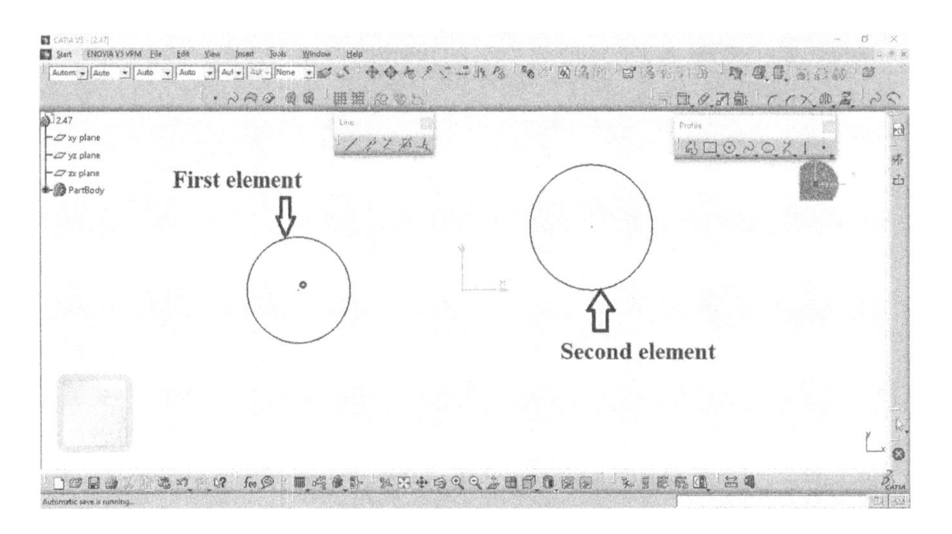

FIGURE 2.49 Constriction of elements for bi-tangent line.

2.3.6.3 Bi-Tangent Line

For drawing bi-tangent lines, first you need to construct two elements as shown in Figure 2.49, then select Bi-tangent tool from Line toolbar as shown in Figure 2.46. After selecting the icon, select the two elements as shown in Figure 2.49 that between which you want to create the line between. Bi-tangent line created between two elements is displayed in Figure 2.50.

2.3.6.4 Bisecting Line

For drawing a bisecting line, first you need to construct two line elements as shown in Figure 2.51, and then select Bisecting line tool from Line toolbar as shown in

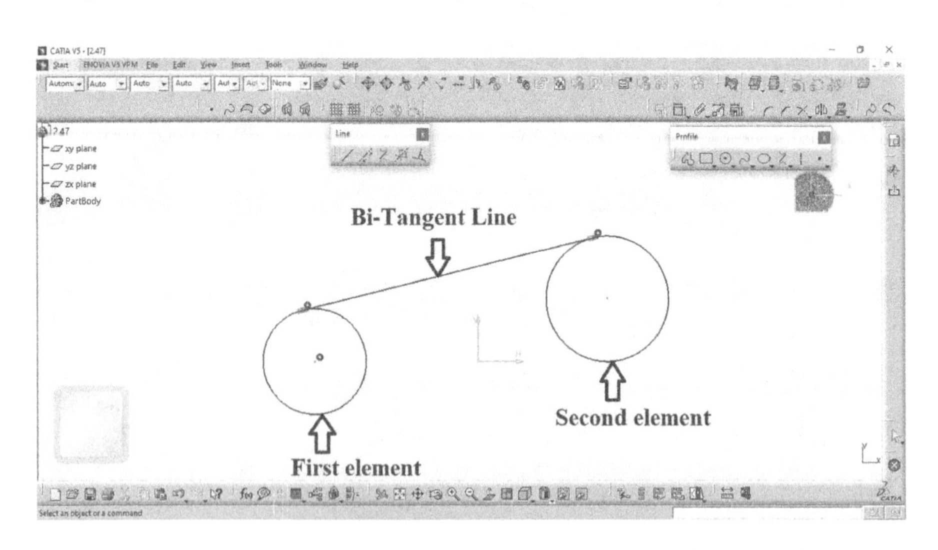

FIGURE 2.50 Constriction of bi-tangent line.

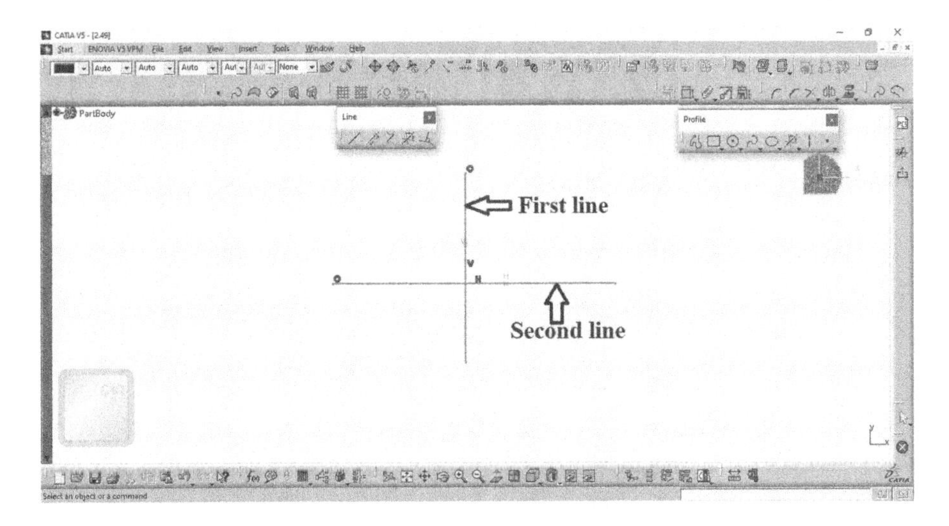

FIGURE 2.51 Constriction of elements for bisecting line.

Figure 2.46. Select the first line and second line as shown in Figure 2.51. Bisecting line created is shown in Figure 2.52.

2.3.6.5 Line Normal to Curve

For drawing a line normal to curve, first you need to construct curve elements as shown in Figure 2.53, and then select Line normal to curve tool from Line toolbar as shown in Figure 2.46. Select the first line point and second curve element as shown in Figure 2.54. The constructed line normal to curve drawn is shown in Figure 2.54.

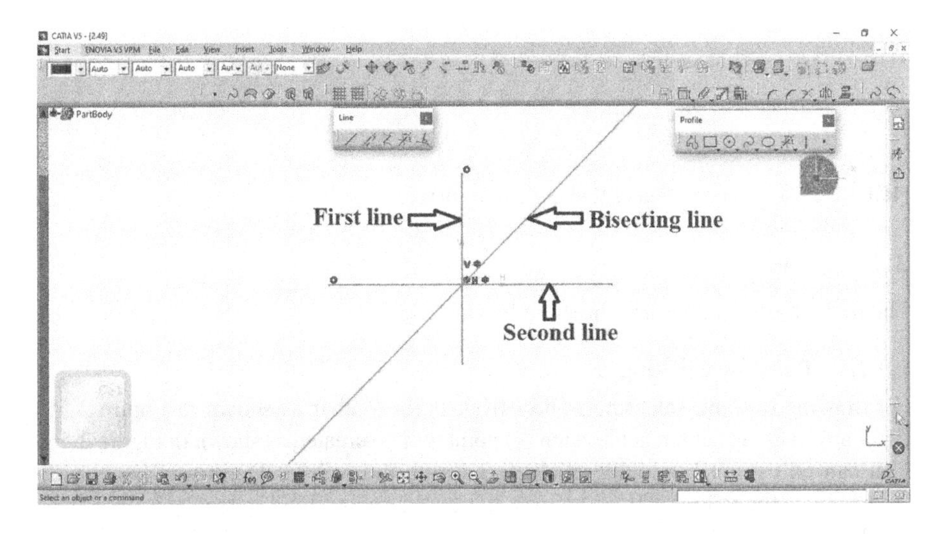

FIGURE 2.52 Constriction of bisecting line.

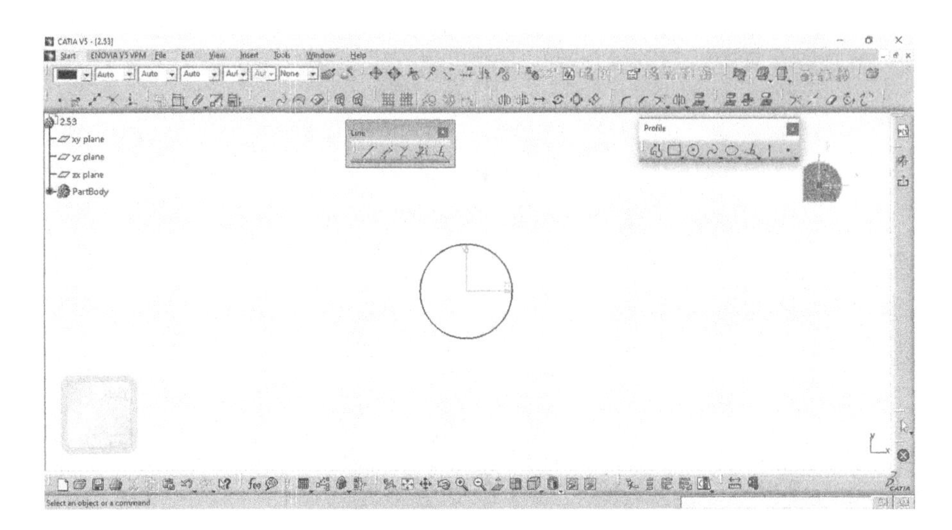

FIGURE 2.53 Constriction of curve for drawing line normal to curve.

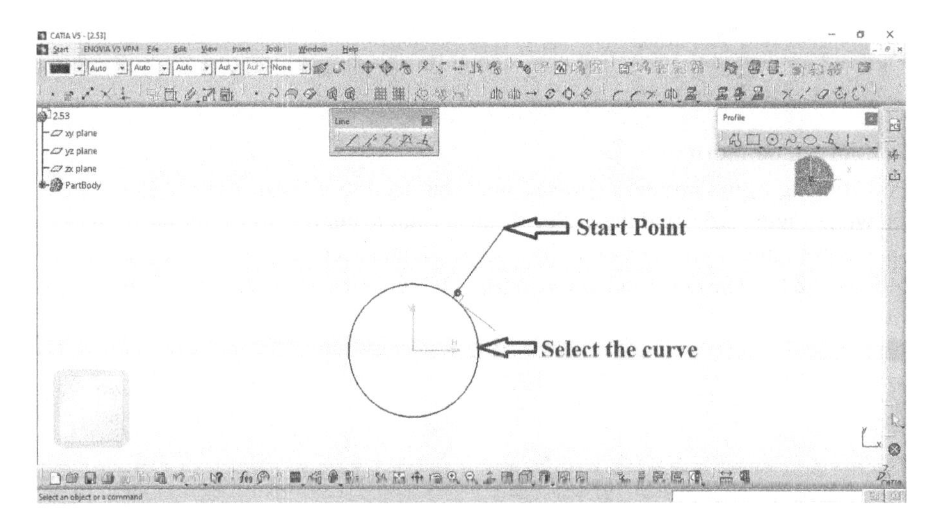

FIGURE 2.54 Constriction of line normal to curve.

2.3.7 Point

The tools in this toolbar are shown in Figure 2.55.

2.3.7.1 Point by Clicking

For drawing a point, select point tool from Point toolbar as shown in Figure 2.55. Click anywhere in screen, a plus sign (+) point will be created as shown in Figure 2.56. You can edit point by double clicking on pint and a Point Definition panel will appear inter the value of H and V in edit panel as shown in Figure 2.57 and select OK button.

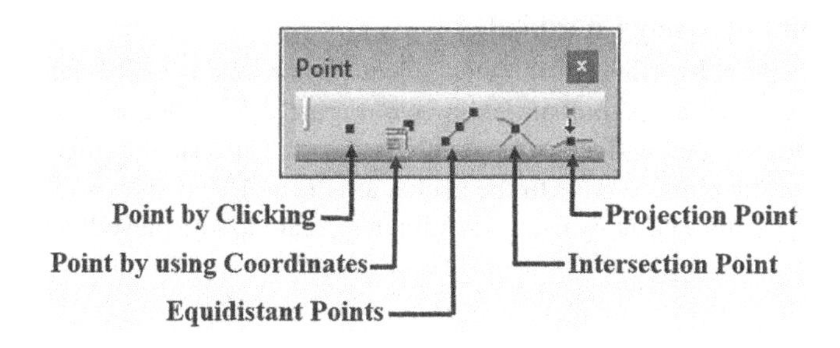

FIGURE 2.55 Point toolbar.

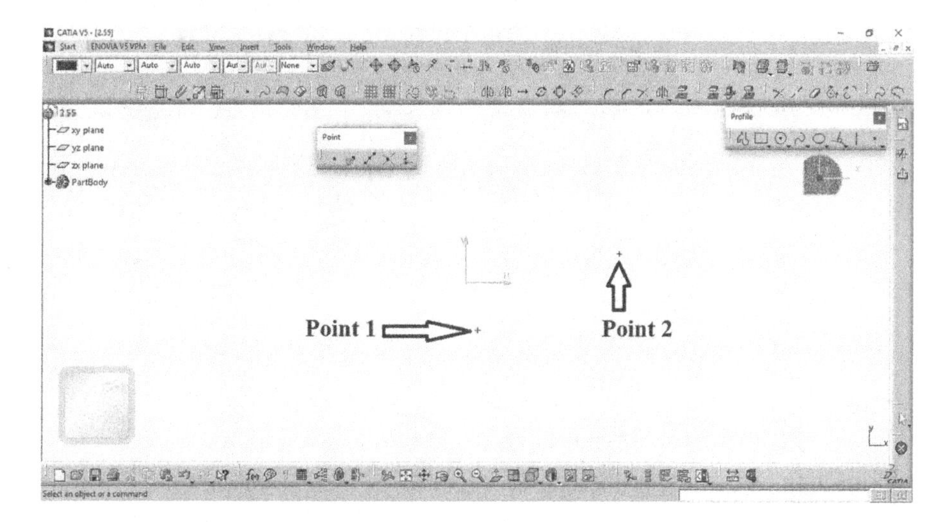

FIGURE 2.56 Constriction of point.

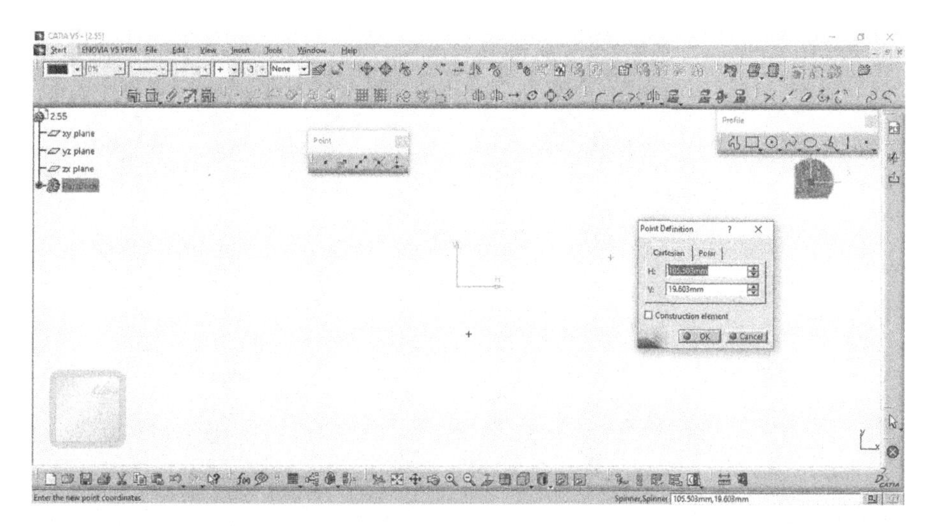

FIGURE 2.57 Constriction of edit point.

2.3.7.2 Point by Using Coordinates

For drawing a point by using coordinates, select Point by using coordinates tool from Point by using coordinates toolbar as shown in Figure 2.55. After selecting this icon, a Point Definition panel will appear and either uses the Cartesian or Polar tab to enter the position of the point relative to the sketch axis H and V as shown in Figure 2.58 and select OK button. Point by using coordinates created as displayed in Figure 2.59.

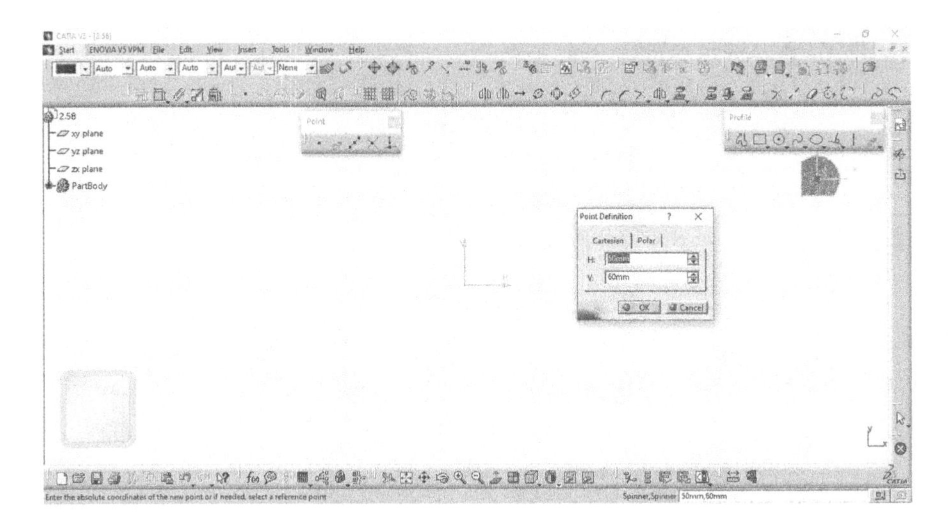

FIGURE 2.58 Value of point for construction of Point by using coordinates.

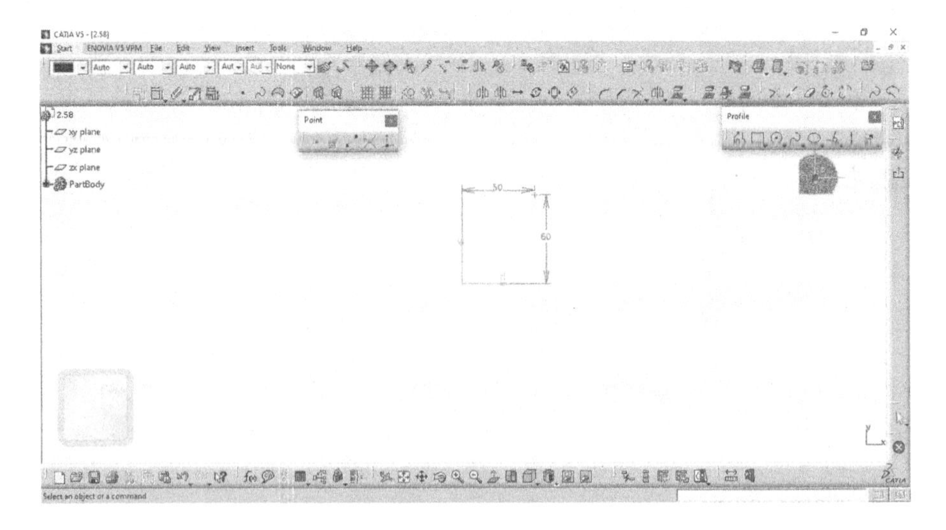

FIGURE 2.59 Constriction of Point by using coordinates.

2.3.7.3 Equidistant Points

For drawing equidistant points, select Equidistant points tool from Point by using Point toolbar as shown in Figure 2.55. After selecting icon followed by the line curve on which you require the points (Figure 2.60), an Equidistant Point Definition panel will appear. Enter the number of point required in the New Points field as shown in Figure 2.61 and select OK button. Equidistant points on line created are displayed in Figure 2.62.

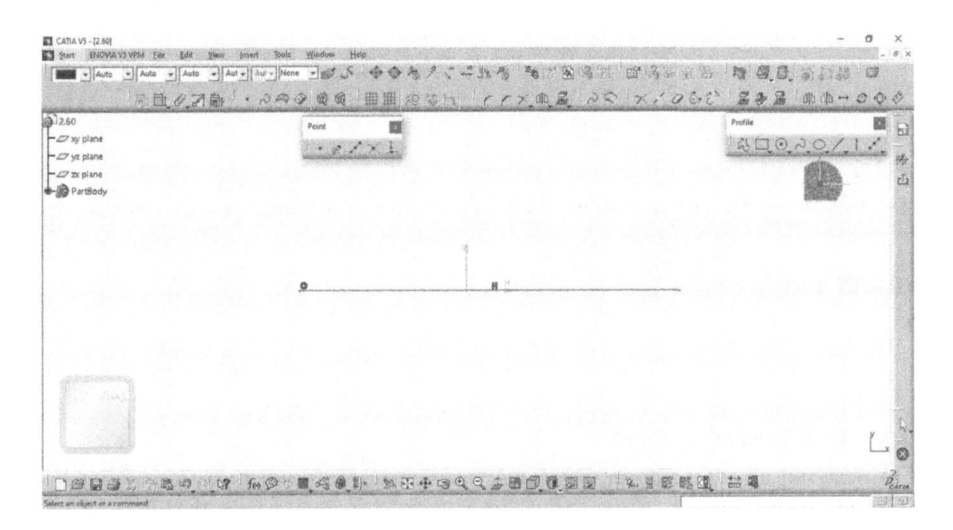

FIGURE 2.60 Constriction of line for equidistant points.

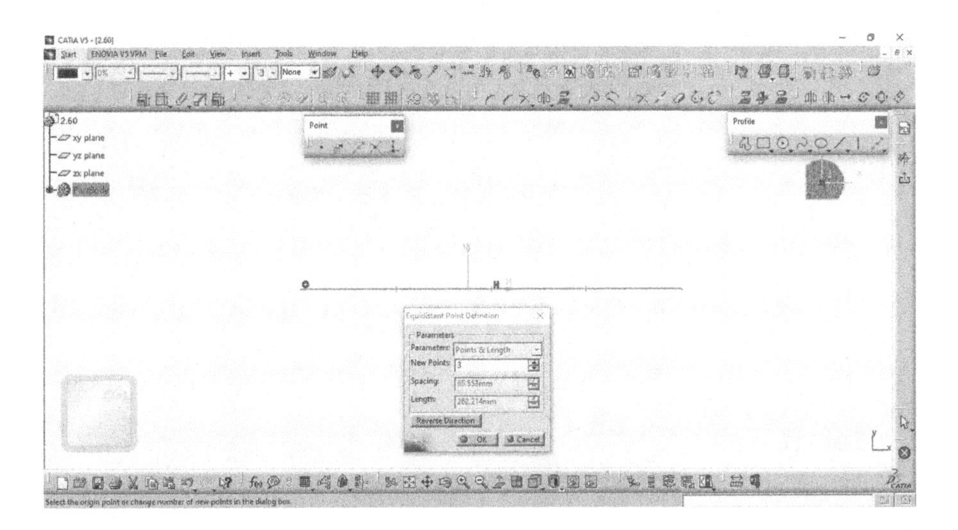

FIGURE 2.61 Entering number of point for equidistant points.

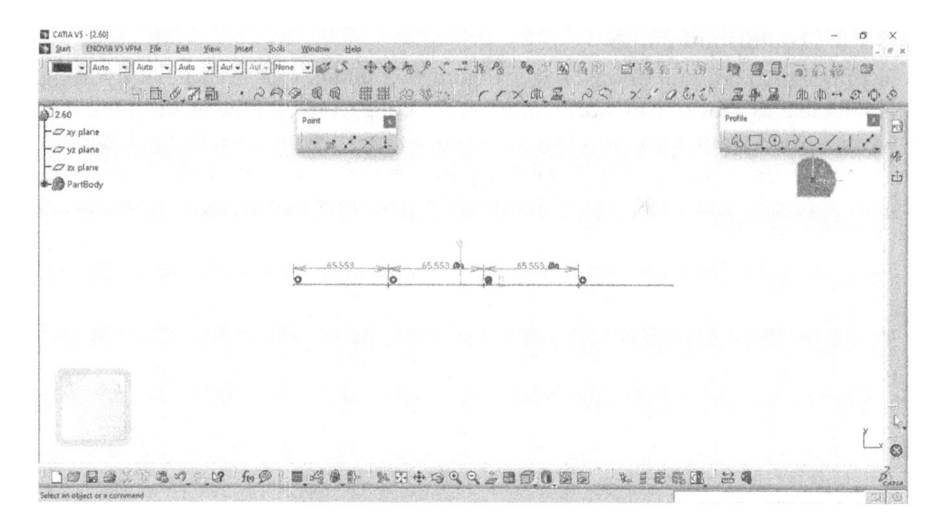

FIGURE 2.62 Constriction of equidistant points.

2.3.7.4 Intersection Point

For drawing an intersection point, select Intersection point tool from Point by using Point toolbar as shown in Figure 2.55. After selecting the icon select (Figure 2.63), the two lines elements to intersect. Intersection on line created is displayed in Figure 2.64.

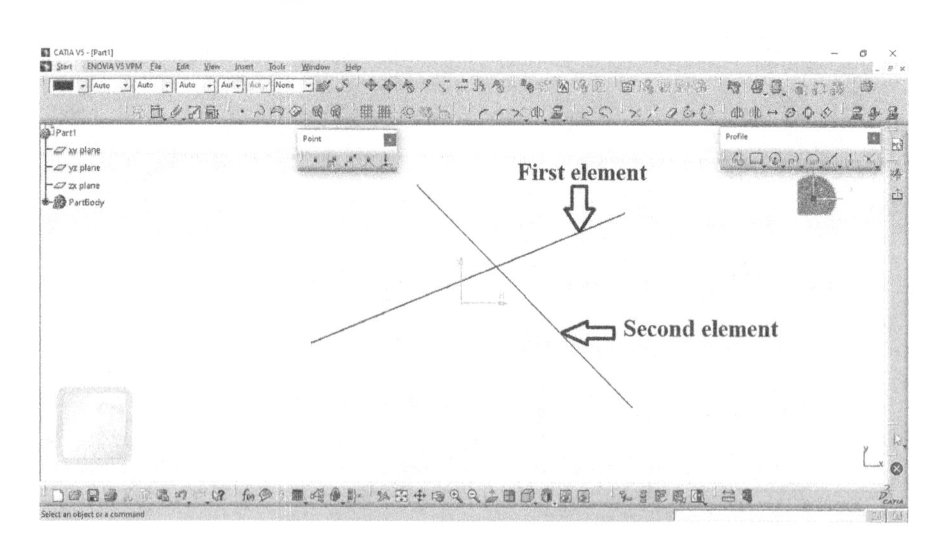

FIGURE 2.63 Constriction of line for intersection point.

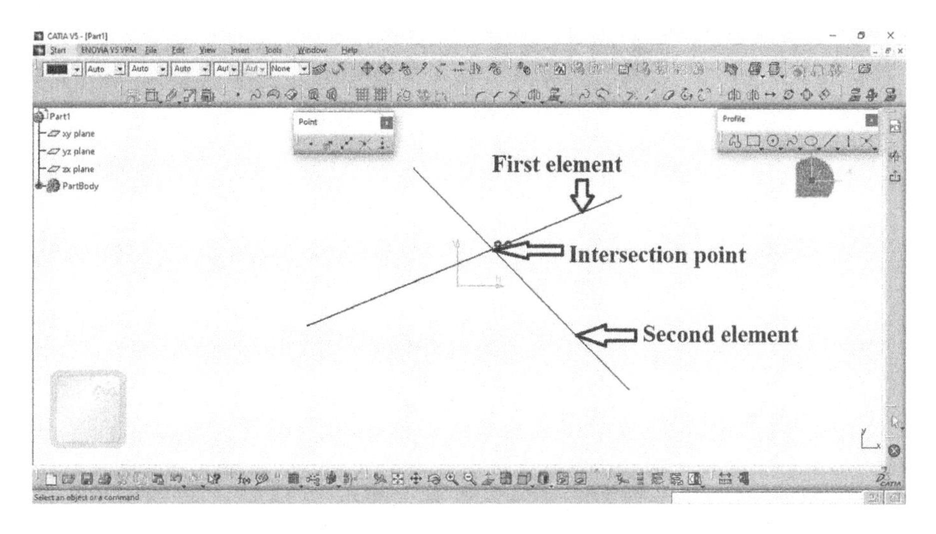

FIGURE 2.64 Constriction of intersection point.

2.3.7.5 Projection Point

For drawing a projection point, select Projection point tool from Point by using Point toolbar as shown in Figure 2.55. After selecting the icon (Figure 2.65), and then proceed to select the point that you wish to project. Finally, select the line on which the points are to be projected. Projection point on line created is displayed in Figure 2.66.

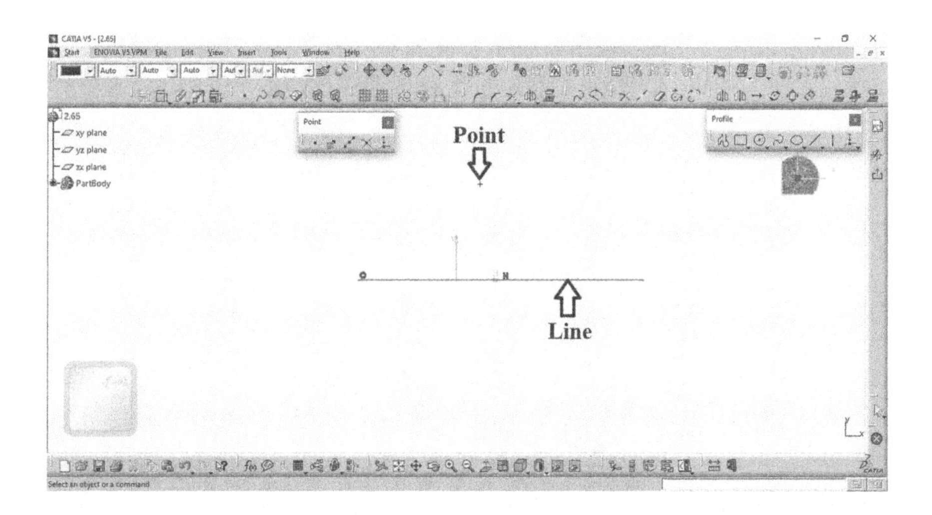

FIGURE 2.65 Constriction of point and line for projection point.

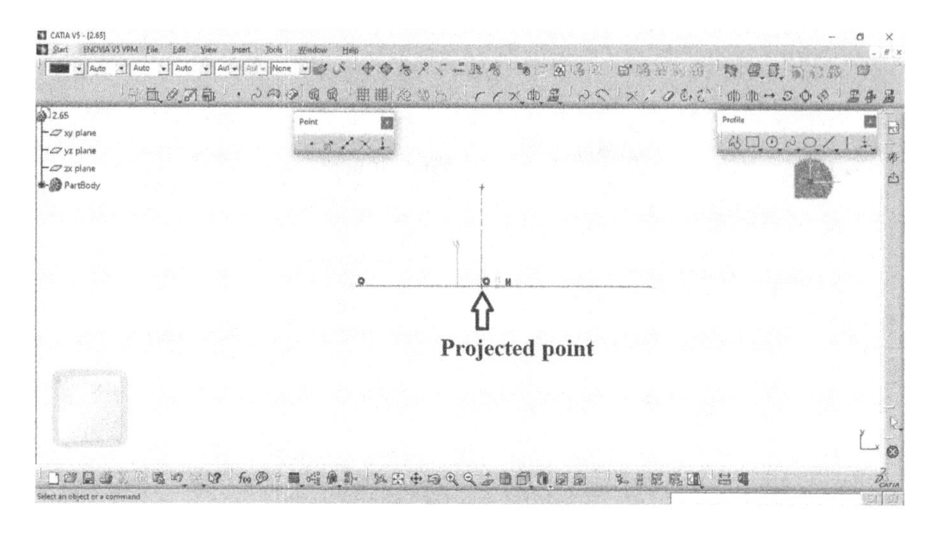

FIGURE 2.66 Constriction of projection point.

3 Operation Toolbar

3.1 INTRODUCTION

In this chapter, users will be exposed to the operation of modifying created products using operation menus like corner, chamfer, mirror, symmetry, translate, rotate, scale and offset rectangle. In this chapter, the above operations would be discussed using screenshots providing the user a hands-on training on various types of operation tools.

3.2 OPERATION TOOLBAR

In order to modify a created sketch, commands such as trim, mirror and chamfer are used. These commands are available in Operation toolbar. The Operation toolbar (Figure 3.1) is utilized to alter the geometry in Sketcher workbench. The tools and toolbar in this toolbar are Corner, Chamfer, Relimitations, Transformation and 3D Geometry.

3.2.1 CORNER

In Sketcher workbench, you are provided with Corner tool under operation toolbar as shown in Figure 3.1 to corner the sketched elements. To draw it, draw two lines that meet or will cross each other as shown in Figure 3.2 after that invoke Corner tool (Figure 3.3).

As appearing in Figure 3.4, Sketch tools toolbar provides various options that are used to create a corner with different type of cornering options.

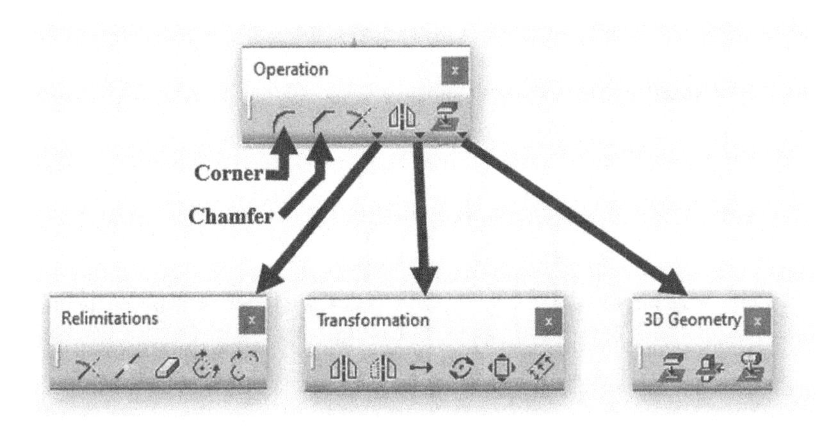

FIGURE 3.1 Operation toolbar.

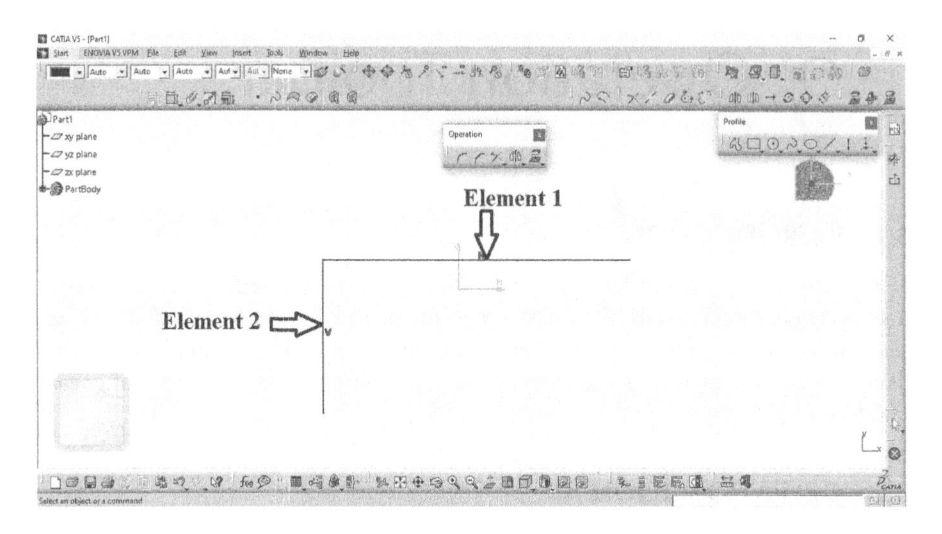

FIGURE 3.2 Constriction of line for corner.

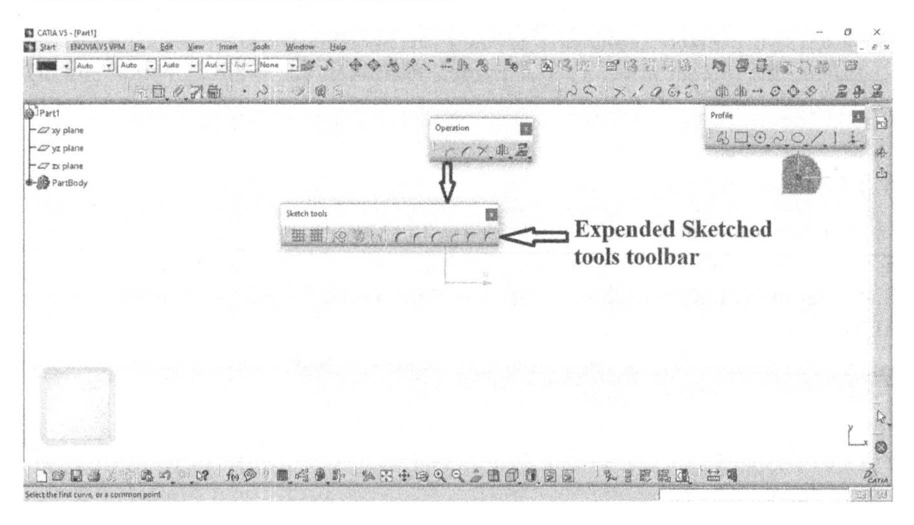

FIGURE 3.3 Expanded Sketched tools toolbar.

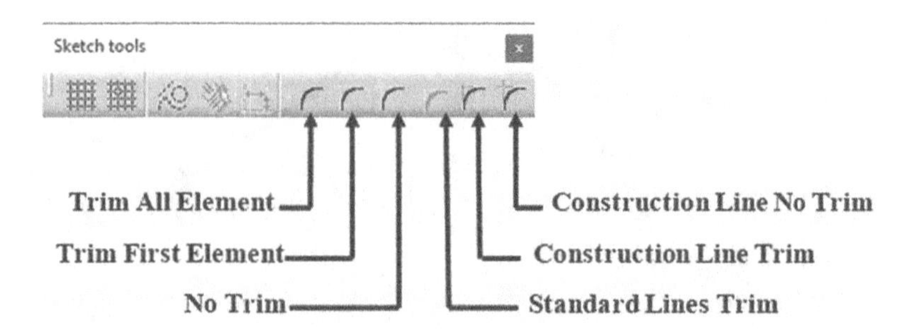

FIGURE 3.4 Options of Expanded Sketched tools toolbar.

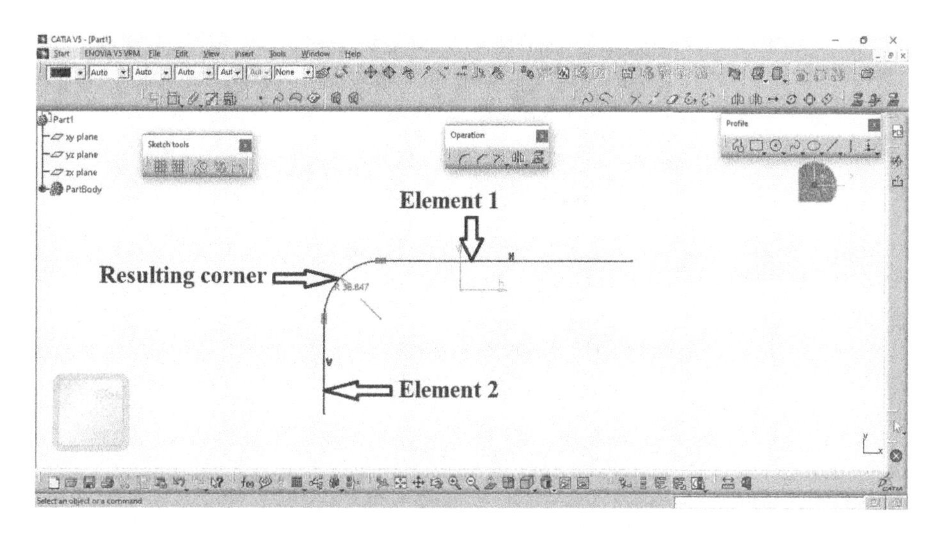

FIGURE 3.5 Constriction of corner by Trim All Element tool.

After selecting Trim All Element button and selecting both entities, both the selected elements are trimmed beyond the filet region, resulting in a corner as shown in Figure 3.5.

In case only one element is required to be trimmed then select Trim First Element button and select the both elements; the resulting corner will be created by trimming only the first element. The second element remains as it is (Figure 3.6).

If you choose No Trim button and select the both elements, the resulting corner will be created by retaining both the selected elements as shown in Figure 3.7.

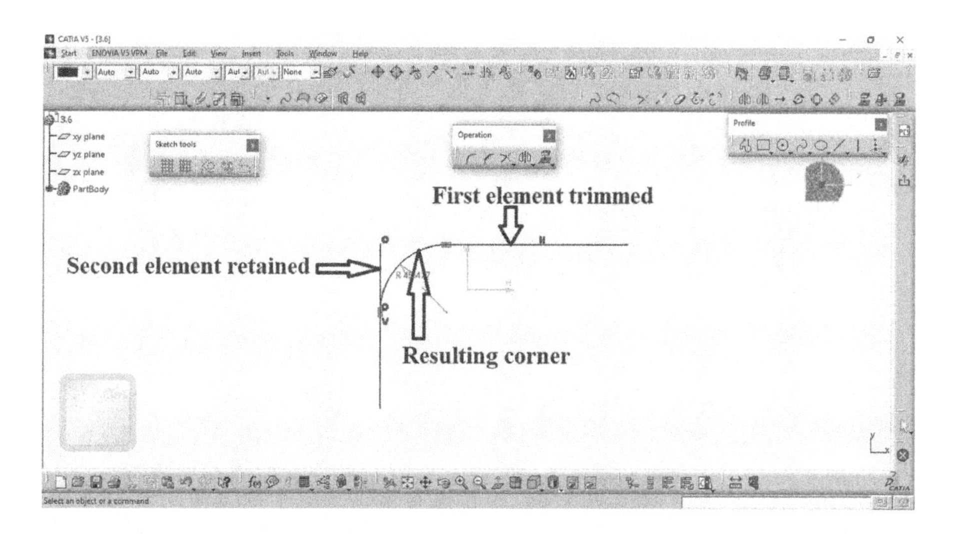

FIGURE 3.6 Constriction of corner by Trim First Element tool.

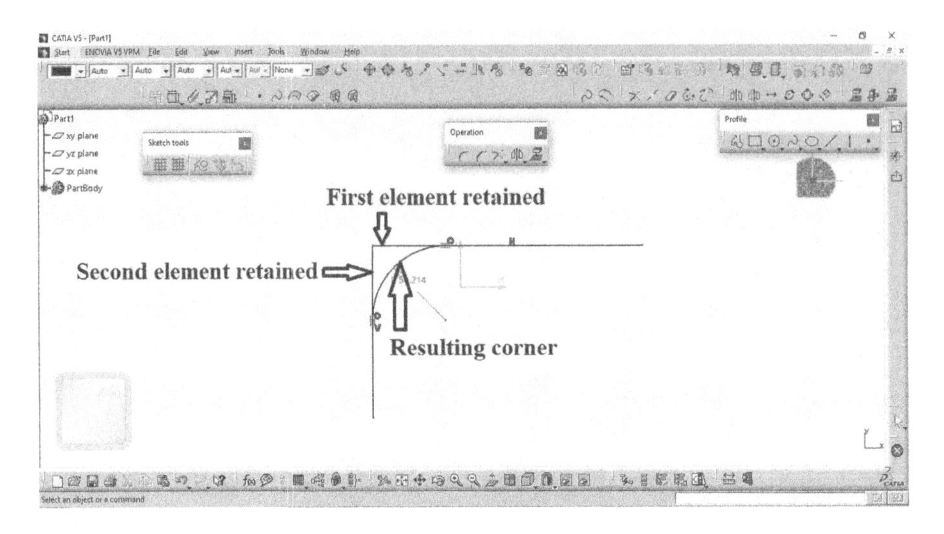

FIGURE 3.7 Constriction of corner by No Trim tool.

If you choose Standard Line Trim button and select the both elements, the result would be,

1. Both the selected lines would be retained.
2. Retained elements will remain as standard elements.
3. And if part of the line is extended beyond the corner (Figure 3.8), the same would be erased (Figure 3.9).

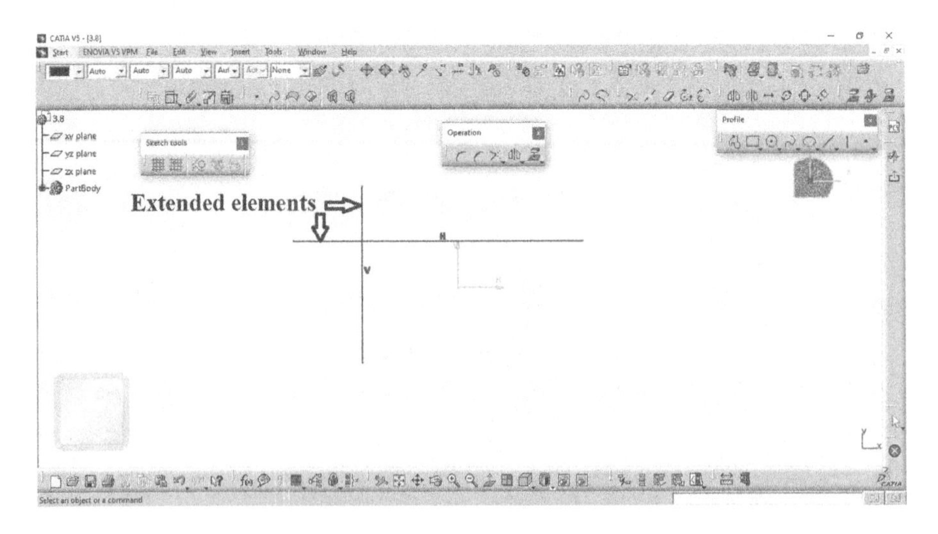

FIGURE 3.8 Constriction of extended element for Standard Line Trim.

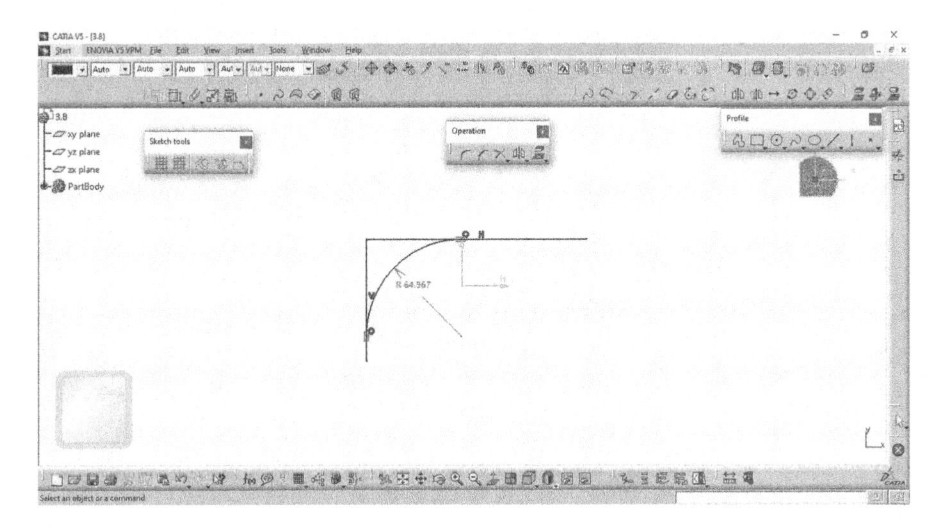

FIGURE 3.9 Constriction of corner by Standard Line Trim.

If you choose Construction Line Trim button and select the both elements, the result would be same as above but the retained part with be as a dotted line/constructional line (Figure 3.10).

If you choose Construction Line No Trim button and select the both elements, the result would be as depicted in Figure 3.11.

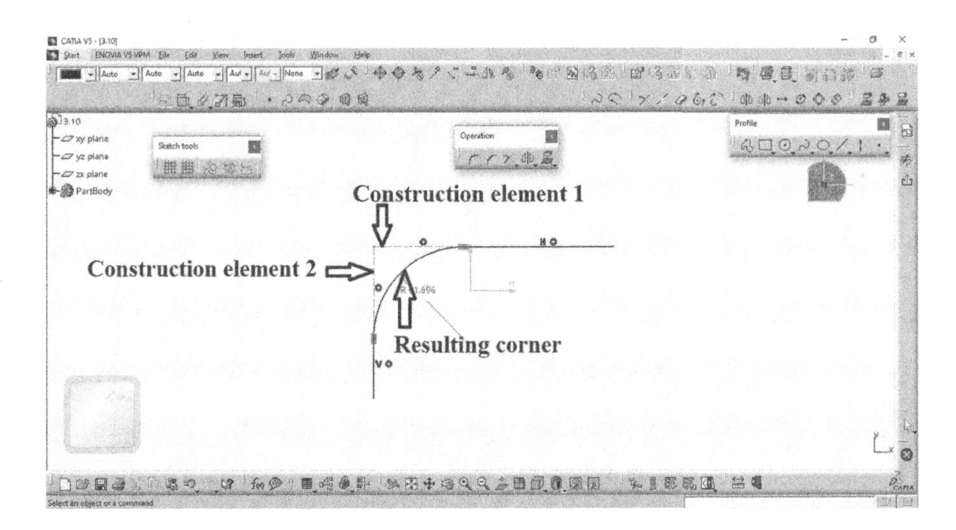

FIGURE 3.10 Constriction of corner by Construction Line Trim.

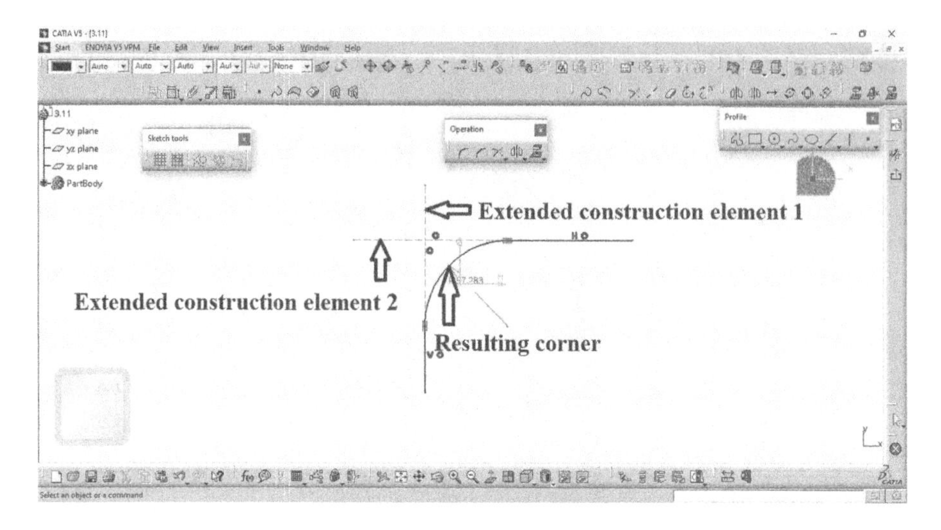

FIGURE 3.11 Constriction of corner by Construction Line No Trim.

3.2.2 Chamfer

In Sketcher workbench, you are provided with Chamfer tool under operation toolbar as shown in Figure 3.1 to Chamfer the sketched elements. To draw it, draw two lines that meet or will cross each other as shown in Figure 3.12 after that invoke Chamfer tool; the Sketched tools toolbar comes up as in Figure 3.13.

As the first and second elements are selected for creation of chamfer, the relevant toolbar provides various different options for creating the same. The various options available are displayed in Figure 3.14.

The descriptive detail of each of the menus has been provided in Figure 3.15.

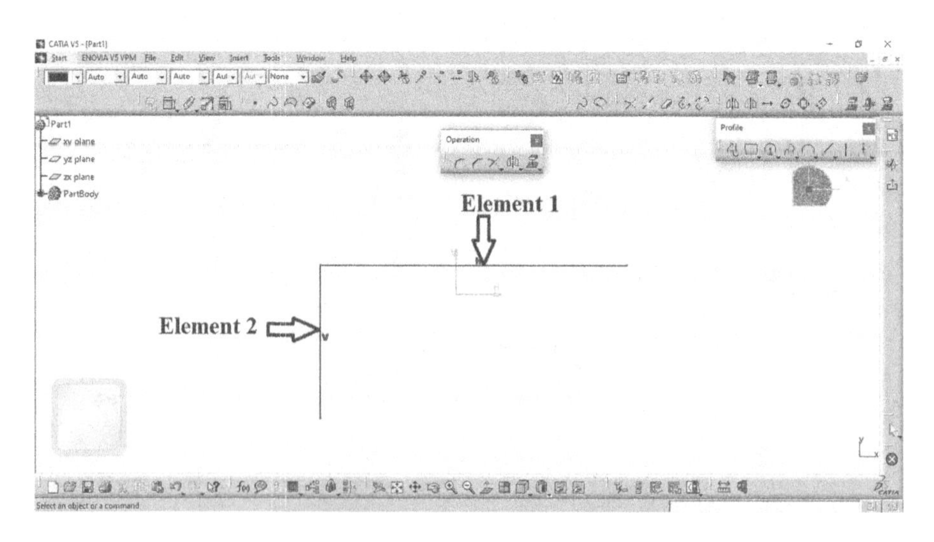

FIGURE 3.12 Constriction of line for chamfer.

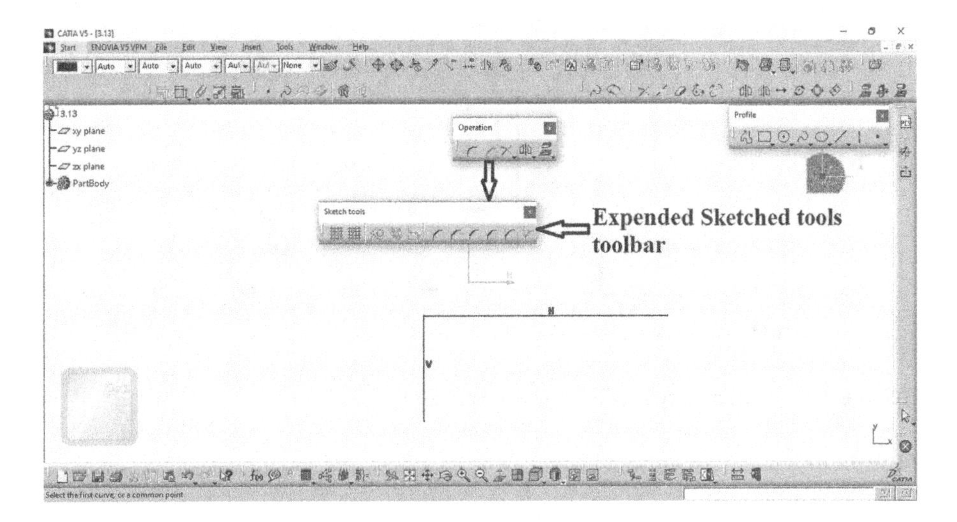

FIGURE 3.13 Expanded Sketched tools toolbar.

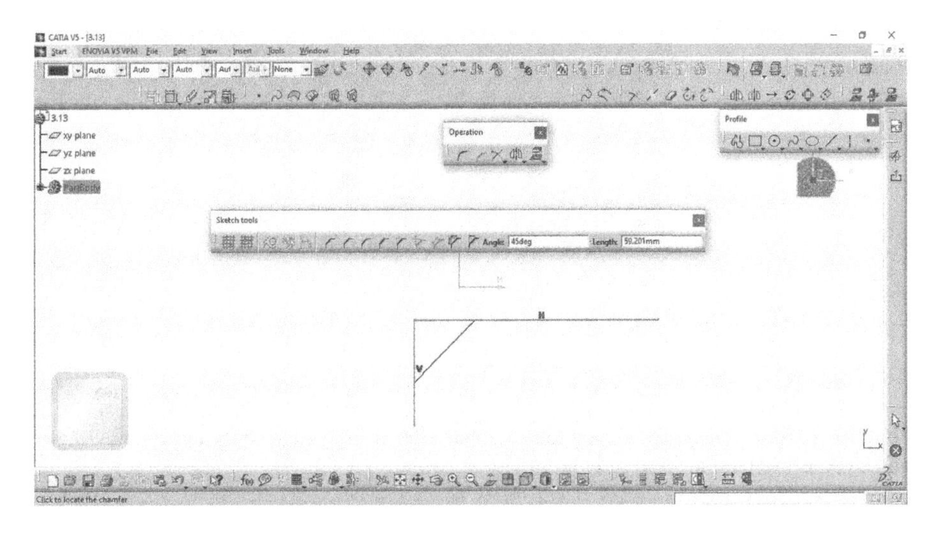

FIGURE 3.14 Expanded Sketched tools toolbar after selecting elements.

In case Angle and Hypotenuse button is selected then the angle and length of the hypotenuse have to be provided in the accompanying box (Figure 3.16), after entering the relevant values, press Enter. The chamfer would be created (Figure 3.17).

If lengths of the two lines are chosen then unlike the earlier case, both the lengths are required to be provided in the relevant boxes (Figure 3.18). After confirmation of the entered data, using Enter button, the resulting chamfer appears (Figure 3.19).

Similarly if Angle and First length button is activated, the relevant values are to be inserted in the boxes (Figure 3.20), and subsequently after confirmation, the same appears in Figure 3.21.

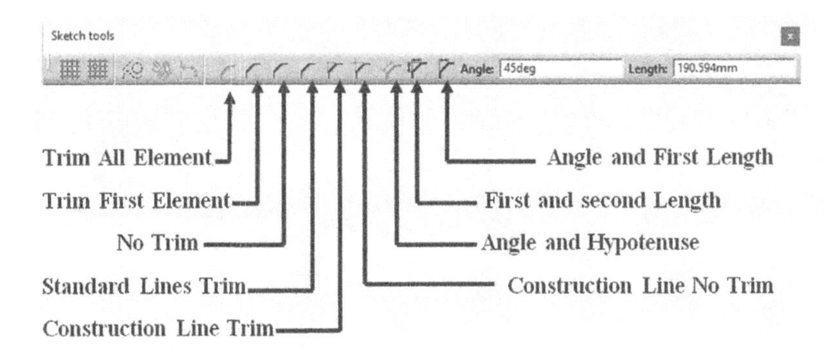

FIGURE 3.15 Different type of chamfering options.

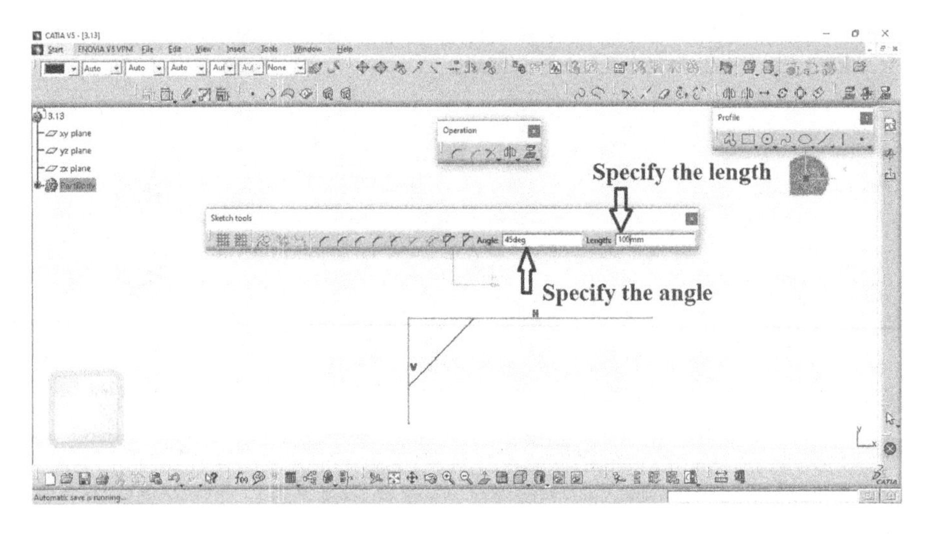

FIGURE 3.16 Expanded Sketched tools toolbar after selecting elements and choose Angle and Hypotenuse button.

In the same manner, trimming or keeping the elements with the aid of other button in the Sketcher tools toolbar can also be done in the same manner as described in the Corner tool.

3.3 RELIMITATIONS

In Relimitations toolbar, editing and/or modification tools of Sketcher workbench are provided. The Relimitations toolbar as shown in Figure 3.22 is used to modify geometry in Sketcher workbench. The tools in this toolbar are Trim, Break, Quick Trim, Close and Complement.

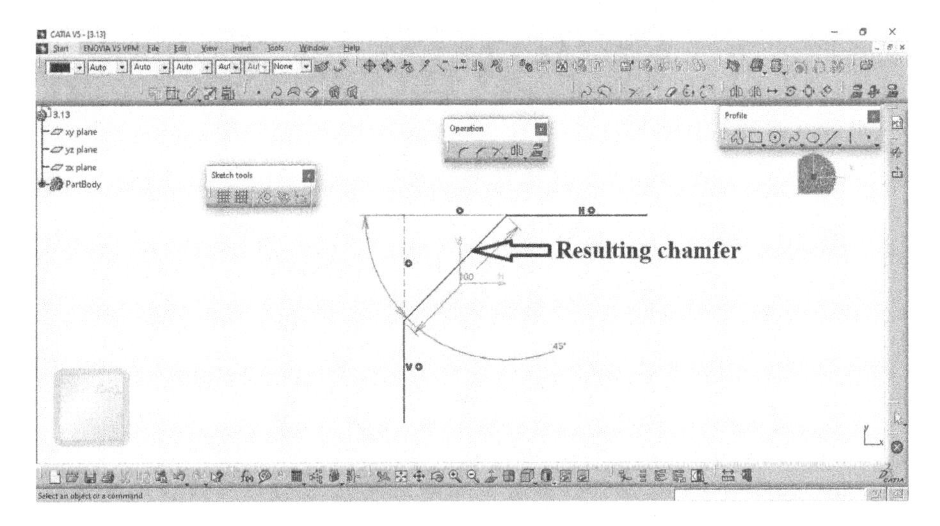

FIGURE 3.17 Constriction of chamfer by choosing Angle and Hypotenuse button.

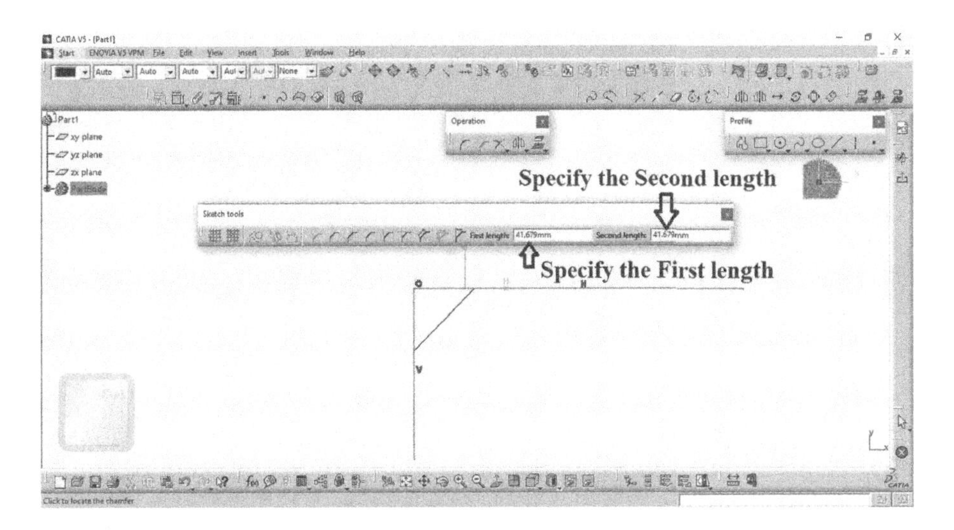

FIGURE 3.18 Expanded Sketched tools toolbar after selecting elements and choose First and Second length button.

3.3.1 Trim

In Sketcher workbench, Trim tool under Relimitations toolbar is provided (Figure 3.22) to trim the sketched elements. To perform it, draw two lines that will cross each other as shown in Figure 3.23 and after that invoke Trim tool; the Sketched tools toolbar expands (Figure 3.24).

Similar to the trimming operation discussed earlier, here also the menu provides various options for the same (Figure 3.25).

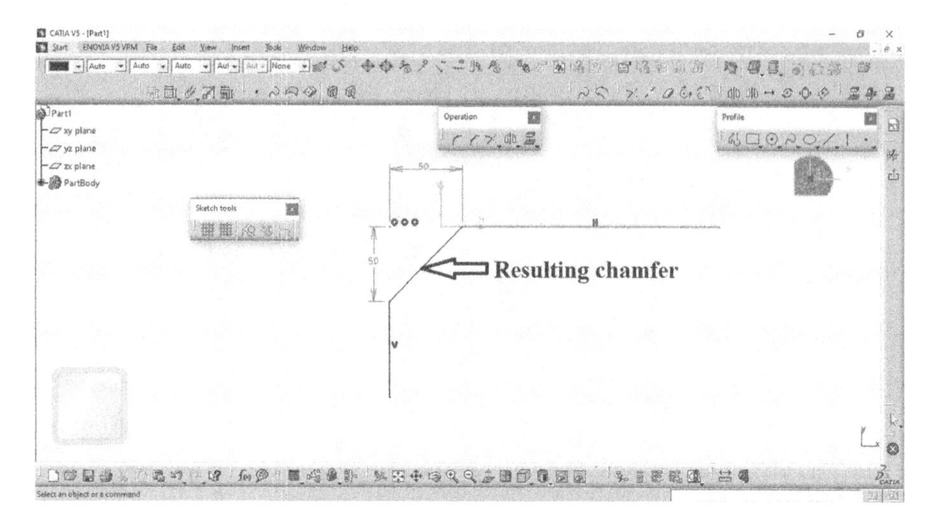

FIGURE 3.19 Constriction of chamfer by choosing First and Second length button.

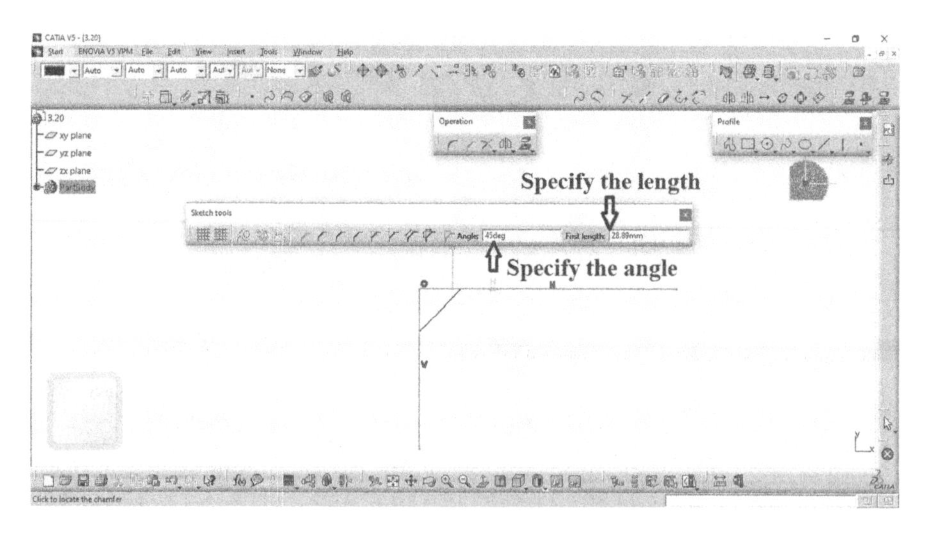

FIGURE 3.20 Expanded Sketched tools toolbar after selecting elements and choose Angle and First length button.

After choosing the Trim icon, ensure the Trim All Elements icon is selected on the Sketch tools toolbar. Then select two elements to trim together by selecting the first element on the portion of it that you wish to keep and then select the second element on the portion you wish to keep. Resulting trim is as shown in Figure 3.26.

After selecting the Trim icon ensures that the Trim First Element icon is selected on the Sketch tools toolbar. Now select the first element on the portion you wish to keep followed by the second element to trim. Resulting trim is as shown in Figure 3.27.

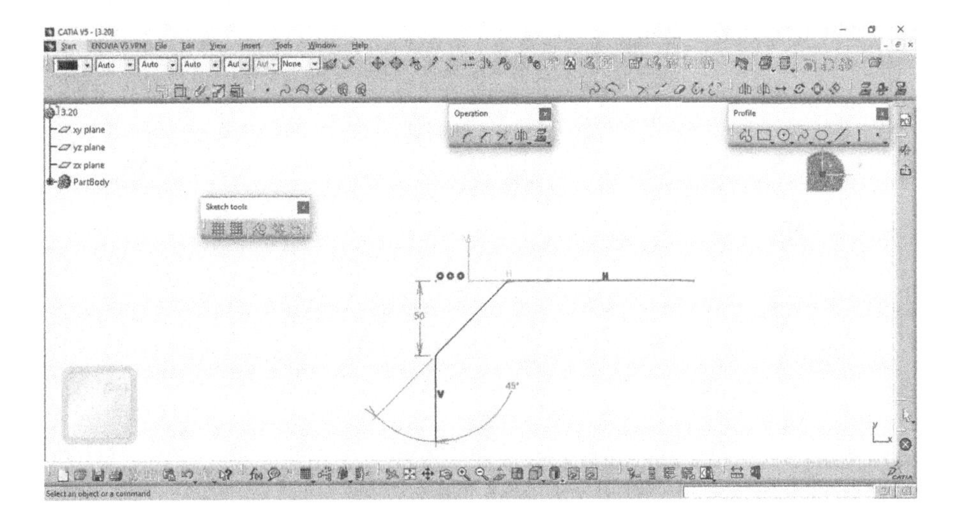

FIGURE 3.21 Constriction of chamfer by choosing Angle and First length button.

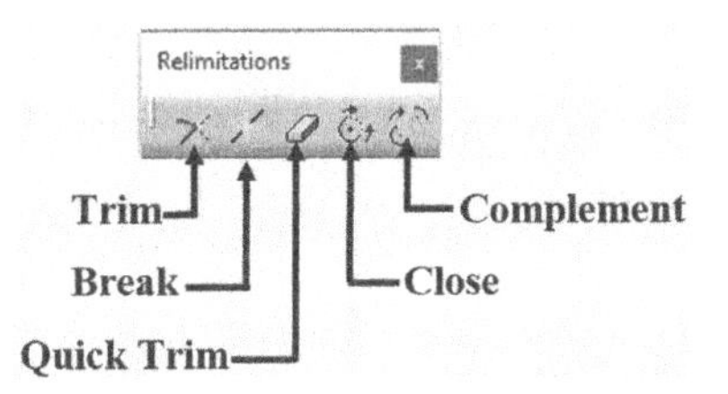

FIGURE 3.22 Relimitations toolbar.

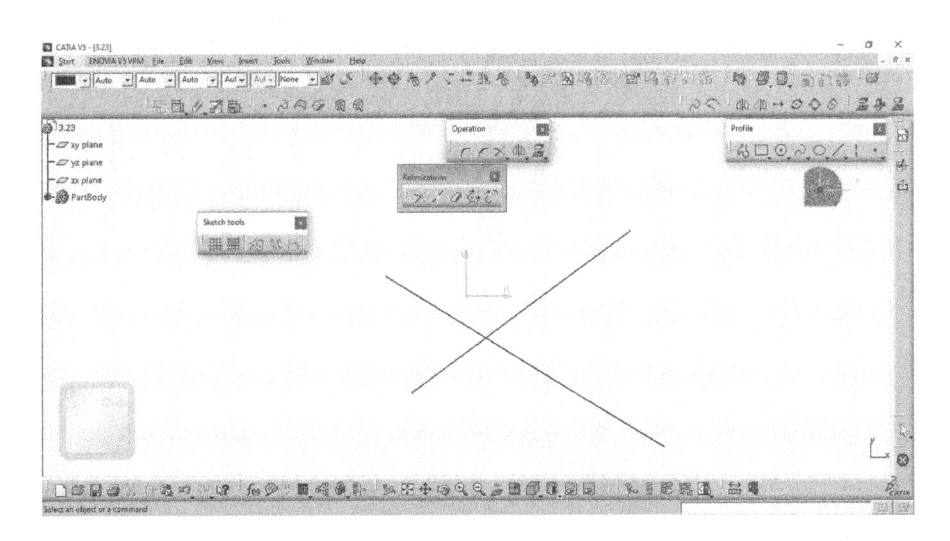

FIGURE 3.23 Constriction of line for Trim.

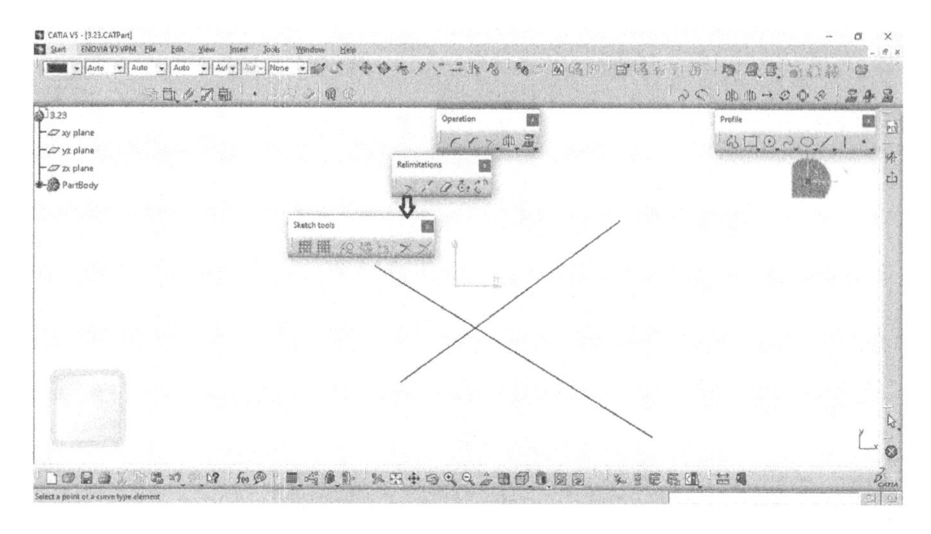

FIGURE 3.24 Expanded Sketched tools toolbar.

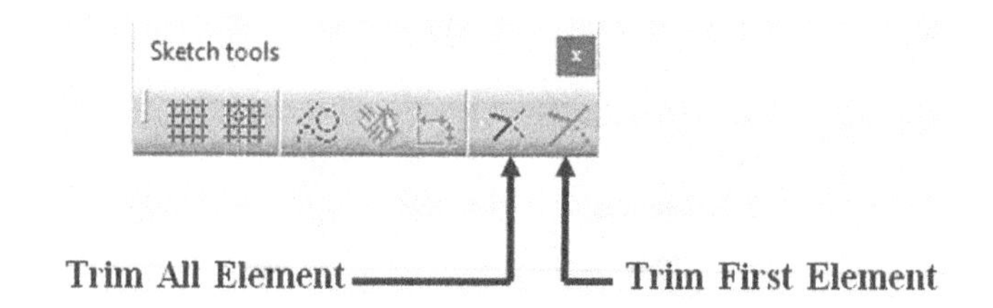

FIGURE 3.25 Options of Expanded Sketched tools toolbar.

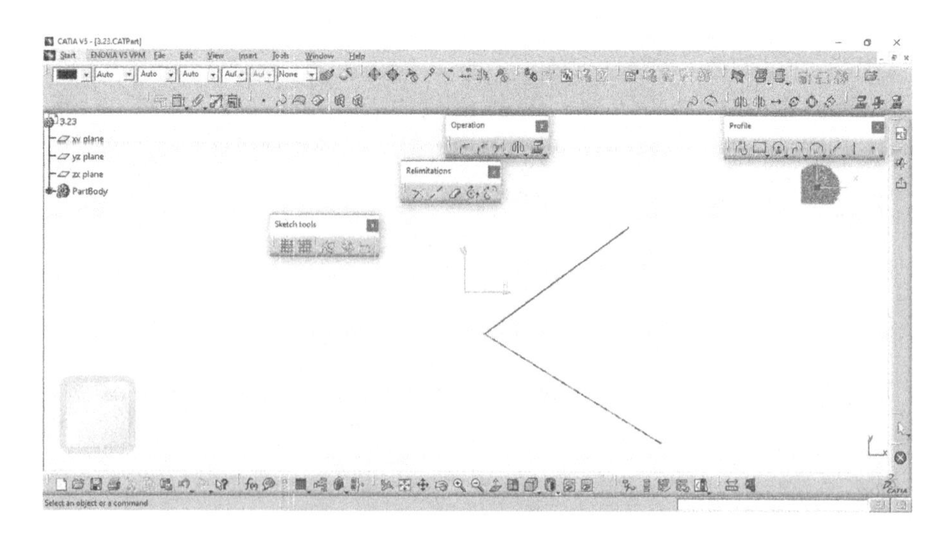

FIGURE 3.26 Resulting Trims by choosing Trim All Elements button.

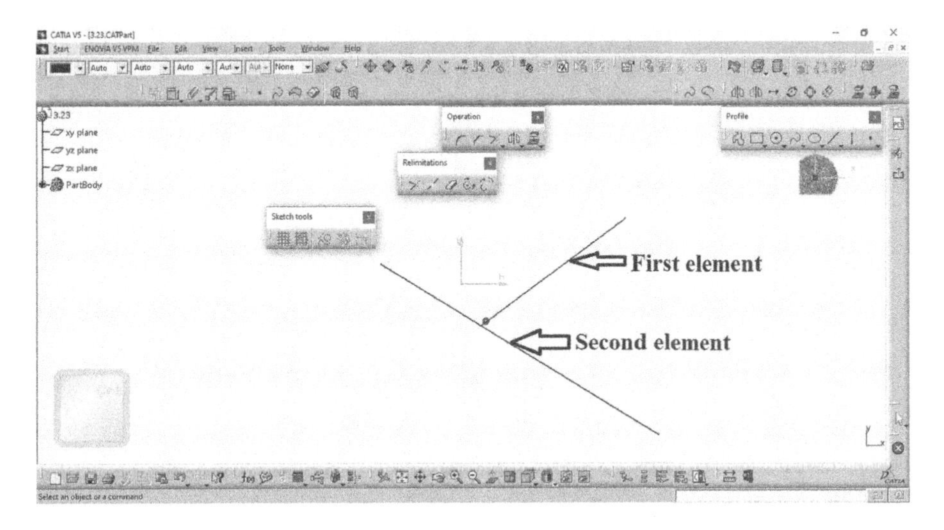

FIGURE 3.27 Resulting Trims by choosing Trim First Element button.

3.3.2 Break

In Sketcher workbench, you are provided with Break tool under Relimitations toolbar as shown in Figure 3.22 to break the sketched elements. To perform it, draw two lines as shown in Figure 3.28.

After selecting the Break icon, select the element (line) to be broken followed by the breaking element or you can select any location along the element to indicate the break point, resulting in a break point as shown in Figure 3.29.

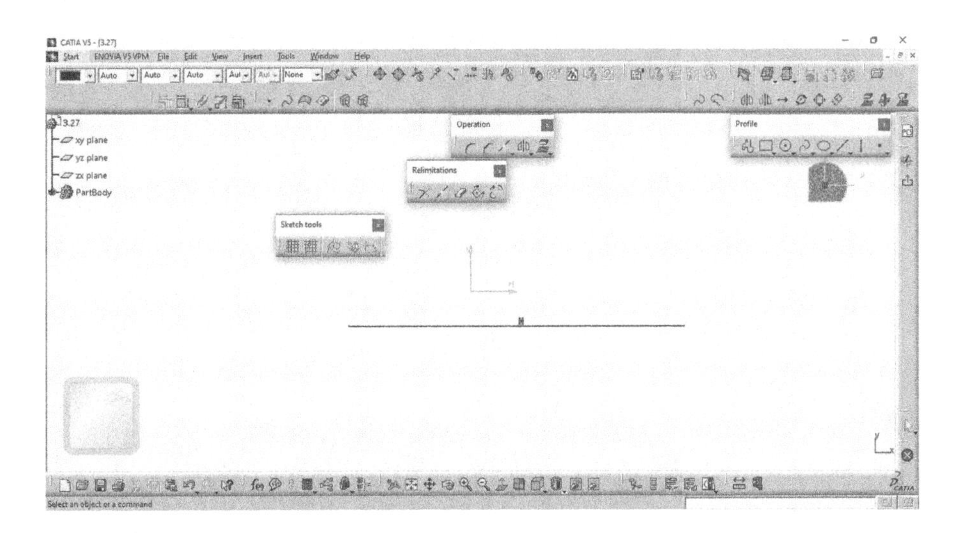

FIGURE 3.28 Constriction of line for Break.

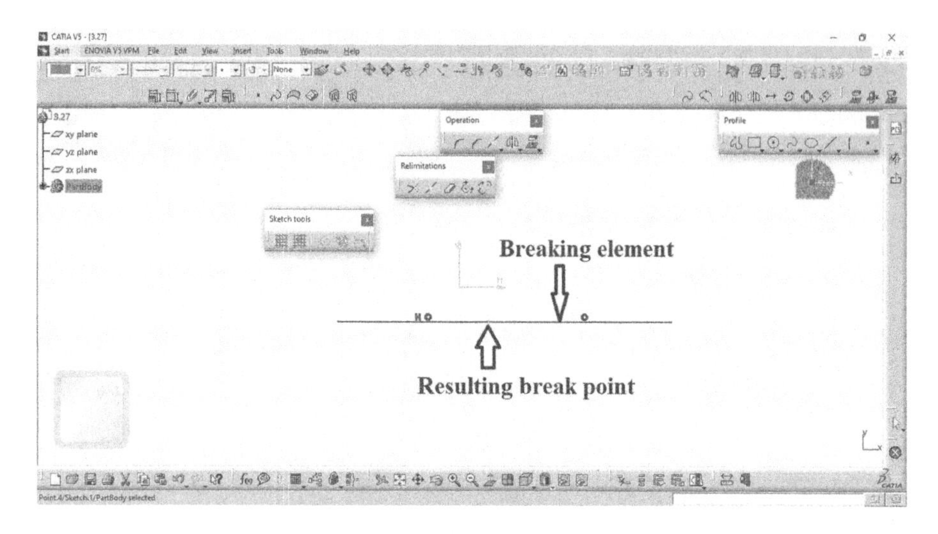

FIGURE 3.29 Constriction of Break Point on line.

3.3.3 Quick Trim

In Sketcher workbench, Quick Trim tool is available under Relimitations toolbar as shown in Figure 3.22 to trim the sketched elements. To perform it, draw a sketch as provided in Figure 3.30.

After selecting the Break icon, select the element to be trimmed as shown in Figure 3.31. This will result in the element being trimmed to the nearest available elements from the location you select on the element. Resulting sketch is as shown in Figure 3.32.

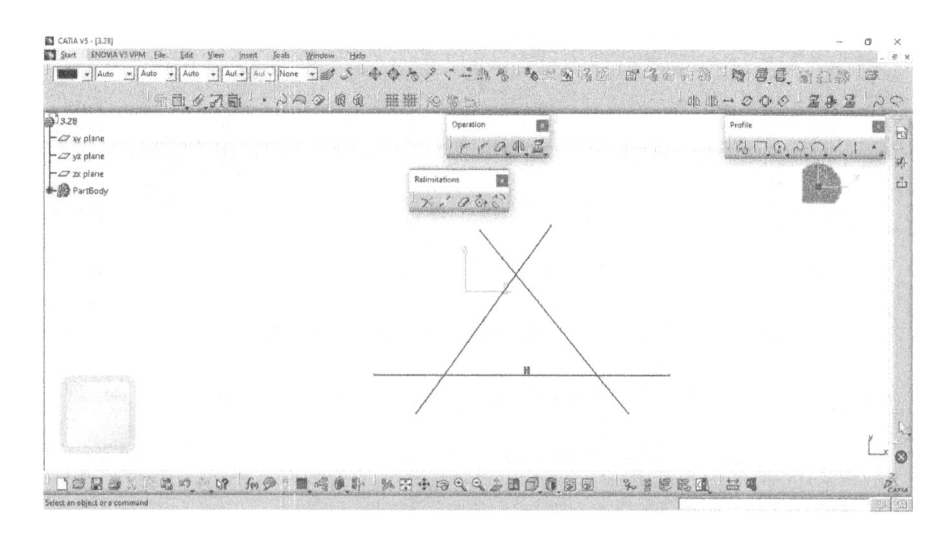

FIGURE 3.30 Constriction of sketch for quick trim.

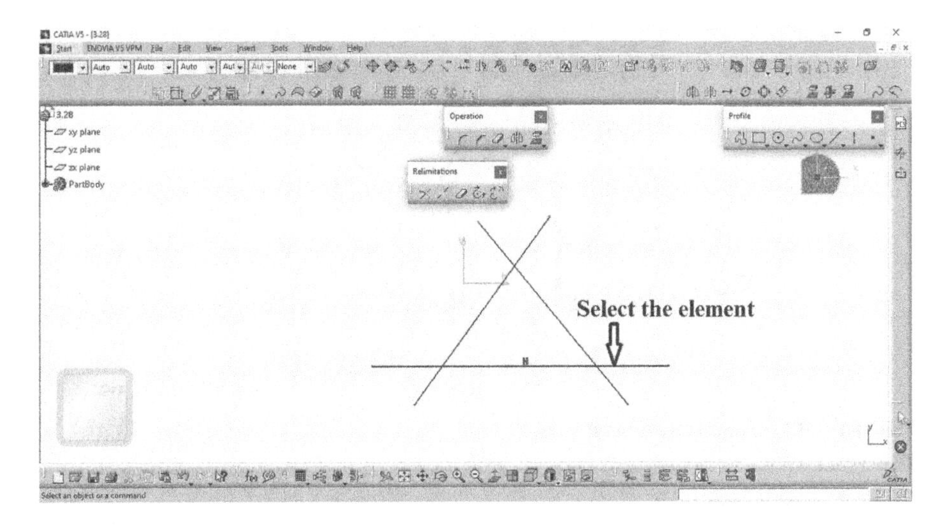

FIGURE 3.31 Selection of element for Quick Trim.

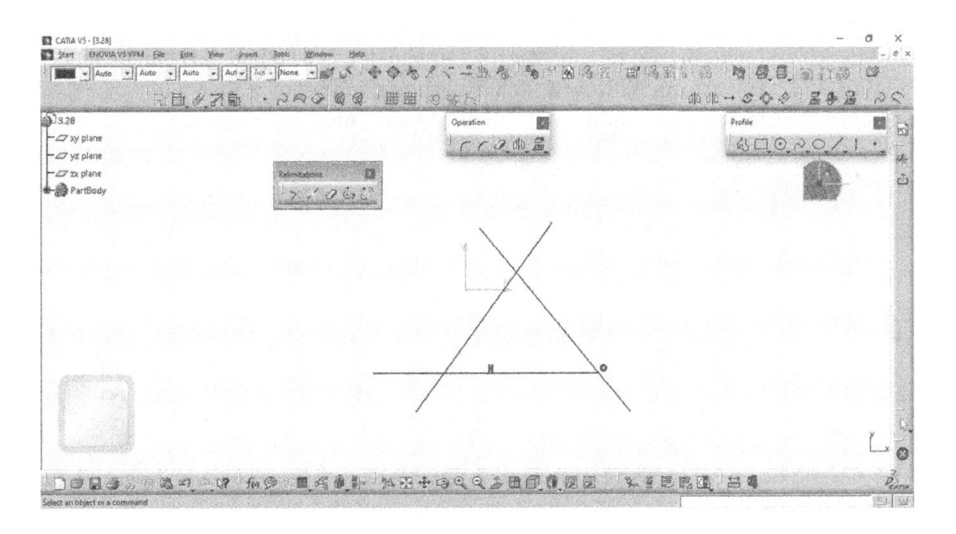

FIGURE 3.32 Resulting sketch after using Quick Trim.

3.4 TRANSFORMATION

The Operation toolbar as shown in Figure 3.33 is used to create geometry in Sketcher workbench. The tools in this toolbar are mirror, symmetry, translate, rotate, scale and offset rectangle.

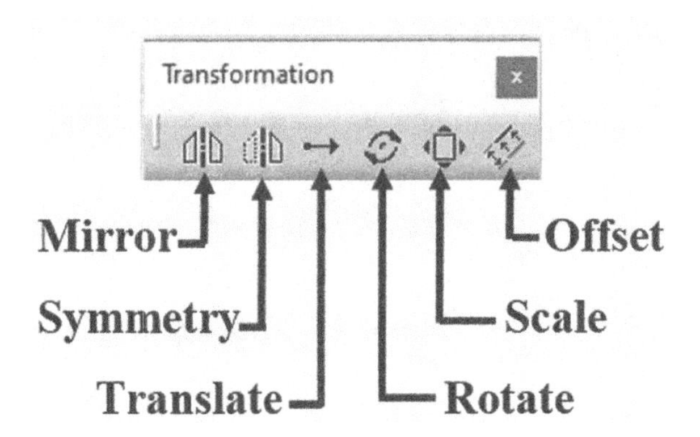

FIGURE 3.33 Transformations toolbar.

3.4.1 Mirror

In the physical world, we can create multiple images using a mirror, similarly in this software using Mirror tool available under Transformation toolbar in Sketcher workbench and with the aid of a mirror line, image of the sketched part can be made (Figure 3.33). To perform it, draw a sketch as shown in Figure 3.34.

Select the elements to be mirrored followed by a line, Construction Line or Axis Line, to indicate the mirror plane. The mirrored elements will now be created as shown in Figure 3.35.

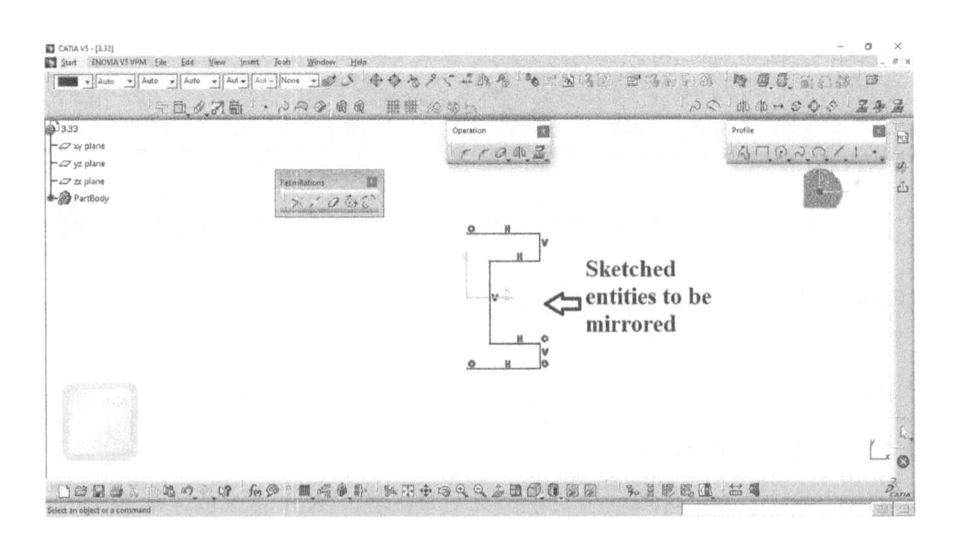

FIGURE 3.34 Sketched entities to be mirrored.

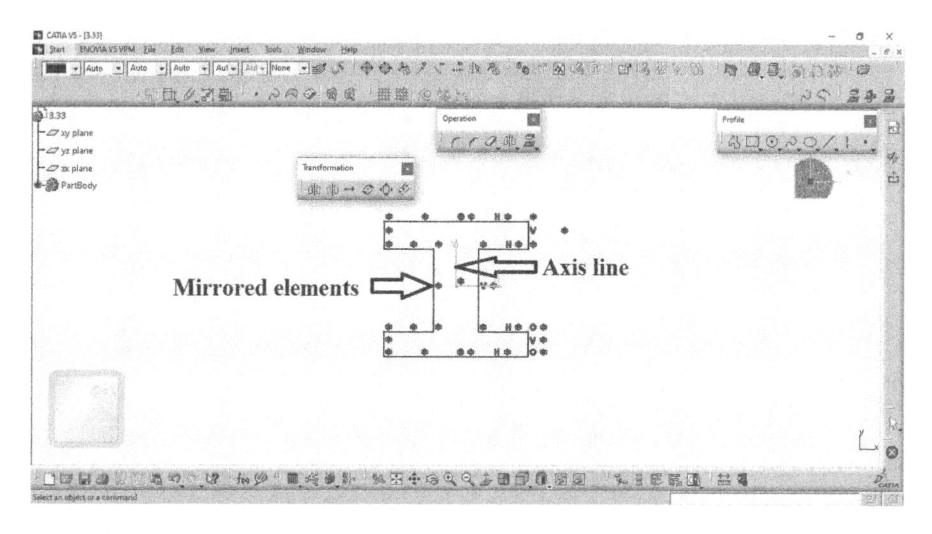

FIGURE 3.35 Mirrored elements.

3.4.2 Symmetry

You can not only make symmetrical sketched elements along the mirror line, but also delete the original element in the Sketcher workbench using Symmetry tool. In Sketcher workbench, you are provided with Symmetry tool under Transformation toolbar as shown in Figure 3.33 to make symmetrical sketched elements. To perform it, draw a sketch as shown in Figure 3.36.

After choosing the Symmetry button in the Transformation toolbar, select the line or axis as shown in Figure 3.36 from which both the sketched part and the mirror part would be at equal distance. On selecting the above line, the selected part will be

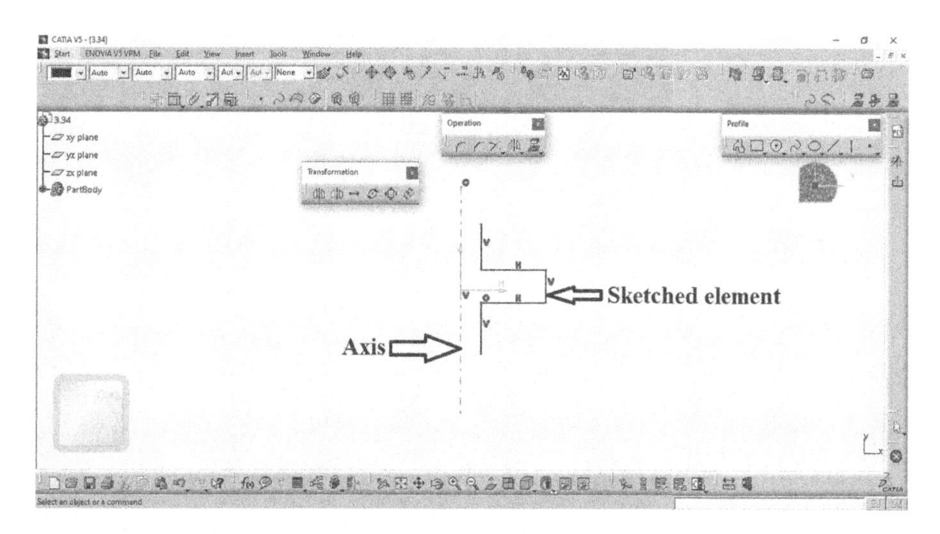

FIGURE 3.36 Sketched entities.

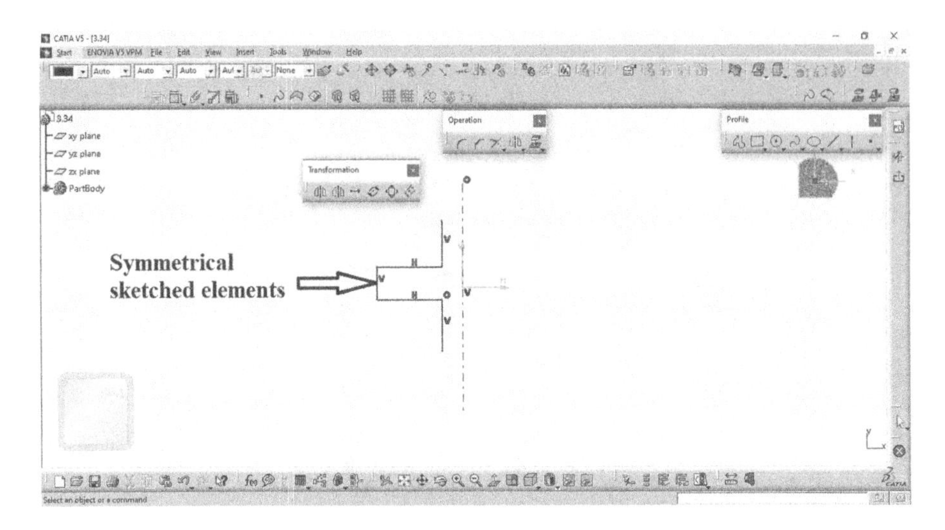

FIGURE 3.37 Symmetrical sketched elements.

mirrored on the other side of the symmetry line, while the original sketched part would get erased. The symmetrical sketched part will now be created as shown in Figure 3.37.

3.4.3 Translate

You can reposition the selected sketched part from its initial position to desired location in the Sketcher workbench using Translate tool. In Sketcher workbench, you are provided with Translate tool under Transformation toolbar as shown in Figure 3.33. To perform it, draw a sketch as shown in Figure 3.38.

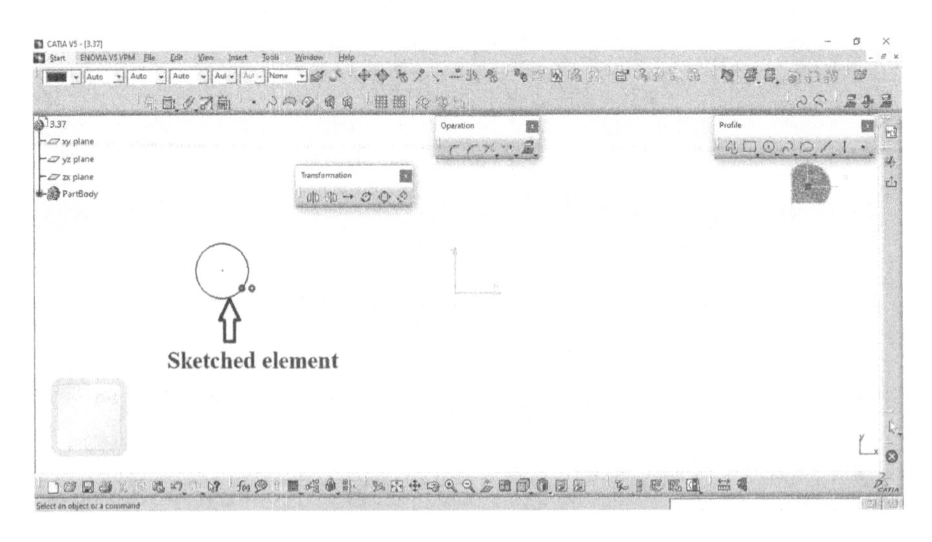

FIGURE 3.38 Sketched entities.

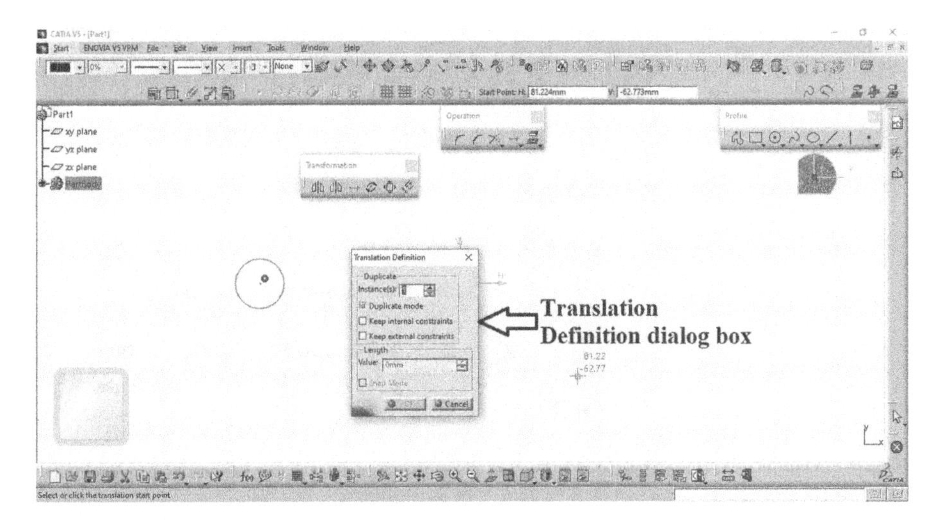

FIGURE 3.39 Translation Definition dialog box.

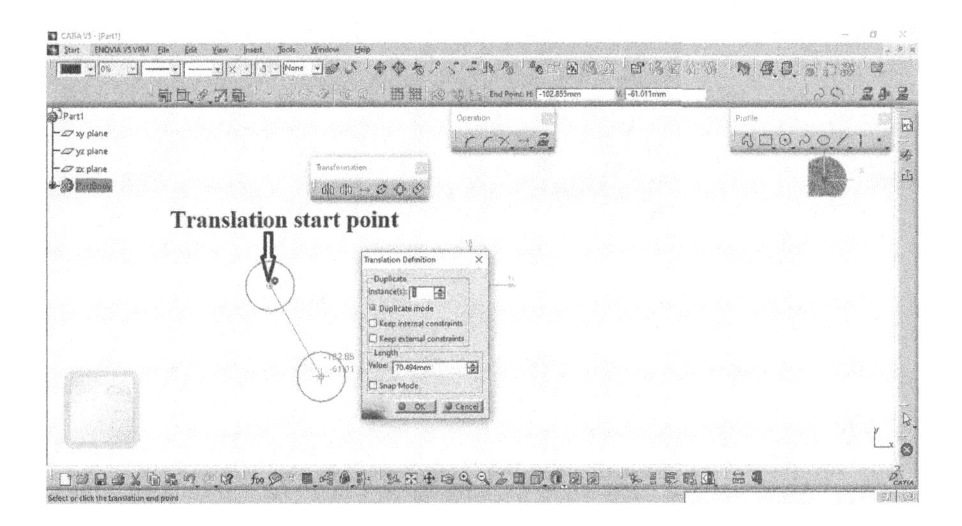

FIGURE 3.40 Selection of translation start point.

To move the sketched elements, as shown in Figure 3.38, select them and then use the Translate icon available in Transformation toolbar (Figure 3.33) and the Translation Definition dialog box appears (Figure 3.39). Also, select the translation start point (Figure 3.40), use the incremental translation distance of 50mm in the space to provide value in length area and the value of number of instance as 4 in the

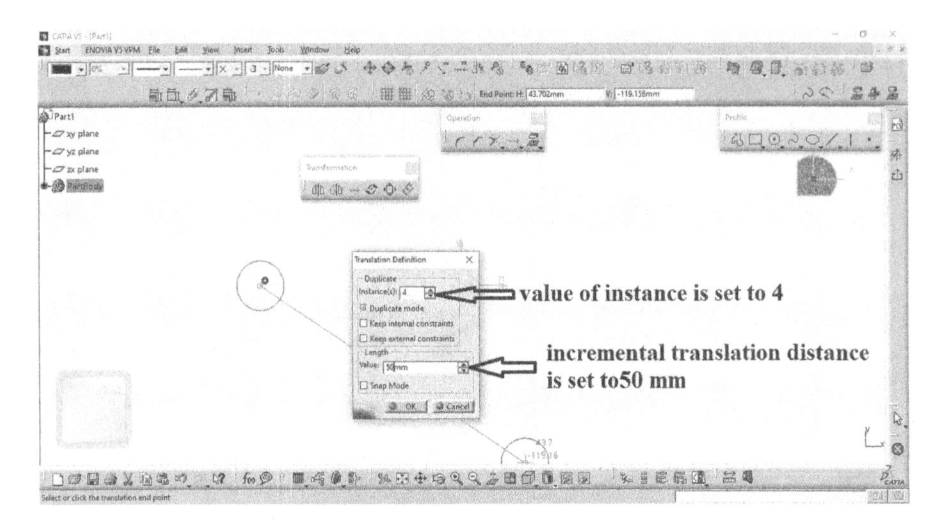

FIGURE 3.41 Value entering.

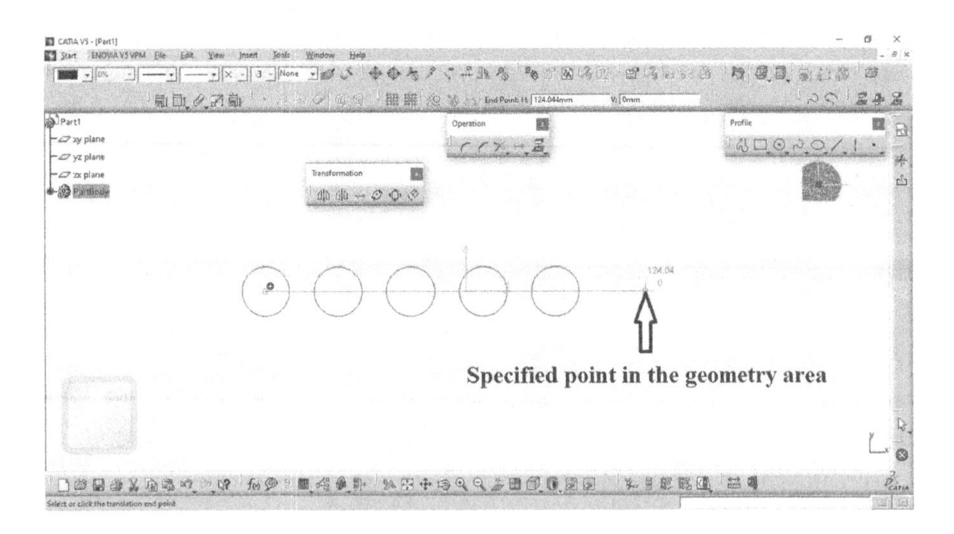

FIGURE 3.42 Specified point in the geometry area.

space in the Duplicate area of the Translation Definition dialog box (Figure 3.41) and press Enter key. You are also required to specify a point in the geometry area to place the sketched part (Figure 3.42). Selected sketched element is created at the specified distance as shown in Figure 3.43.

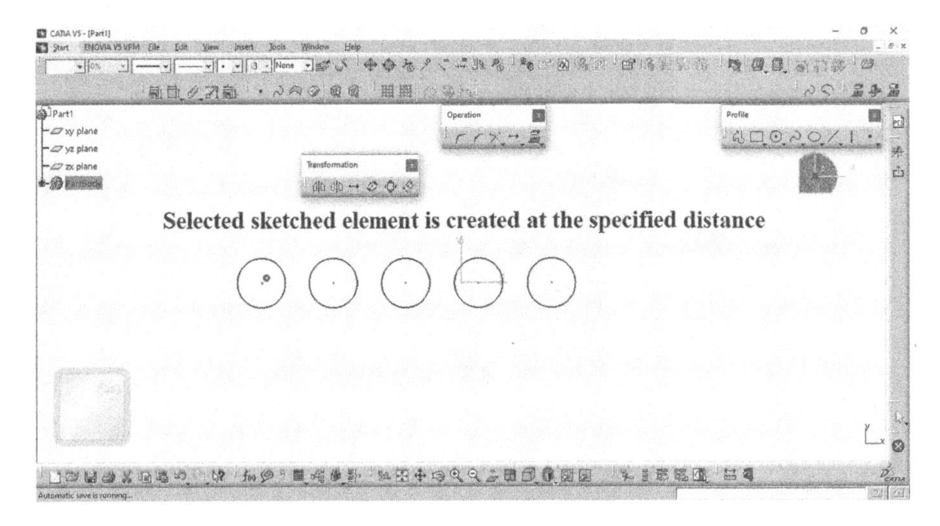

FIGURE 3.43 Created elements using Translation tool.

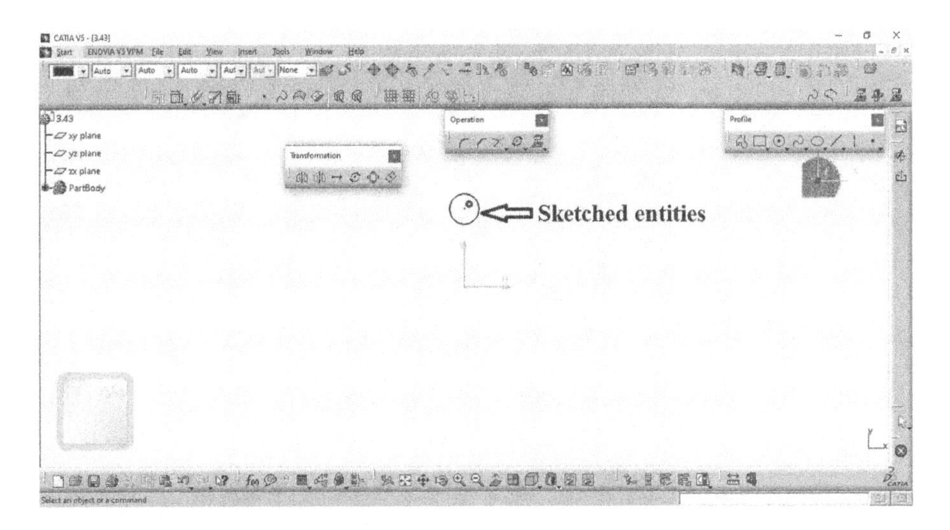

FIGURE 3.44 Sketched entities.

3.4.4 Rotate

This tool is used to rotate the selected sketched part about a rotation center point. In Sketcher workbench, you are provided with Rotate tool under Transformation toolbar as shown in Figure 3.33. To perform it, draw a sketch as shown in Figure 3.44.

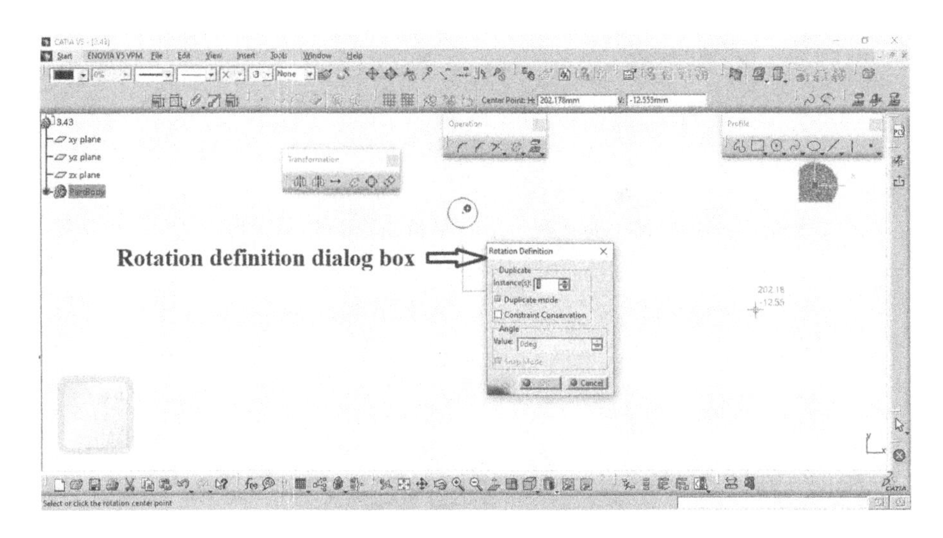

FIGURE 3.45 Rotation definition dialog box.

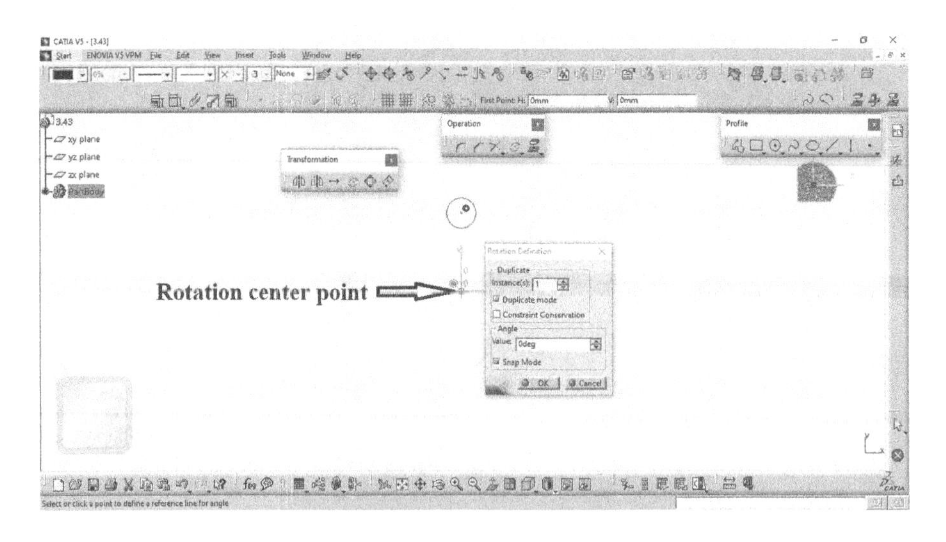

FIGURE 3.46 Selection of rotation center point.

Select the sketched part (Figure 3.44) and also select Rotation button from transformation toolbar, as shown in Figure 3.33. In the Rotation definition dialog box displayed (Figure 3.45), select the rotation center point about which the selected sketched part will be rotated (Figure 3.46). Next, to define a reference line for angle, select a point (Figure 3.47). Set the angle as 30deg in the value providing space in

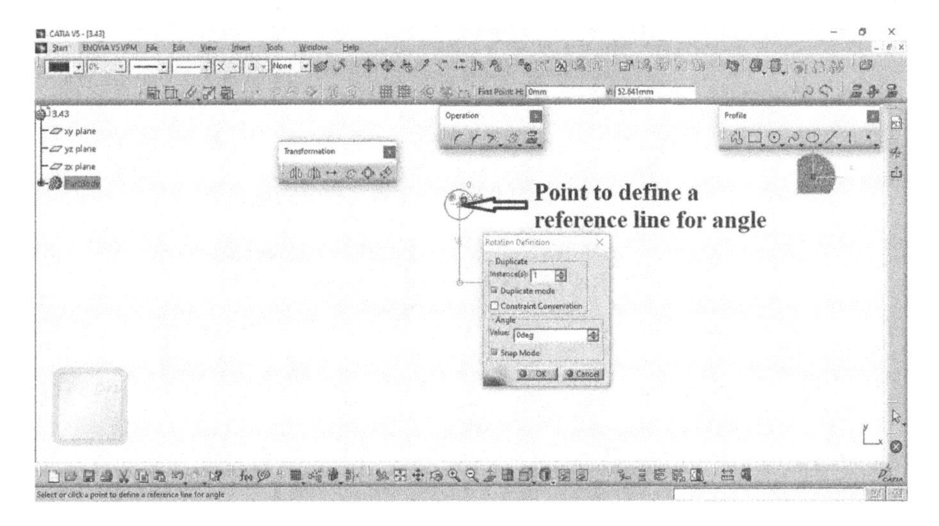

FIGURE 3.47 Selection of point to define a reference line for angle.

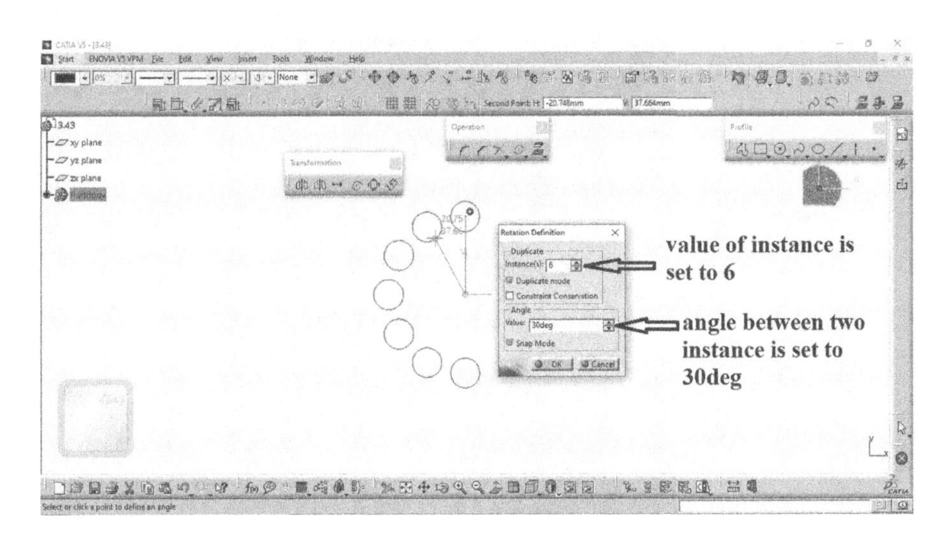

FIGURE 3.48 Value entering.

the angle area and the number of instance as 6 in the instances providing space in the Duplicate area of the Rotation definition dialog box (Figure 3.48) and press Enter key. Selected sketched element is created at the specified rotated instance as shown in Figure 3.49.

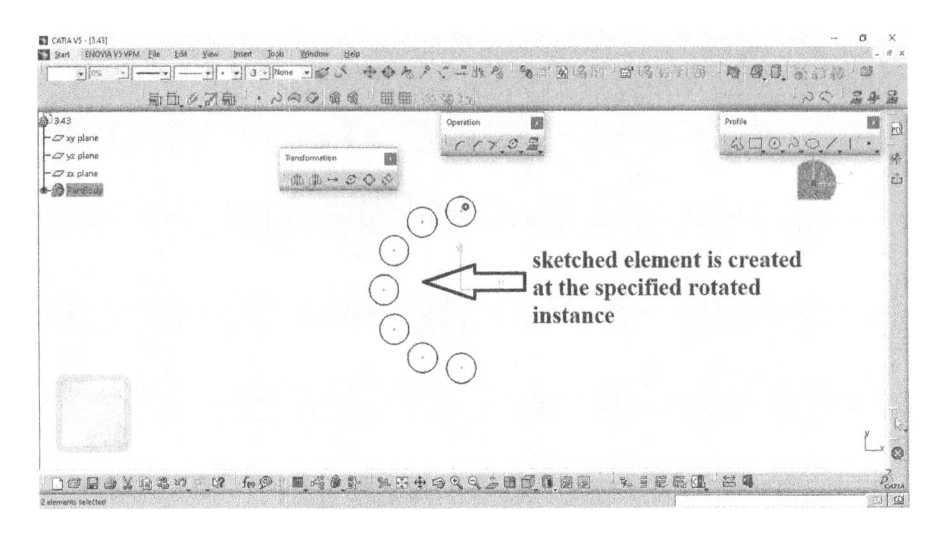

FIGURE 3.49 Created elements using Rotate tool.

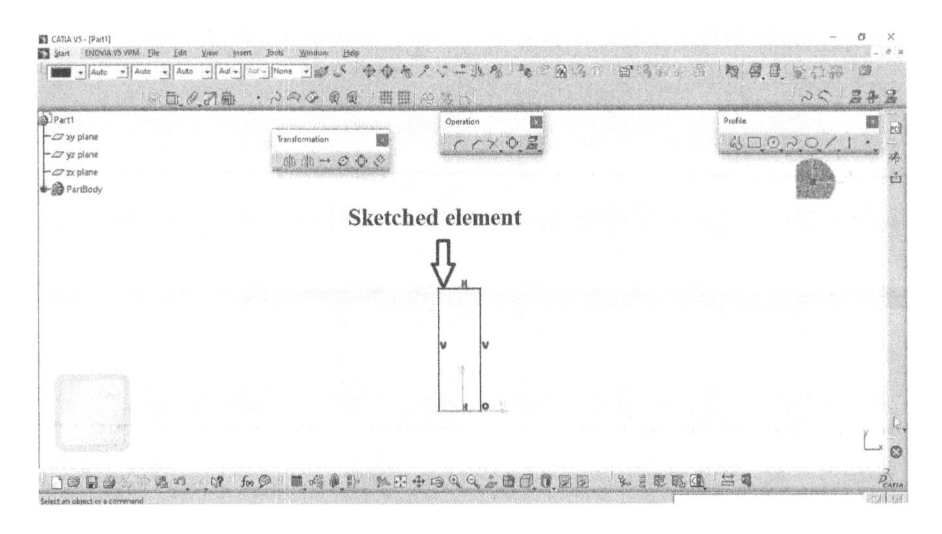

FIGURE 3.50 Sketched element for Scale tool.

3.4.5 Scale

This tool is used to increase and decrease size of the selected sketched elements. In Sketcher workbench, you are provided with Scale tool under Transformation toolbar as shown in Figure 3.33. To perform it, draw a sketch as shown in Figure 3.50.

For scale the sketched elements, after selecting the relevant part, choose the Scale button from the Transformation toolbar. You would find that Scale

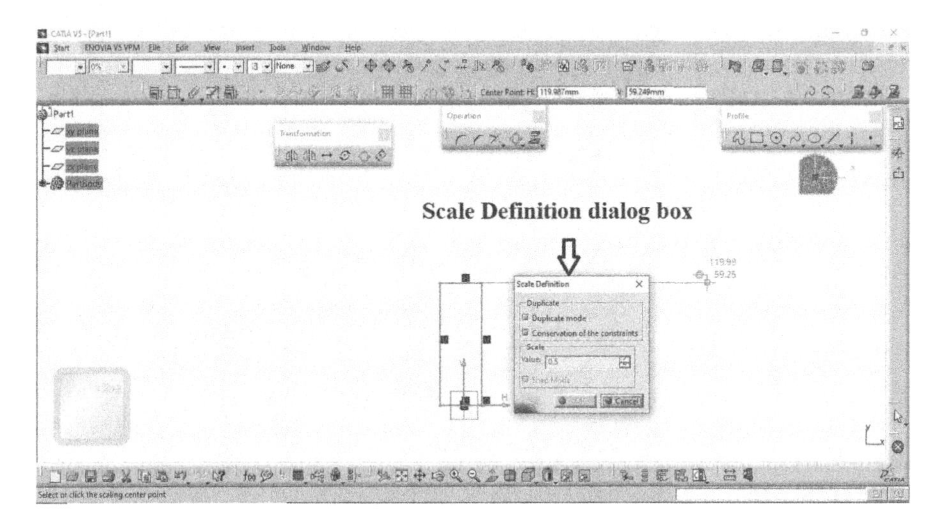

FIGURE 3.51 Scale definition dialog box.

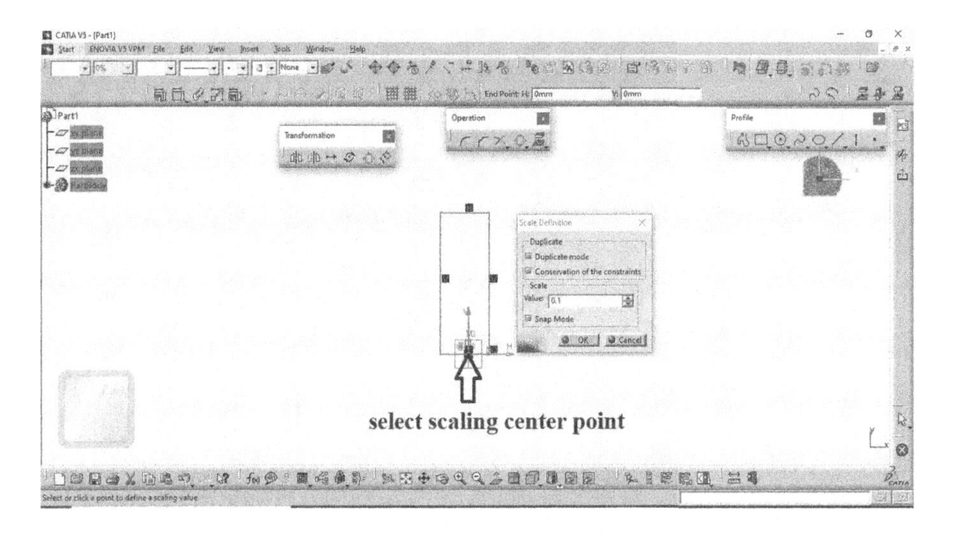

FIGURE 3.52 Select scaling center point for scaling.

Definition dialog box gets displayed in the window (Figure 3.51). Similar to the earlier, select scaling center point in the drawing window (Figure 3.52). Again provide the scaling value as 2 in the relevant space (Figure 3.53) and press Enter key. Selected sketched element is created at the specified scaled value as shown in Figure 3.54.

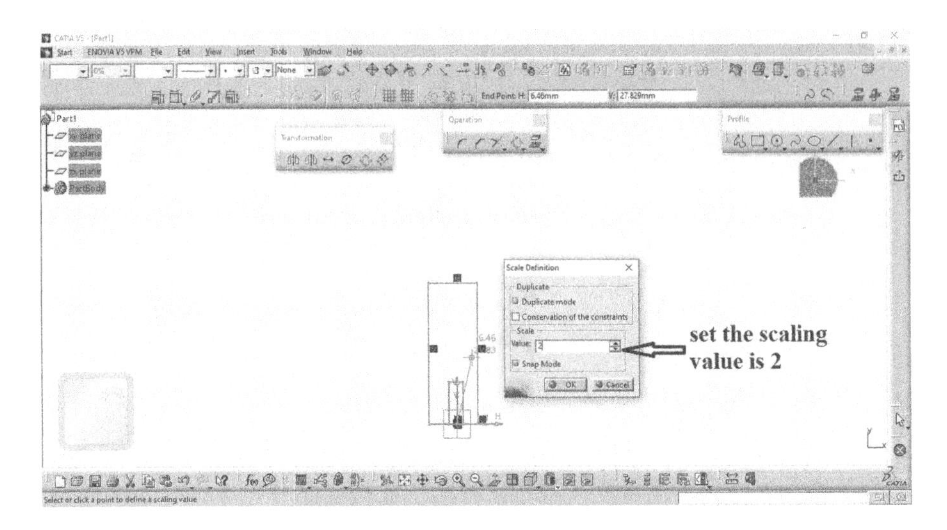

FIGURE 3.53 Set the scaling value entering.

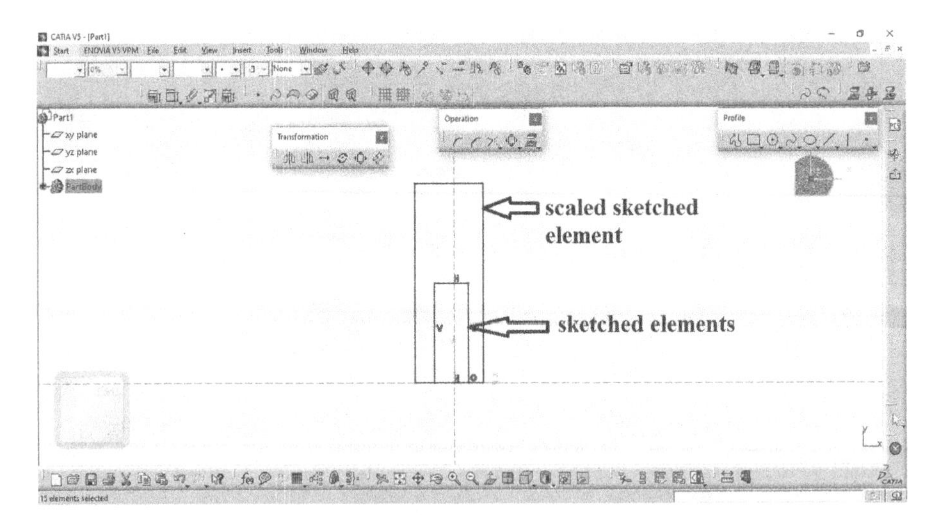

FIGURE 3.54 Created elements using Scale tool.

3.4.6 Offset

In Sketcher workbench, you are provided with Offset tool under Transformations toolbar as shown in Figure 3.33 to offset the sketched elements. To perform it, draw the Sketch element as shown in Figure 3.55 and after that invoke Offset tool; the Sketched tools toolbar expands as shown in Figure 3.56.

The Sketch tools toolbar provides the various options of offset as provided in Figure 3.57.

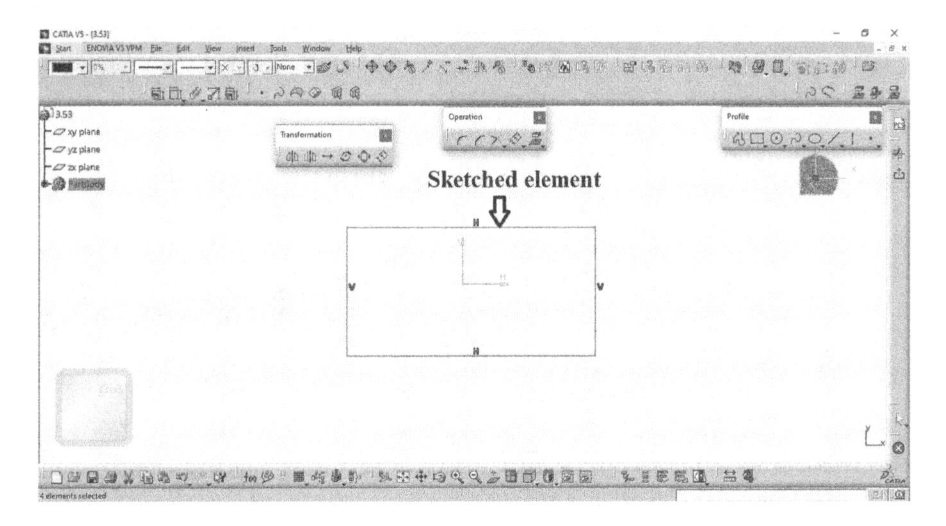

FIGURE 3.55 Sketched element for Offset tool.

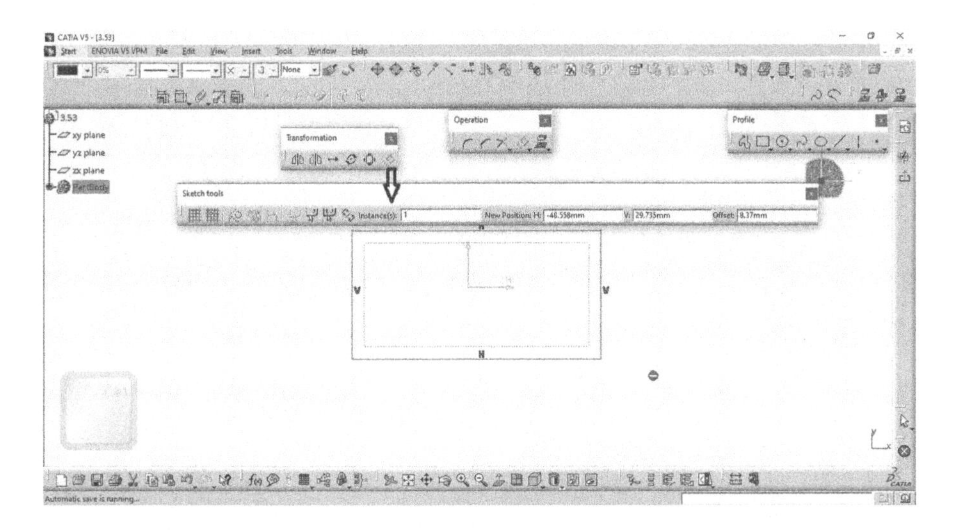

FIGURE 3.56 Expanded Sketched tools toolbar.

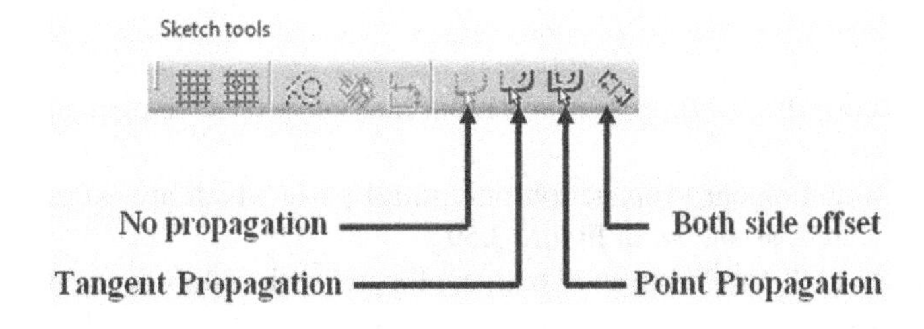

FIGURE 3.57 Options of Expanded Sketched tools toolbar.

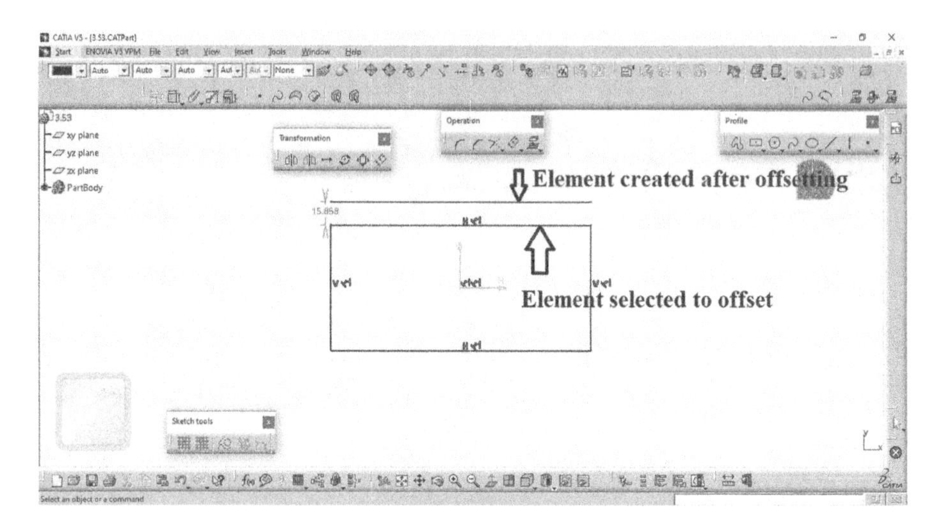

FIGURE 3.58 Element created after offsetting using No Propagation button.

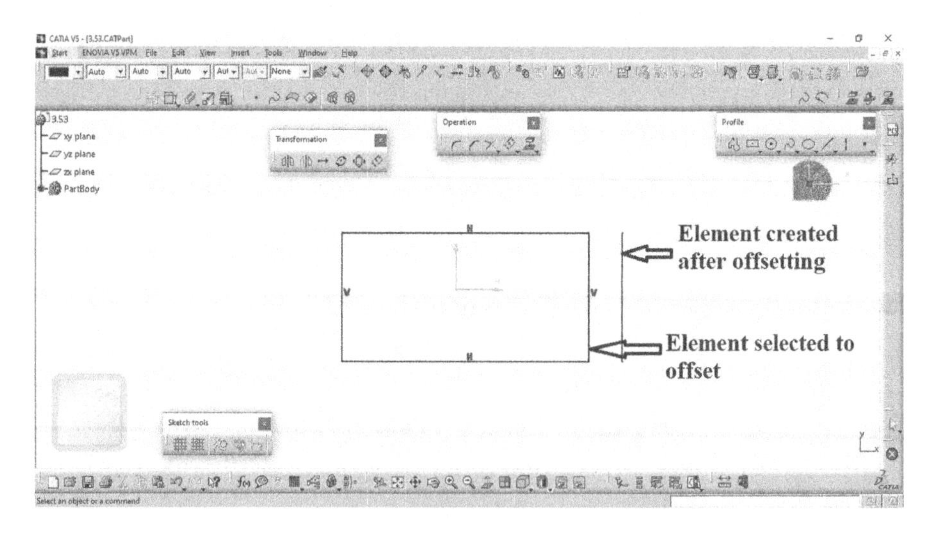

FIGURE 3.59 Element created after offsetting using Tangent Propagation button.

If you choose the No Propagation button, then only selected element is offset as shown in Figure 3.58.

On selecting Tangent Propagation option, all parts which are tangent are automatically selected, as shown in Figure 3.59.

Similarly with Point Propagation button, all parts with end-to-end connection with the selected part get automatically connected and form a closed loop (Figure 3.60).

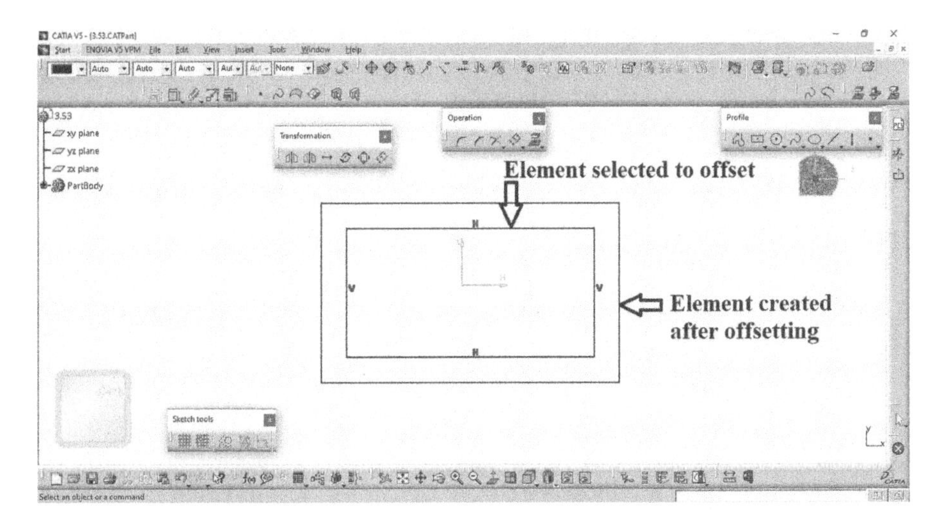

FIGURE 3.60 Element created after offsetting using Point Propagation button.

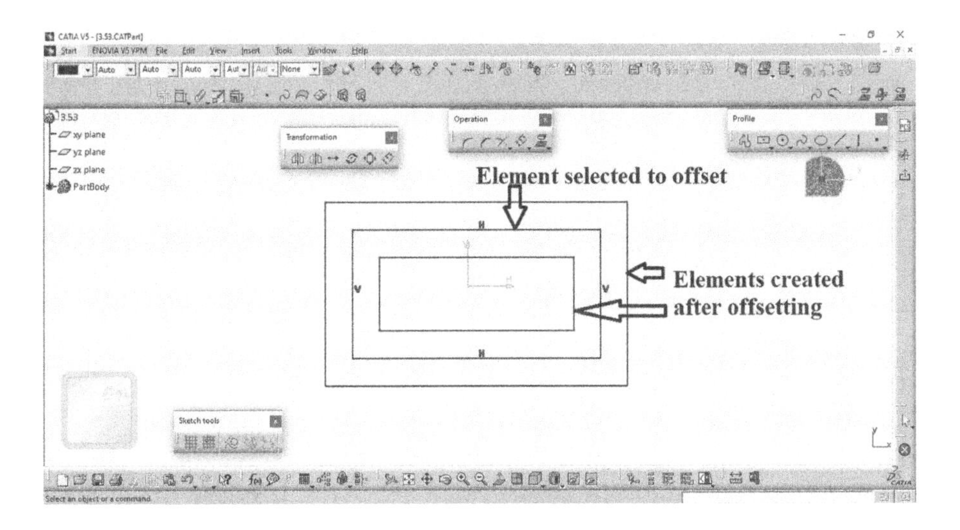

FIGURE 3.61 Element created after offsetting using Both Side Offset button.

For creating offset on both the sides, Both Side offset button should be used (Figure 3.61).

4 Constraints Toolbar

4.1 INTRODUCTION

In this chapter, the usage of constraining the sketched component is discussed. The operation is essential to restrict the figures from getting deformed or distorted by creating zero degrees of freedom. So in this chapter, with the aid of screenshots, different constraint tools would be discussed and a hands-on training would be provided.

4.2 CONSTRAINTS

Constraints are used to control an element size, position and its relationship to other elements.

There are two types of constraint:

1. *Geometric*–Controls the relationship between elements, i.e., Tangency, Parallelism, Coincidence, etc. and may also be used to control the orientation of an element, i.e., Vertical and Horizontal.
2. *Dimensional*–Controls the size of an element, i.e., the length of a line, the radius of an arc, etc. They can also control the distance and angle between elements.

Following graphic symbols represents geometric and dimensional constraints, as shown in Figure 4.1.

4.3 CONSTRAINT AND ELEMENT COLORS

Constraints and elements can be display in the following colors, which depict different states, as shown in Figure 4.2.

4.4 CONSTRAINTS TOOLBAR

The following toolbar is used to apply constraints in sketcher. Generally you can either select elements to be constraint and then the command icon or vice versus. Figure 4.3 shows constraints toolbar.

4.5 GEOMETRIC CONSTRAINT

Generally the Constraint applied automatically to the sketch is not enough to make it Constraints. But for absolute stability manual application of additional Constraints to the sketch are required to be provided with the help of

Geometric		
Symbols	Constraints	Description
	Fix	Holds the elements in a fix position.
	Horizontal	Holds the element parallel to the sketch axis horizontal.
	Vertical	Holds the element parallel to the sketch axis vertical.
	Perpendicular	Holds two elements perpendicular to each other.
	Coincidence	Applies coincidence between two elements.
	Parallelism	Applies parallelism between two elements.
Dimensional		
	Angular	Represent angle between two elements.
	Radial	Represent radius of an arc
	Diametric	Represent diameter of a circle
	linear	Represent length of a line

FIGURE 4.1 Graphic symbols geometric and dimensional constraints representation.

Symbols	Description
	Unconstrained or partially constrained elements are displayed in white.
	A green constraint indicates the constraint is valid and up to date.
	Green elements indicate that they are fully constrained.
	Brown constraints and elements indicate over-defined or inconsistent constraints, which can be resolved by deleting the relevant constraint.
	Purple constraints and elements indicate that the elements are over-constrained. Remove the unnecessary constraint to resolve the problem.
	Red constraints and elements indicate that the at least one of the constraints has to be changed.

FIGURE 4.2 Element colors representation.

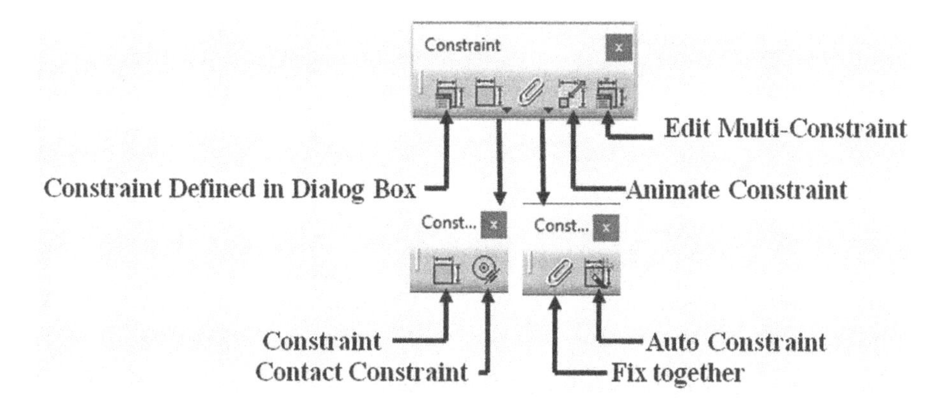

FIGURE 4.3 Constraints toolbar.

Constraints Definition toolbar. To apply additional Constraints, select the Sketched elements. You can use the CTRL key to select more than one element. Next choose the Constraints Definition dialog box button from the Constraint toolbar, as shown in Figure 4.3, then the Constraint Definition dialog box comes out (Figure 4.4).

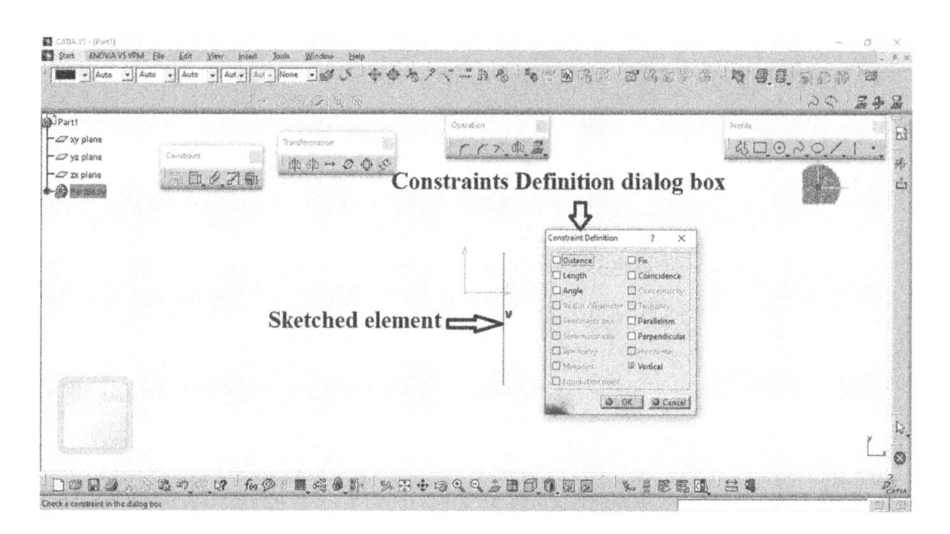

FIGURE 4.4 Constraint Definition dialog box is displayed.

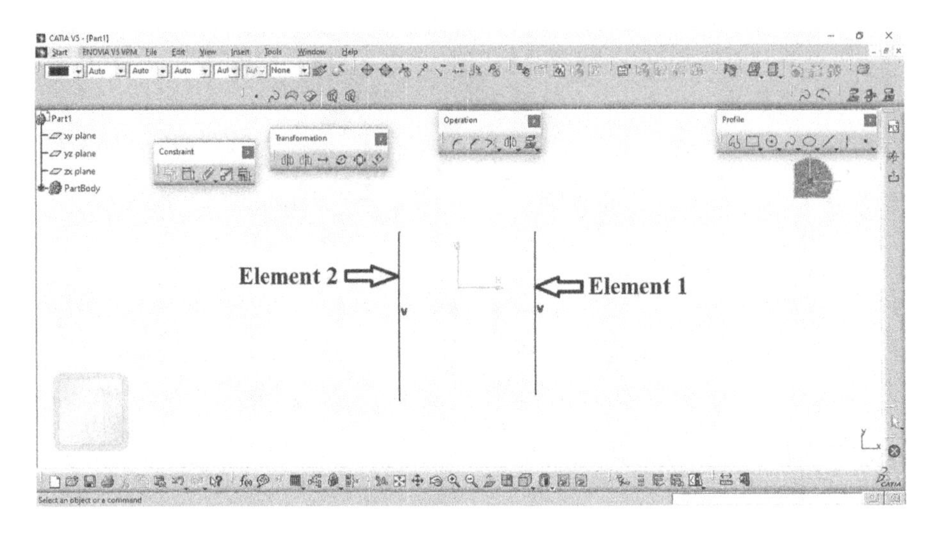

FIGURE 4.5 Sketched element for Distance Constraint.

After selecting sketched elements as shown in Figure 4.5, if you choose Distance check box from Constraint Definition dialog box, and press the OK button as shown in Figure 4.6, it will add a distance dimension between two selected elements, as shown in Figure 4.7.

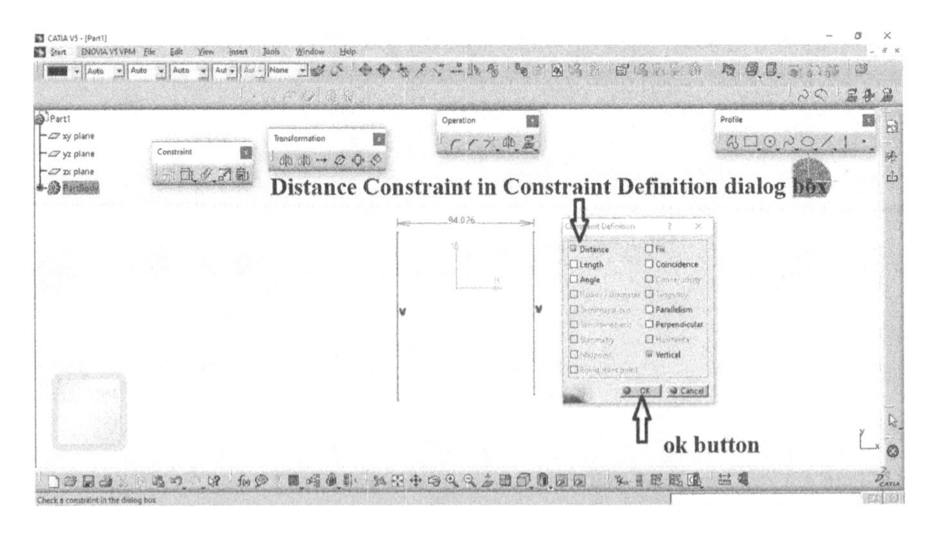

FIGURE 4.6 Selection of Distance Constraint in Constraint Definition dialog box.

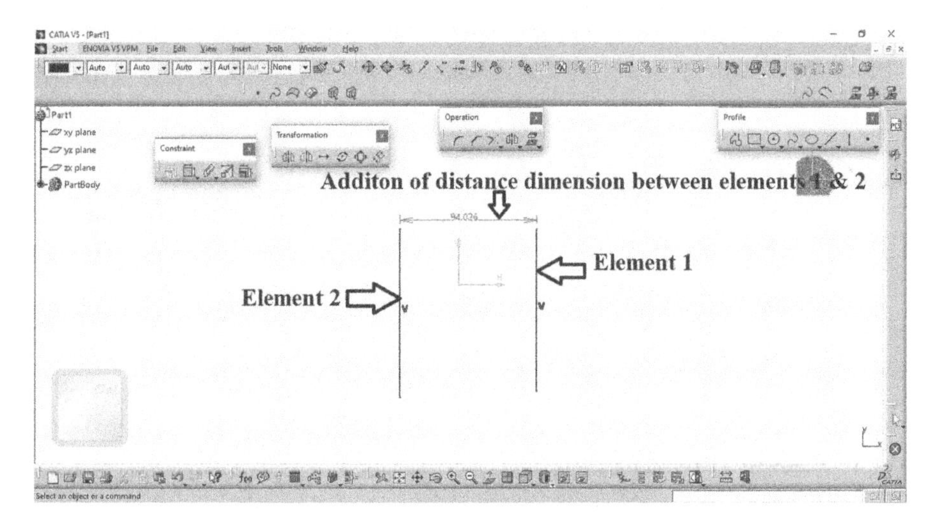

FIGURE 4.7 Constraining to elements using Distance Constraint.

After selecting sketched elements as shown in Figure 4.8, if you choose Length check box from Constraint Definition dialog box, and press the OK button as shown in Figure 4.9, a linear dimensions is displayed on selected elements, as shown in Figure 4.10. The Length Constraint is always applied on linear dimension to the selected line.

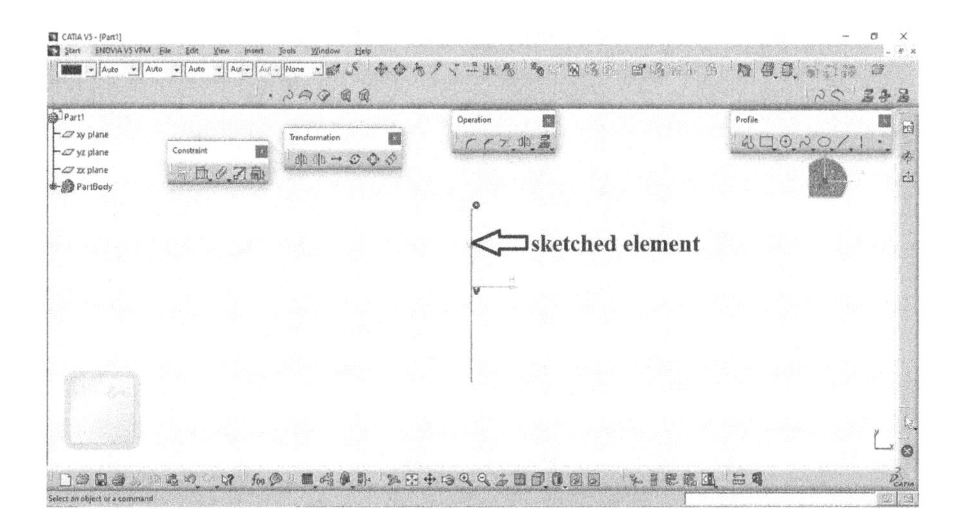

FIGURE 4.8 Sketched element for Length constraint.

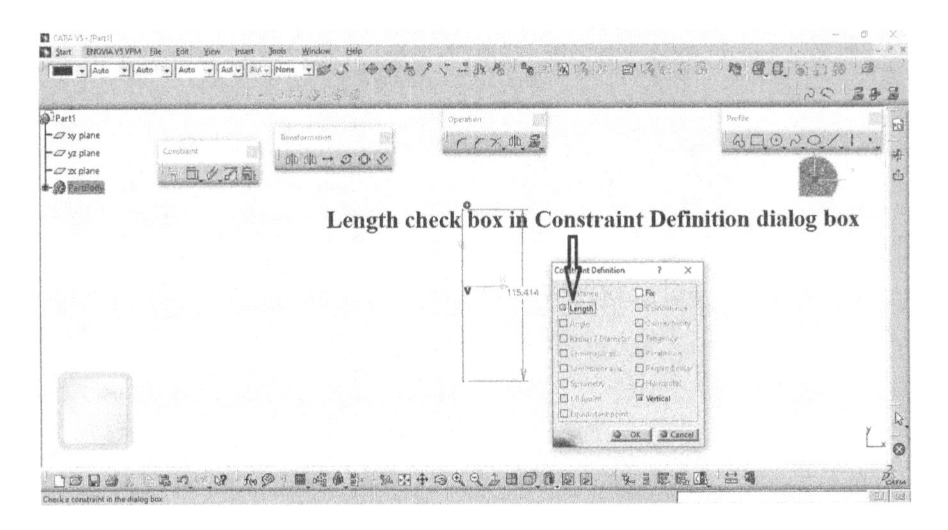

FIGURE 4.9 Selection of Length Constraint in Constraint Definition dialog box.

After selecting sketched elements as shown in Figure 4.11, if you choose Angle check box from Constraint Definition dialog box, and press the OK button as shown in Figure 4.12, an angular dimension is displayed on selected elements, as shown in Figure 4.13. The Angle Constraint is always applied on angular dimension between two selected lines or between a line and an existing edge of a sketch.

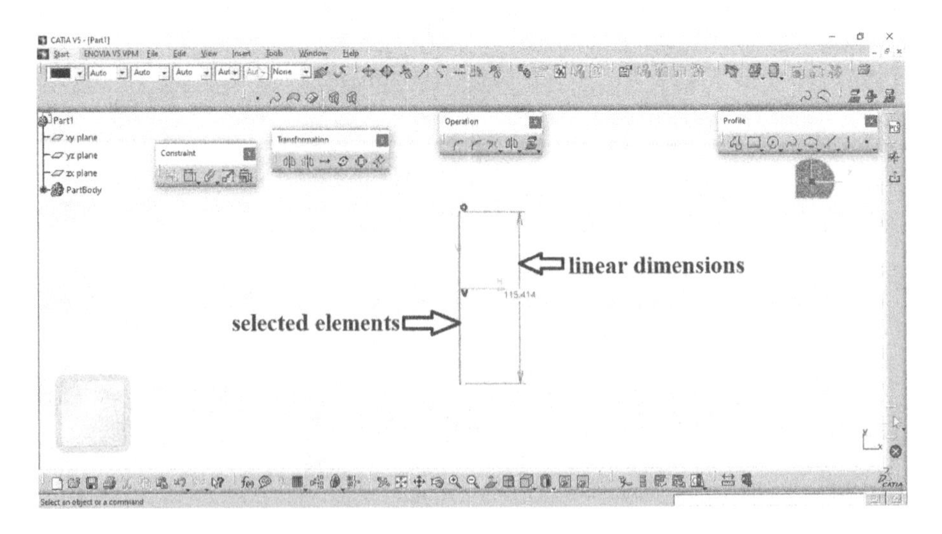

FIGURE 4.10 Constraining to elements using Length Constraint.

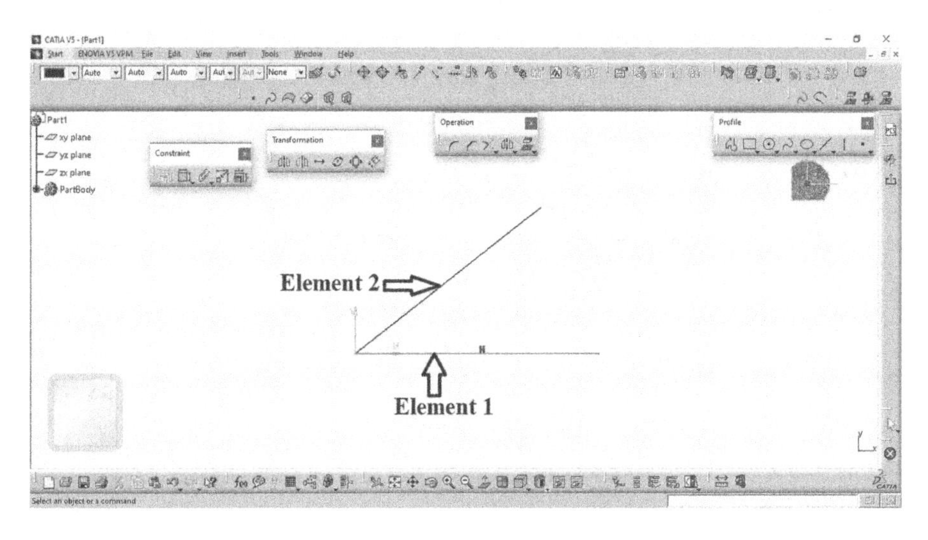

FIGURE 4.11 Sketched element for Angle Constraint.

After selecting sketched elements as shown in Figure 4.14, if you choose Radius/ Diameter check box from Constraint Definition dialog box, and press the OK button as shown in Figure 4.15. If the selected elements are an arc, the radius dimension is applied. If the selected element is circle, the diameter dimension is applied as shown in Figure 4.16. The Radius/Diameter Constraint is always applied on radius or diameter dimension to the selected are or circle.

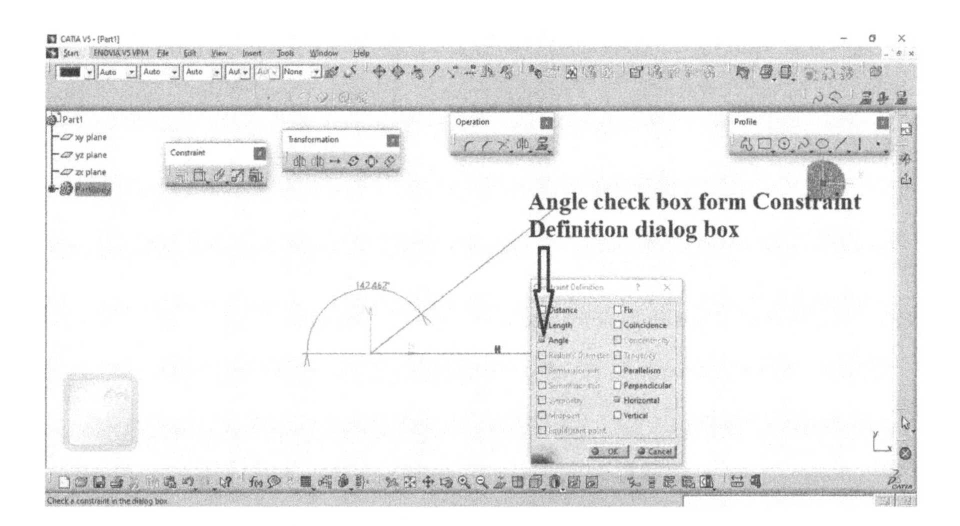

FIGURE 4.12 Selection of Angle Constraint in Constraint Definition dialog box.

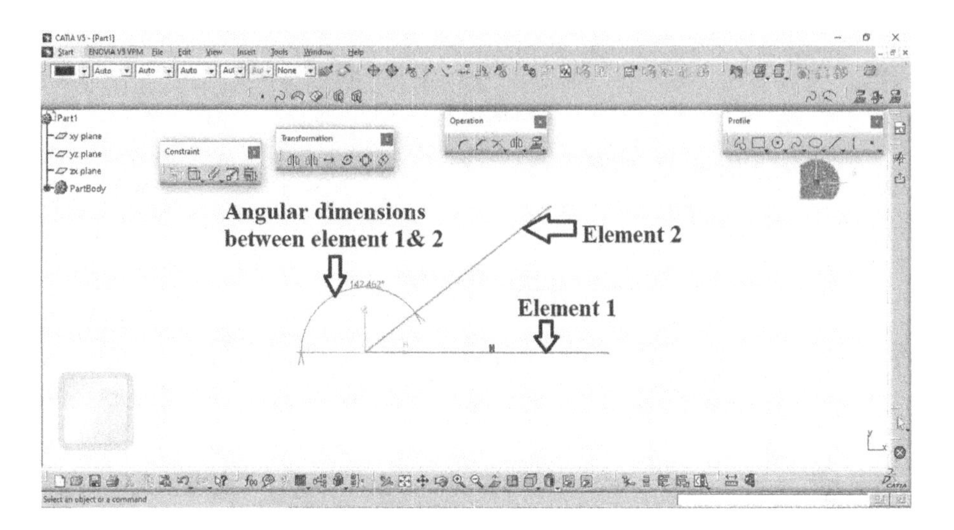

FIGURE 4.13 Constraining elements using Angle Constraint.

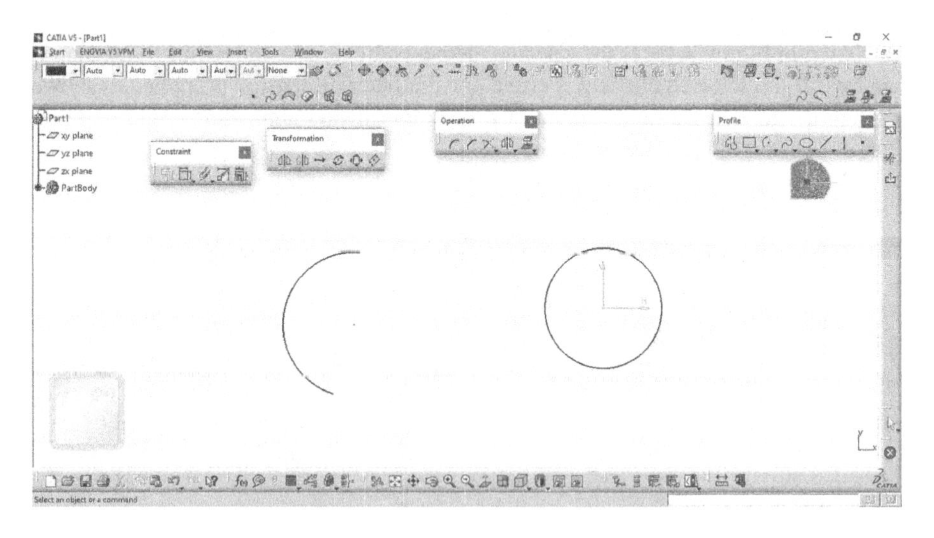

FIGURE 4.14 Sketched element for Radius/Diameter Constraint.

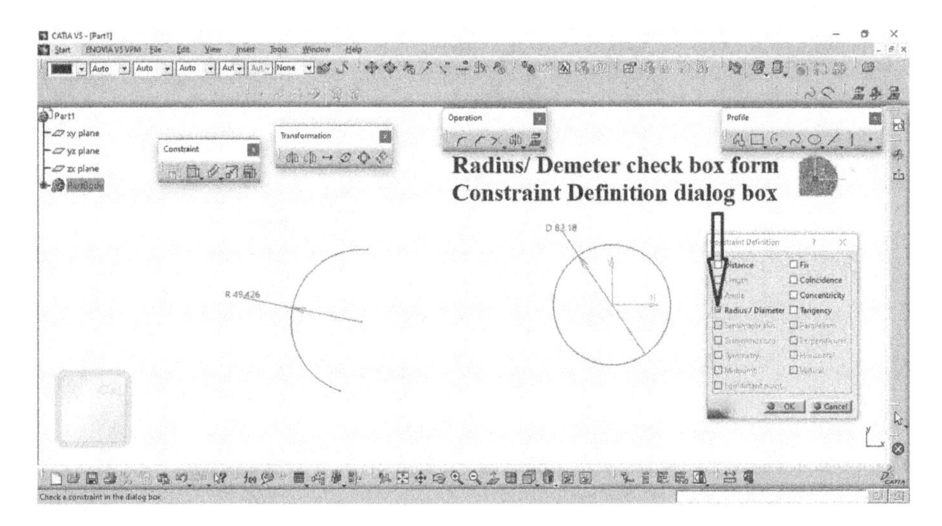

FIGURE 4.15 Selection of Angle Constraint in Constraint Definition dialog box.

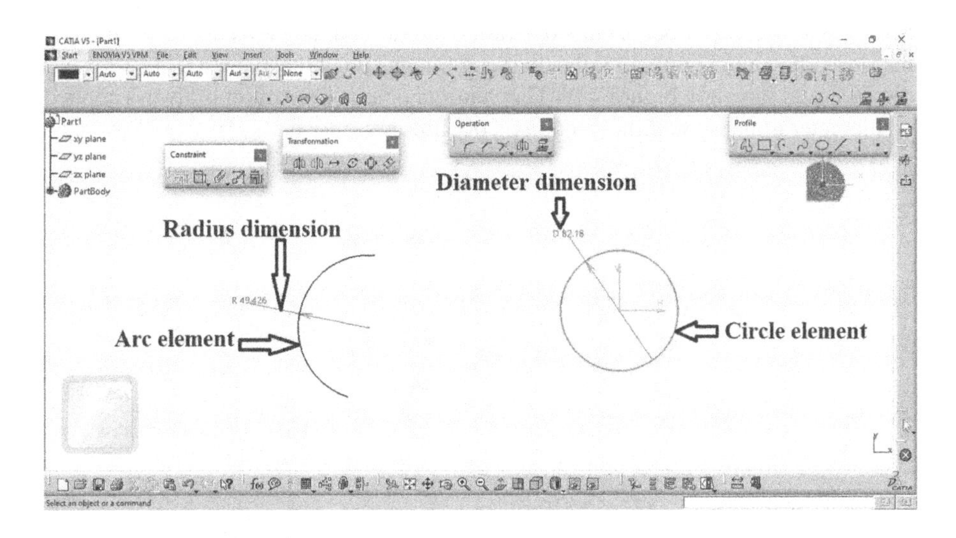

FIGURE 4.16 Constraining elements using Angle Constraint.

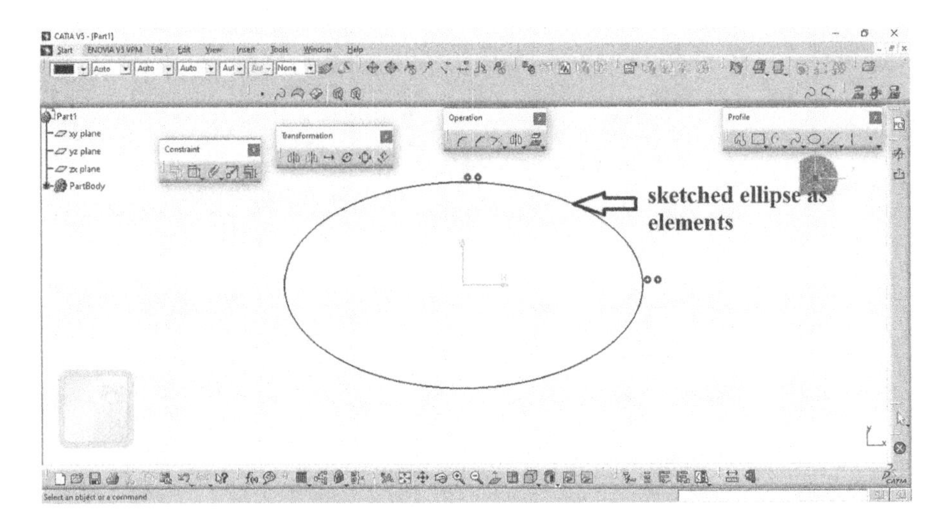

FIGURE 4.17 Sketched element for Semimajor axis Constraint.

After selecting sketched elements as shown in Figure 4.17, if you choose Semimajor axis check box from Constraint Definition dialog box, and press the OK button as shown in Figure 4.18, the diameter dimension of the major axis of ellipse is displayed, as shown in Figure 4.19. The Semimajor axis Constraint is always applied on diameter dimension of the major axis of ellipse.

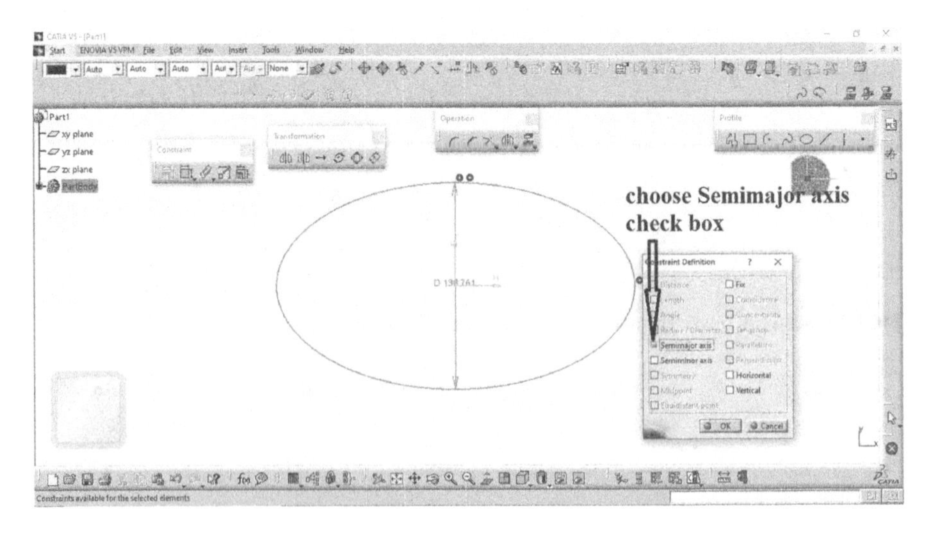

FIGURE 4.18 Selection of Semimajor axis Constraint in Constraint Definition dialog box.

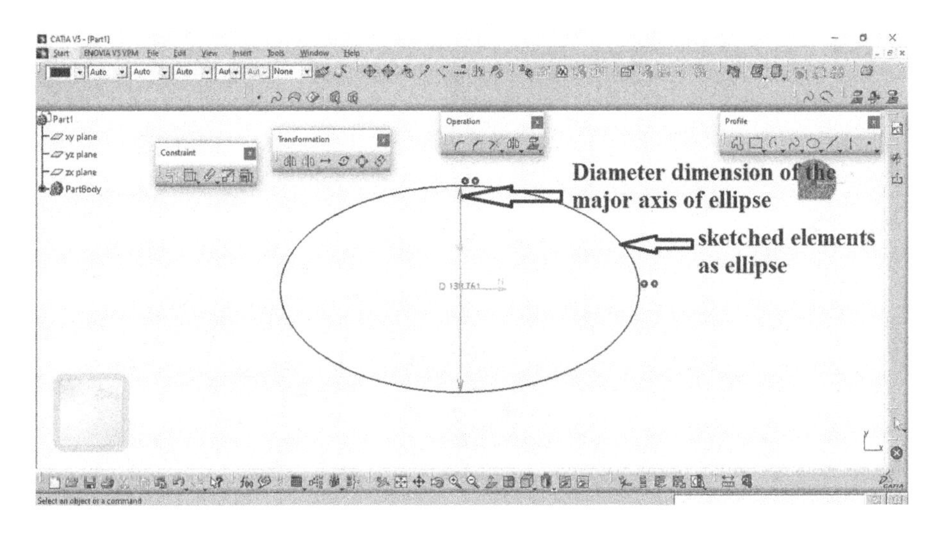

FIGURE 4.19 Constraining elements using Semimajor axis Constraint.

After selecting sketched elements as shown in Figure 4.20, if you choose Semiminor axis check box from Constraint Definition dialog box, and press the OK button as shown in Figure 4.21, the diameter dimension of the Semiminor axis of ellipse is displayed, as shown in Figure 4.22. The Semiminor axis Constraint is always applied on diameter dimension of the Semiminor axis.

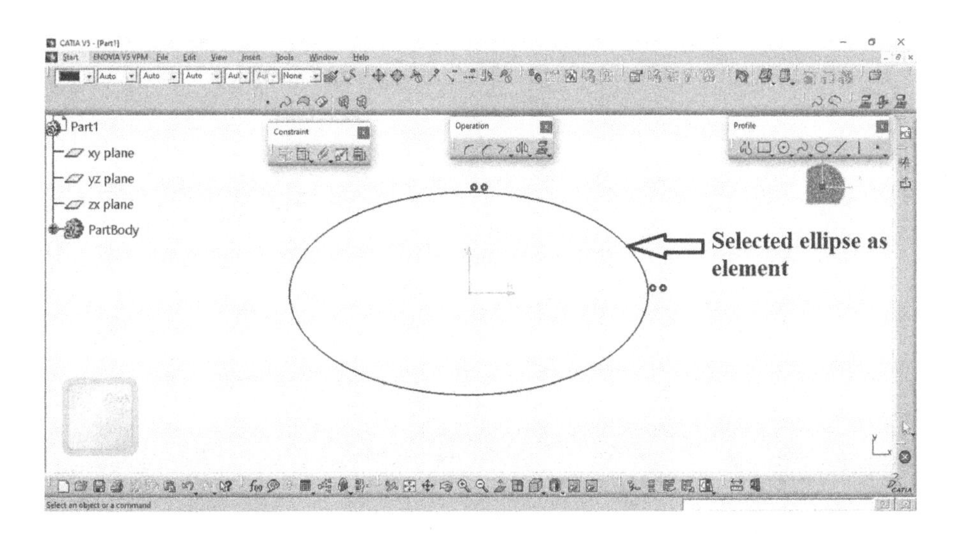

FIGURE 4.20 Sketched element for Semiminor axis Constraint.

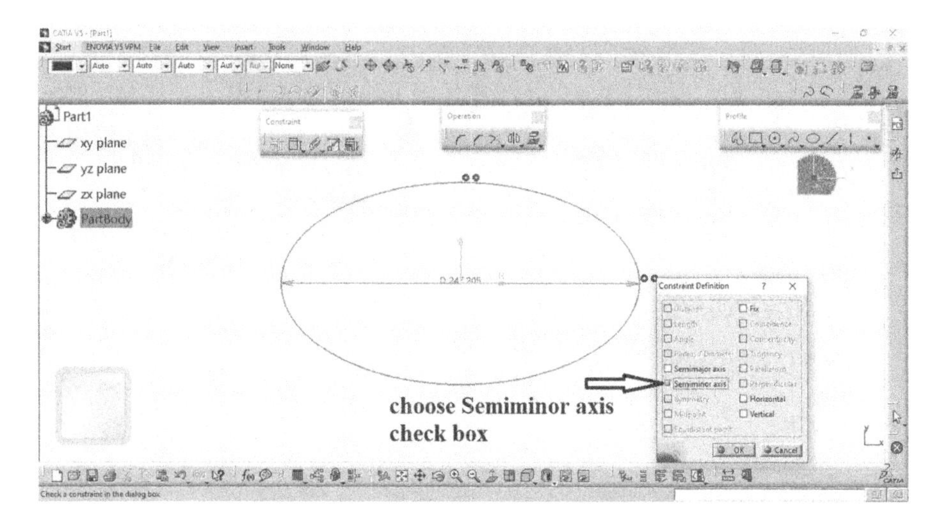

FIGURE 4.21 Selection of Semiminor axis Constraint in Constraint Definition dialog box.

After selecting sketched elements and the symmetry, which can be a line, axis, or the vertical and horizontal axis of the coordinate system in proper sequence as shown in Figure 4.23, if you choose Symmetry check box from Constraint Definition dialog box, and press the OK button as shown in Figure 4.24, the selected

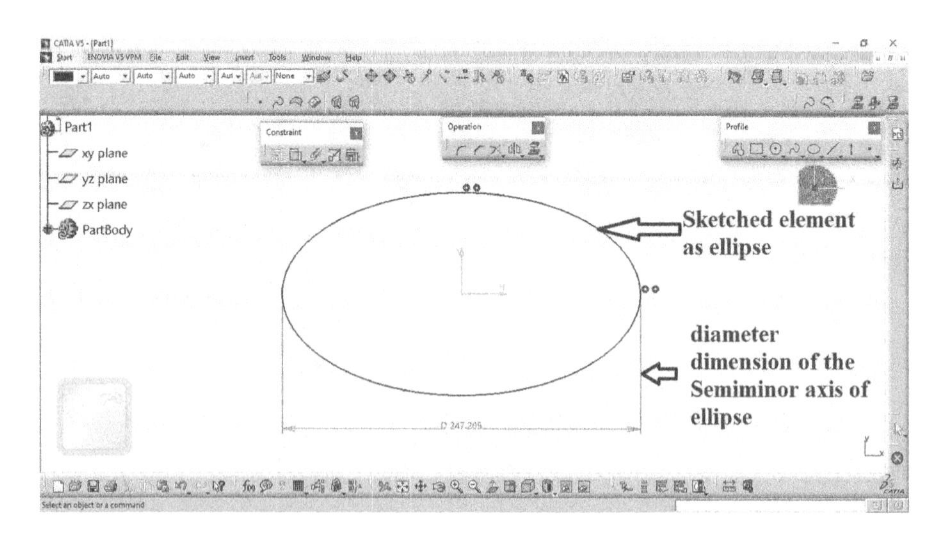

FIGURE 4.22 Constraining elements using Semiminor axis Constraint.

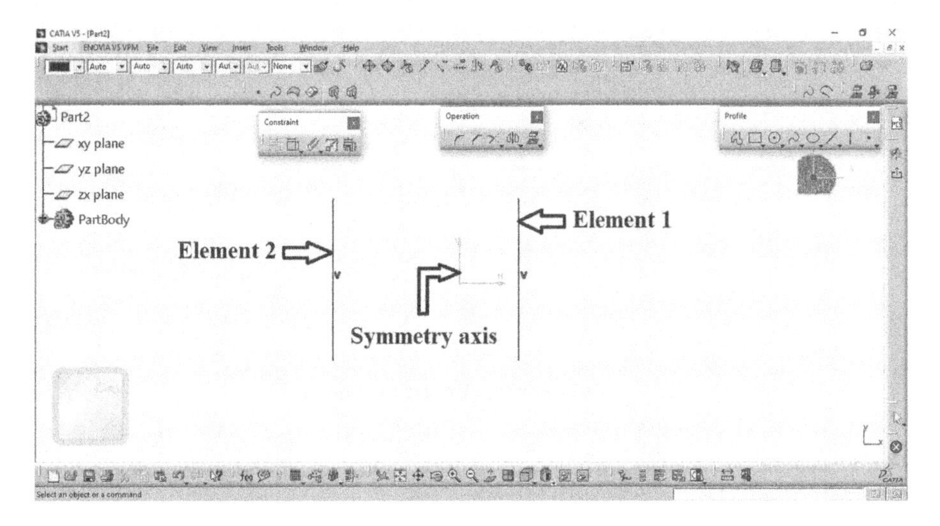

FIGURE 4.23 Sketched element for Symmetry Constraint.

sketched elements are symmetric along the selected symmetry axis and the symmetric symbol is displayed on the entities, as shown in Figure 4.25. The Symmetry Constraint always forces the selected elements to maintain an equal distance about a symmetry axis.

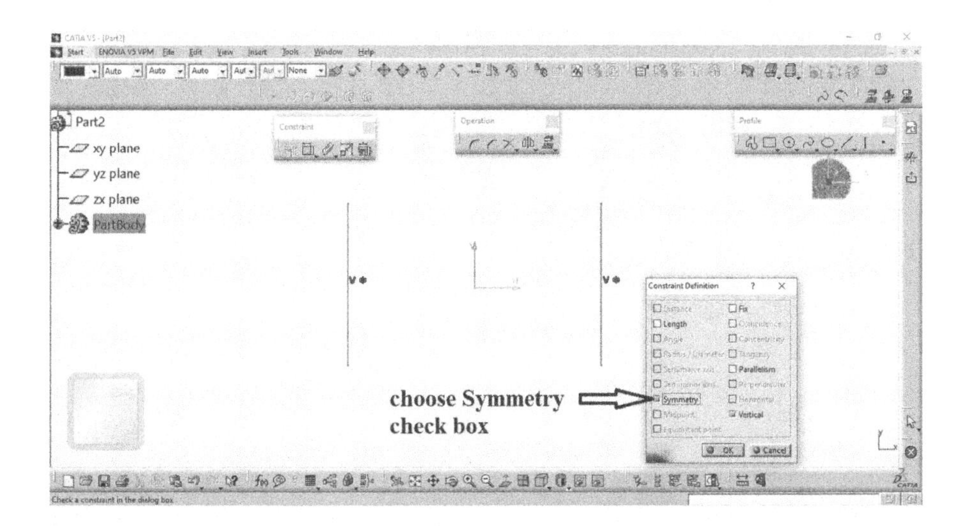

FIGURE 4.24 Selection of Symmetry Constraint in Constraint Definition dialog box.

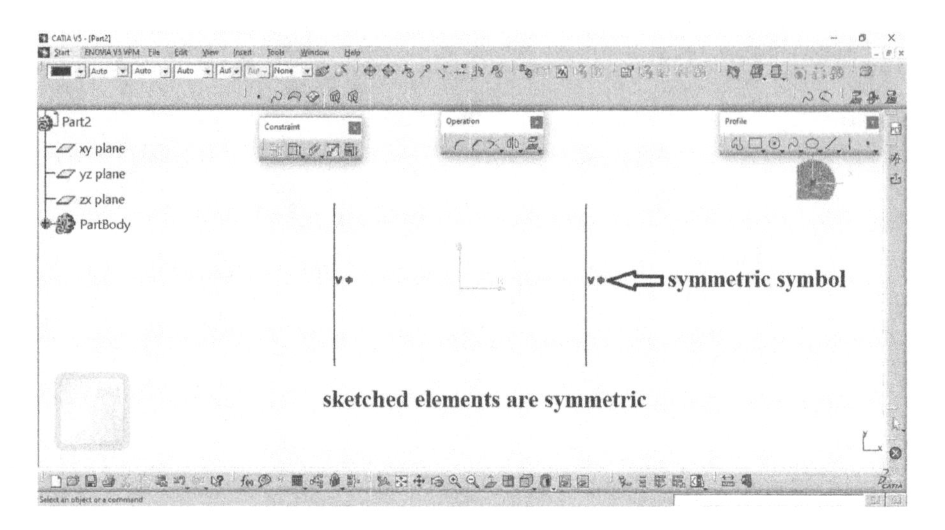

FIGURE 4.25 Constraining elements using Symmetry Constraint.

After selecting sketched elements line and point as shown in Figure 4.26, if you choose Midpoint check box from Constraint Definition dialog box, and press the OK button as shown in Figure 4.27, the selected point is placed in the middle of the selected line as shown in Figure 4.28. The Midpoint Constraint always forces the selected point to be placed on the middle of the selected line.

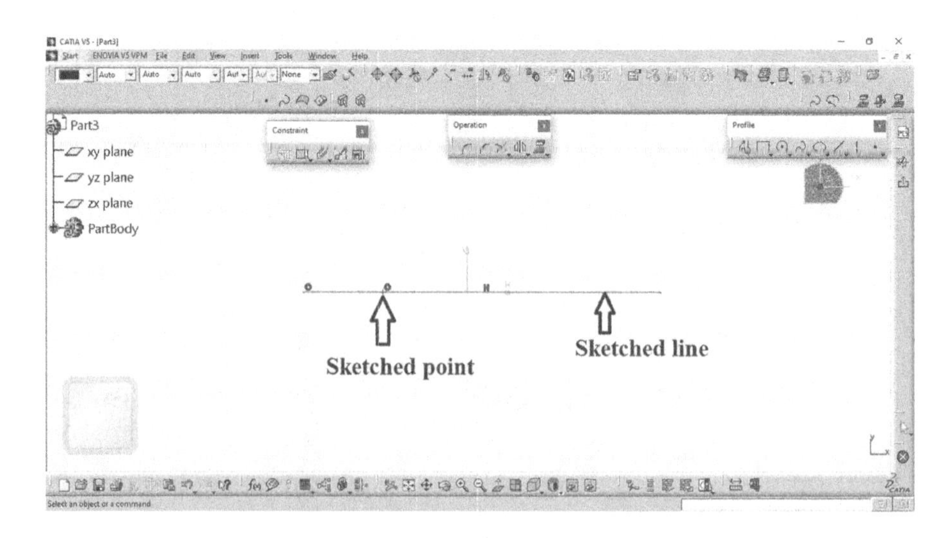

FIGURE 4.26 Sketched element for Midpoint Constraint.

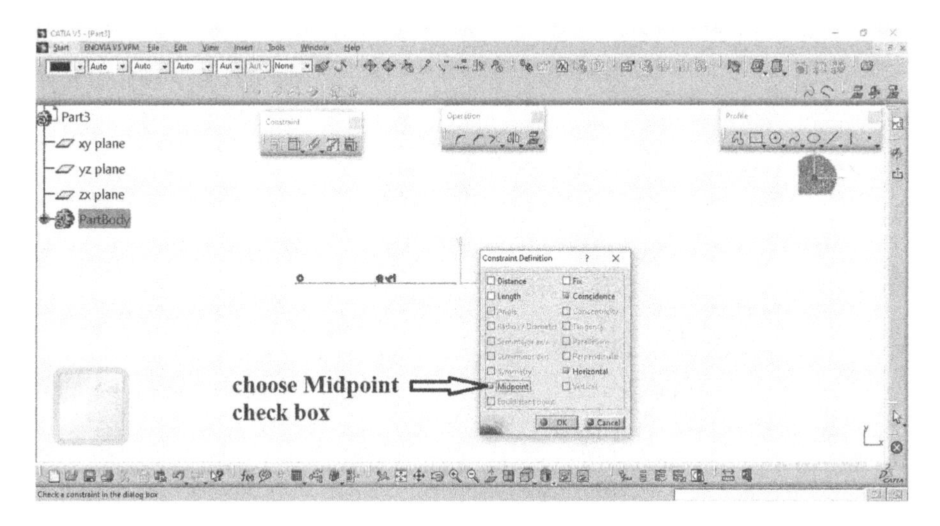

FIGURE 4.27 Selection of Midpoint Constraint in Constraint Definition dialog box.

After selecting sketched point-1, 2 and 3 in proper sequence as shown in Figure 4.29, if you choose Equidistant point check box from Constraint Definition dialog box, and press the OK button as shown in Figure 4.30, the two selected point will be placed at an equal distance from the third point, as shown in Figure 4.31. The Equidistant point Constraint always forces the selected points to maintain an equal distance from third point.

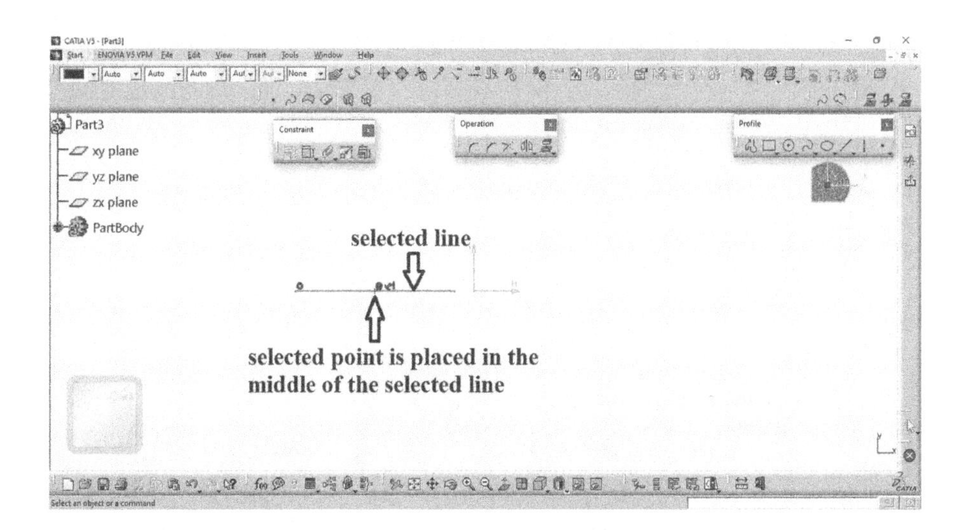

FIGURE 4.28 Constraining elements using Midpoint Constraint.

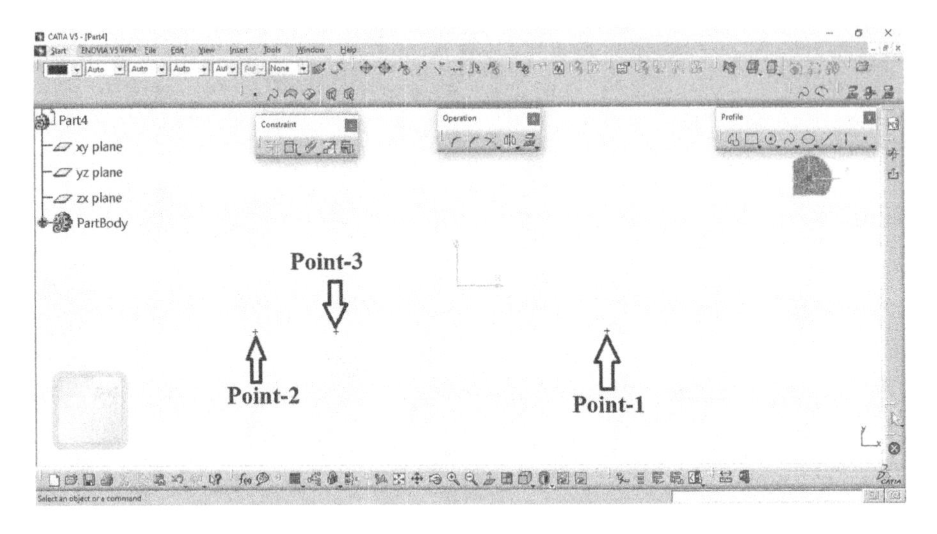

FIGURE 4.29 Sketched element for Equidistant point Constraint.

After selecting sketched elements as shown in Figure 4.32, if you choose Fix check box from Constraint Definition dialog box, and press the OK button as shown in Figure 4.33, the element is fixed as shown in Figure 4.34. The Fix Constraint always forces the selected entity to be locked at its location.

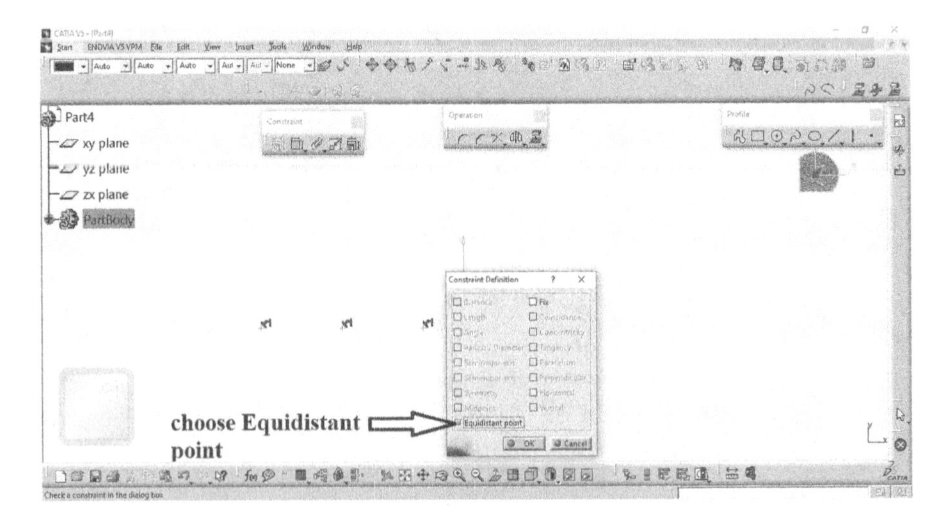

FIGURE 4.30 Selection of Equidistant point Constraint in Constraint Definition dialog box.

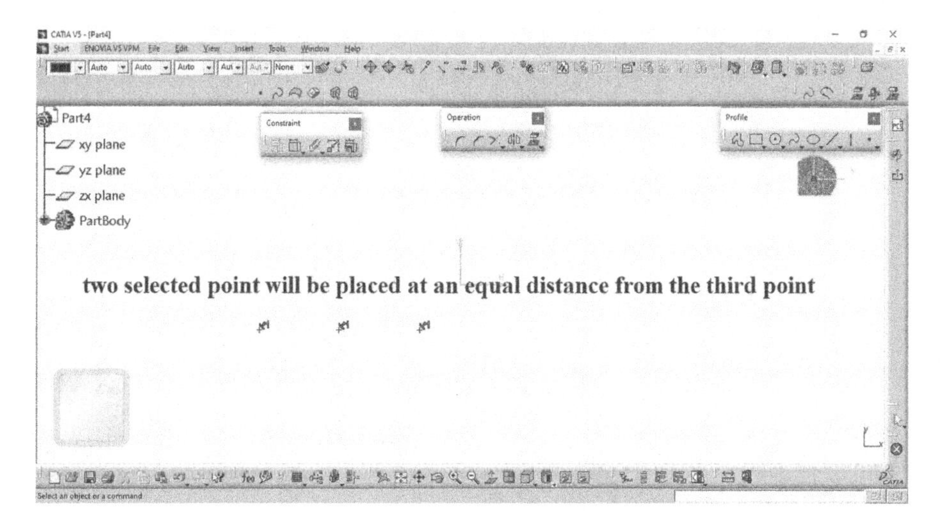

FIGURE 4.31 Constraining elements using Equidistant point Constraint.

After selecting sketched elements as shown in Figure 4.35, if you choose Coincident check box from Constraint Definition dialog box, and press the OK button as shown in Figure 4.36, the selected elements will coincide with each other as shown in Figure 4.37. The Coincident Constraint always forces a selected entity to coincide with each other.

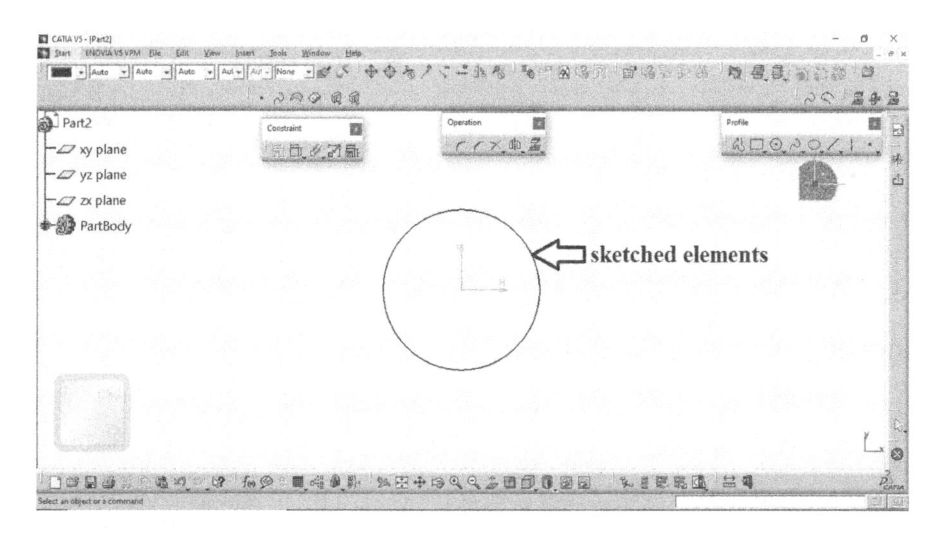

FIGURE 4.32 Sketched element for Fix Constraint.

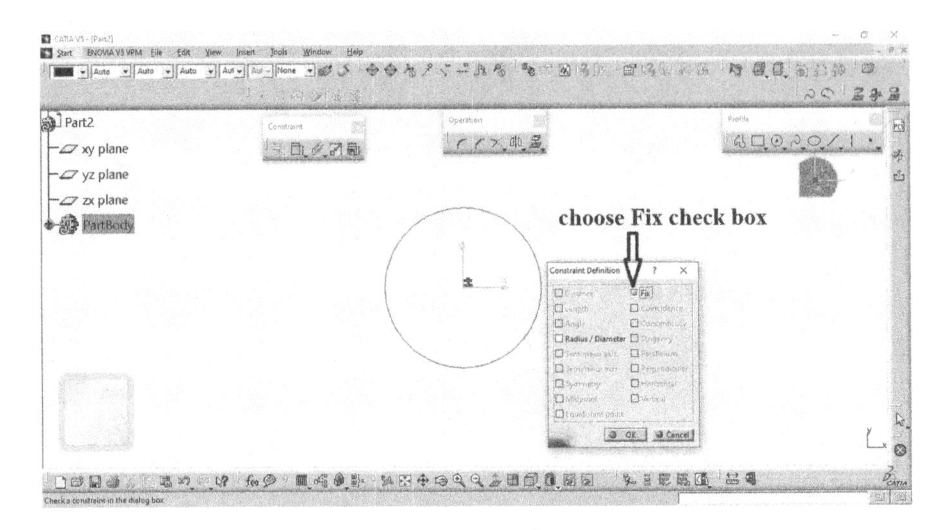

FIGURE 4.33 Selection of Fix Constraint in Constraint Definition dialog box.

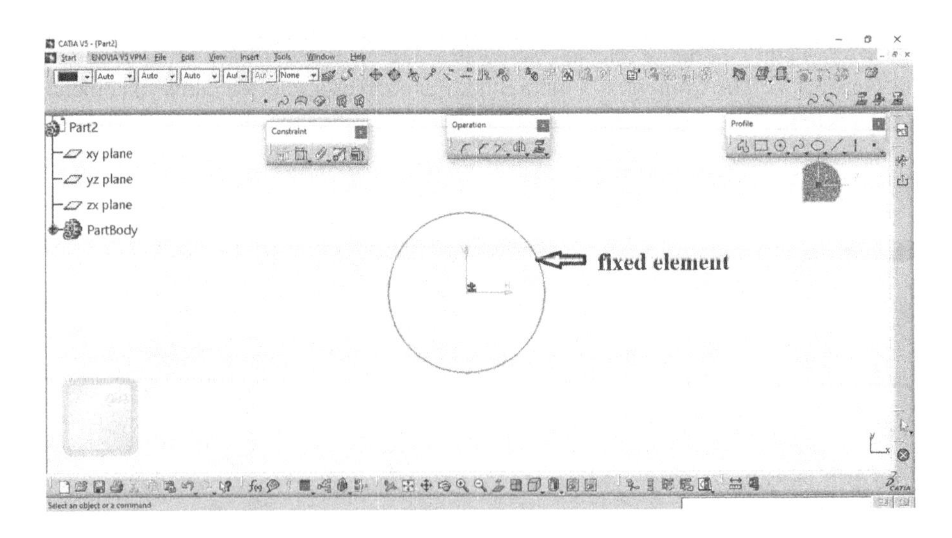

FIGURE 4.34 Constraining elements using Fix Constraint.

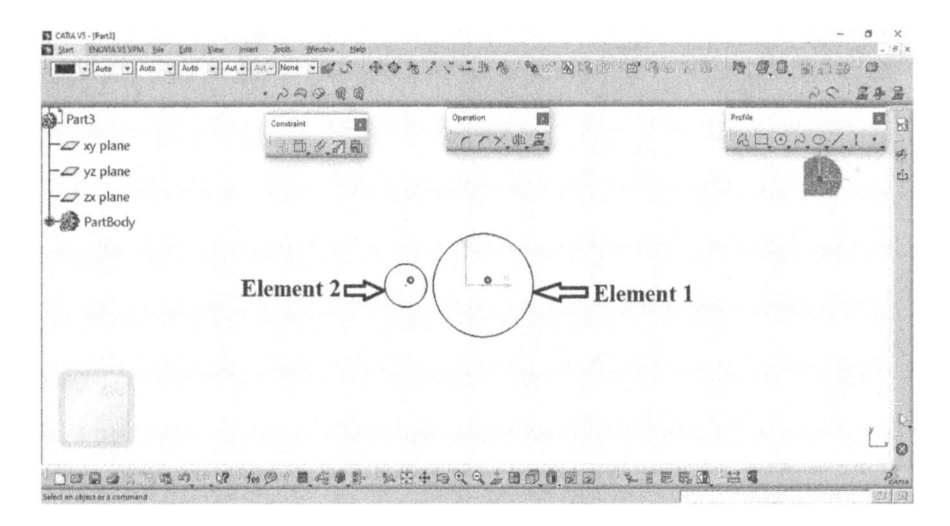

FIGURE 4.35 Sketched element for Coincident Constraint.

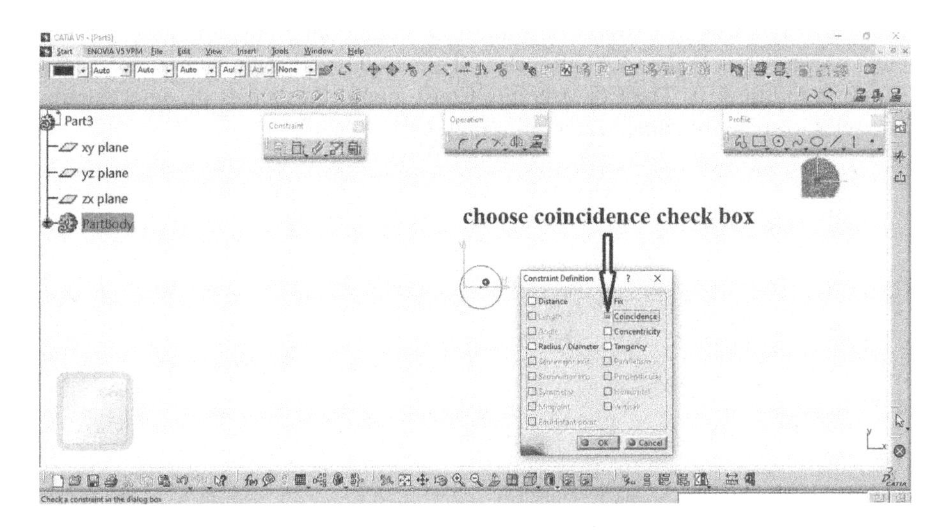

FIGURE 4.36 Selection of Coincident Constraint in Constraint Definition dialog box.

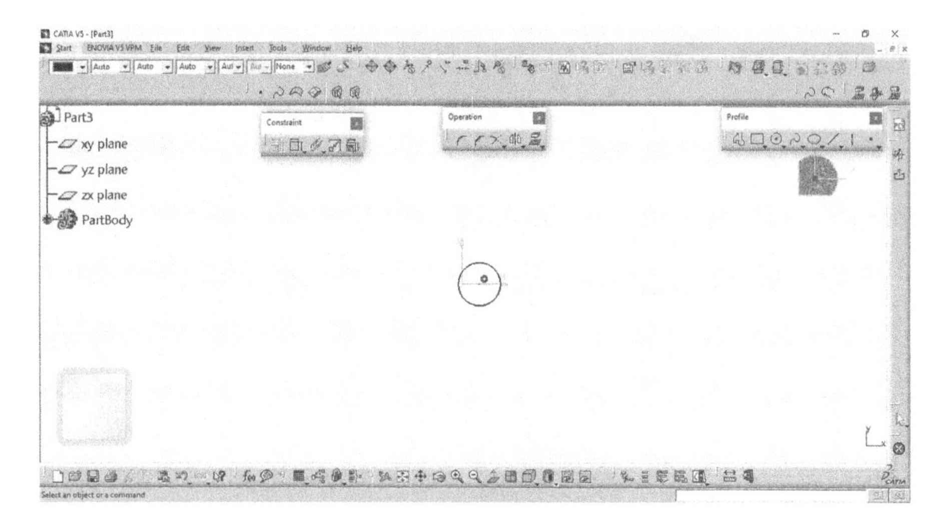

FIGURE 4.37 Constraining elements using Coincident Constraint.

After selecting sketched element circles as shown in Figure 4.29, if you choose Concentricity check box from Constraint Definition dialog box, and press the OK button as shown in Figure 4.30, the selected circles with coincide with each other as shown in Figure 4.31. The Concentricity Constraint always forces the selected arcs or circles to share the same center point Figures 4.38–4.40.

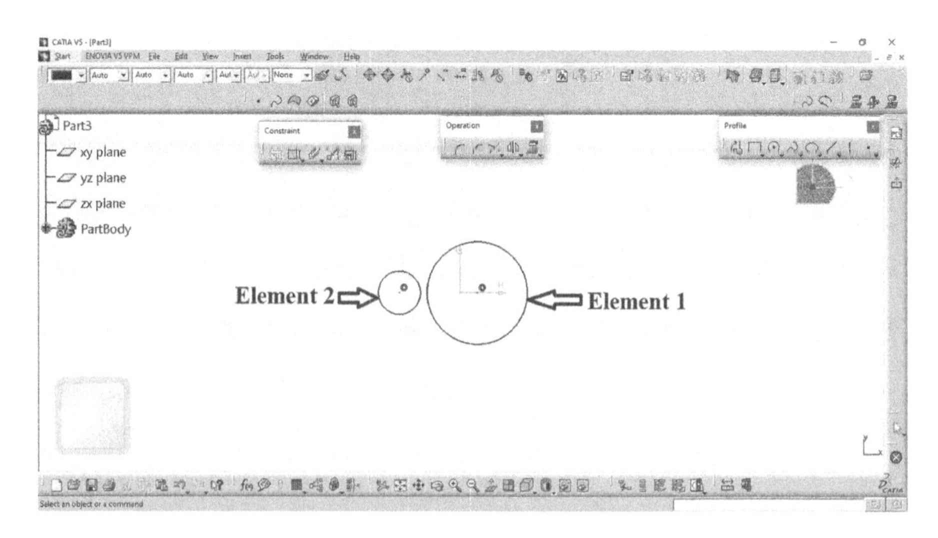

FIGURE 4.38 Sketched element for Concentricity Constraint.

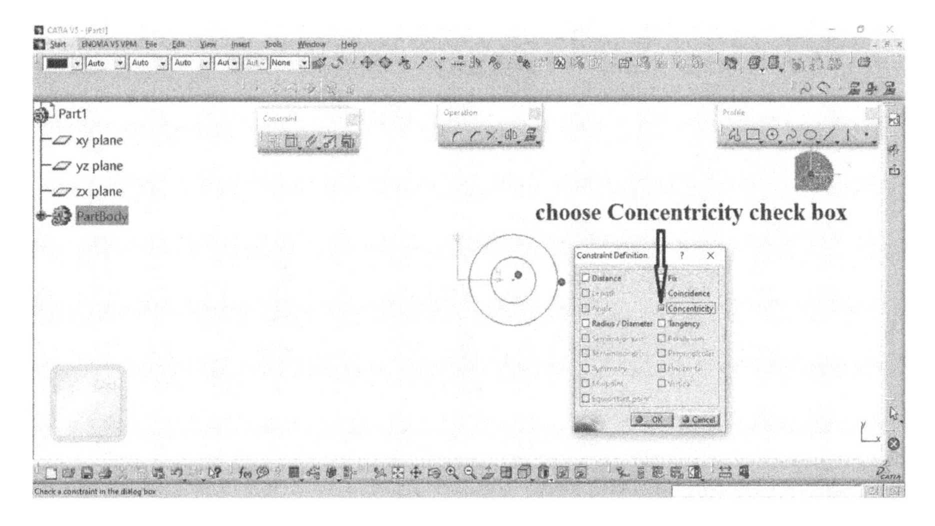

FIGURE 4.39 Selection of Concentricity Constraint in Constraint Definition dialog box.

After selecting sketched elements line and circle as shown in Figure 4.41, if you choose Tangency check box from Constraint Definition dialog box, and press the OK button as shown in Figure 4.42, the selected line and circle become tangent to each other as shown in Figure 4.43. The Tangency Constraint always forces the selected elements to become tangent to each other.

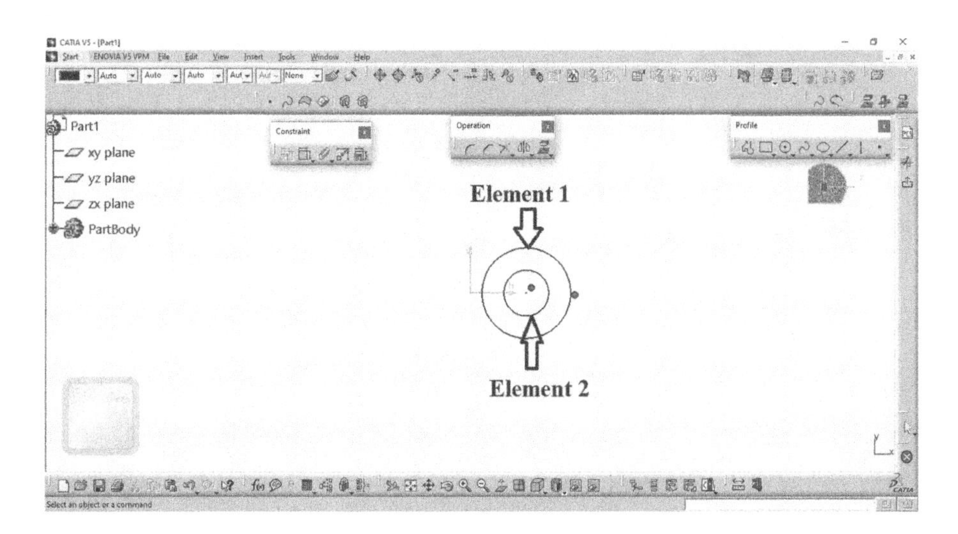

FIGURE 4.40 Constraining elements using Concentricity Constraint.

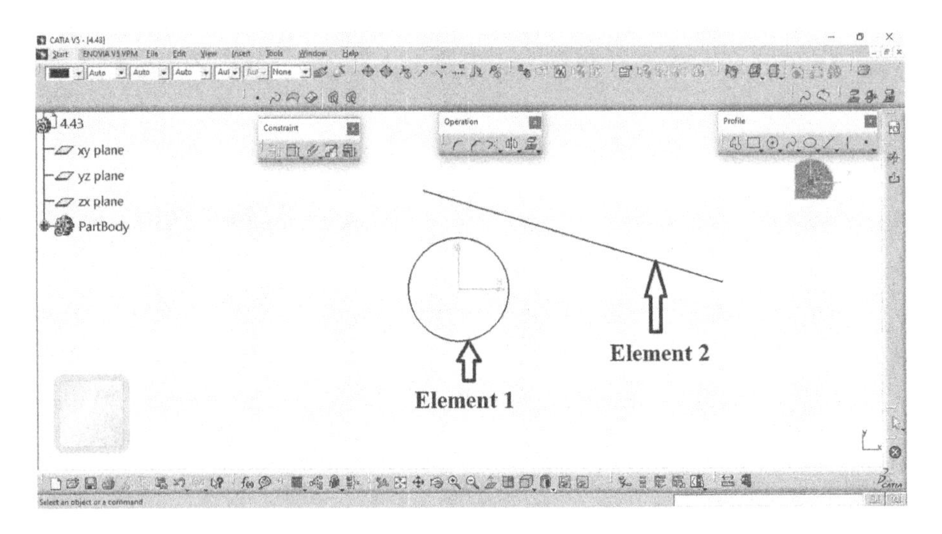

FIGURE 4.41 Sketched element for Tangency Constraint.

After selecting sketched two lines as shown in Figure 4.44, if you choose Parallelism check box from Constraint Definition dialog box, and press the OK button as shown in Figure 4.45, the selected two lines to be parallel to each other as shown in Figure 4.46. The Parallelism Constraint always forces the selected lines to be parallel to each other.

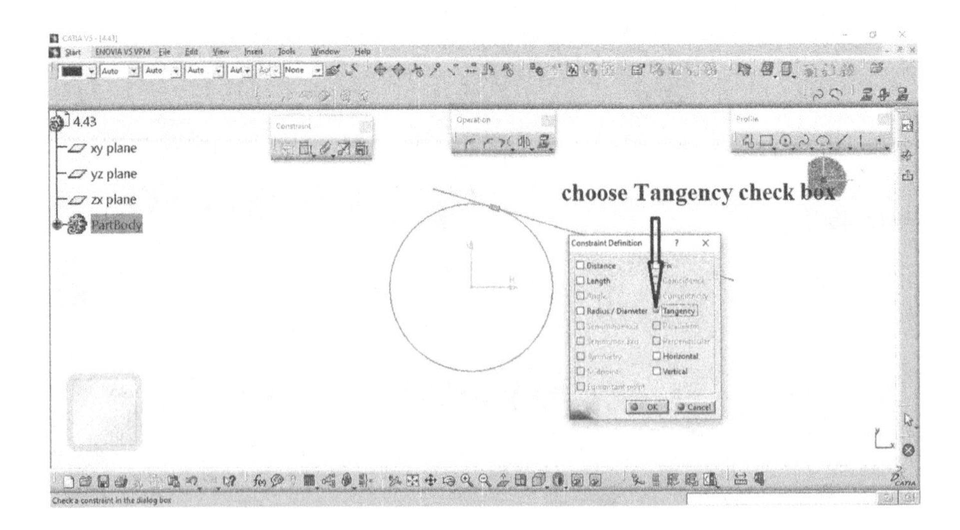

FIGURE 4.42 Selection of Tangency Constraint in Constraint Definition dialog box.

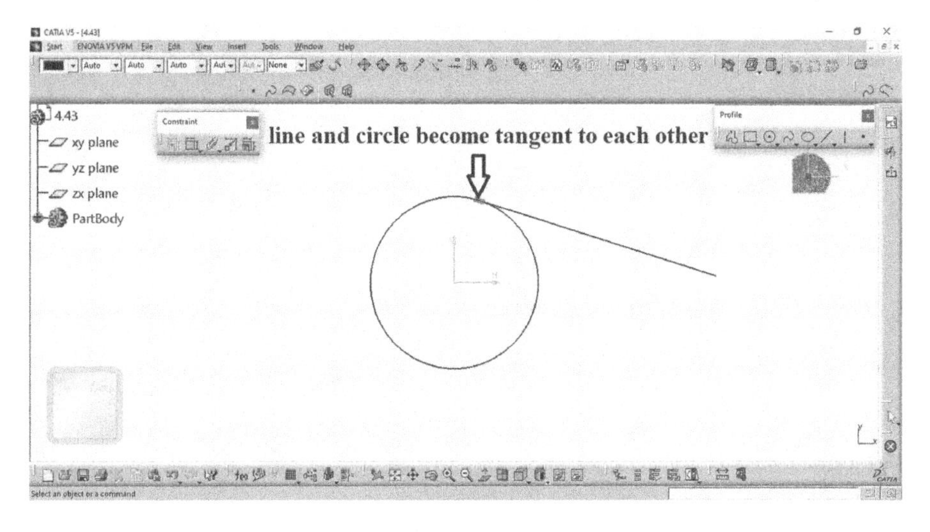

FIGURE 4.43 Constraining elements using Tangency Constraint.

After selecting sketched two lines as shown in Figure 4.47, if you choose Perpendicular check box from Constraint Definition dialog box, and press the OK button as shown in Figure 4.48, the selected two lines have to be perpendicular to each other as shown in Figure 4.49. The Perpendicular Constraint always forces the selected lines to be perpendicular to each other.

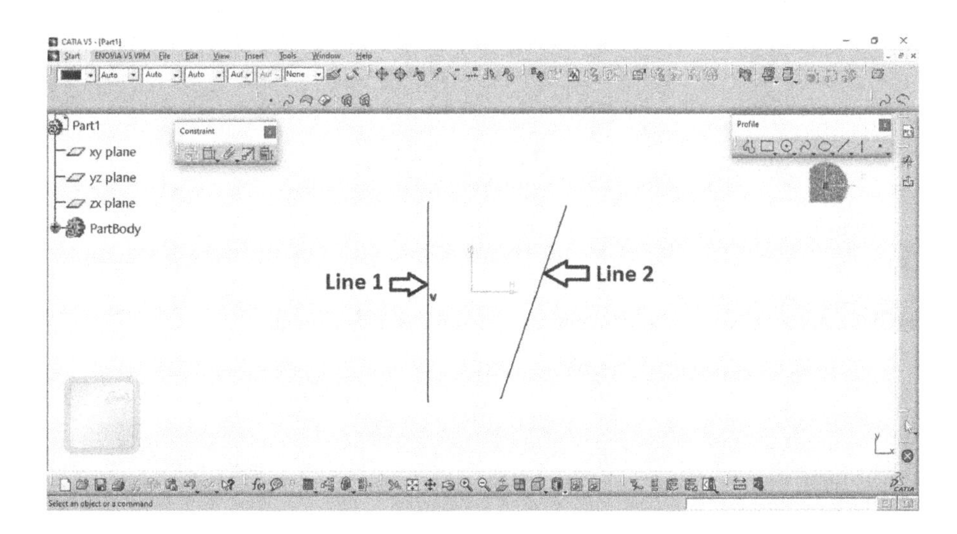

FIGURE 4.44 Sketched element for Parallelism Constraint.

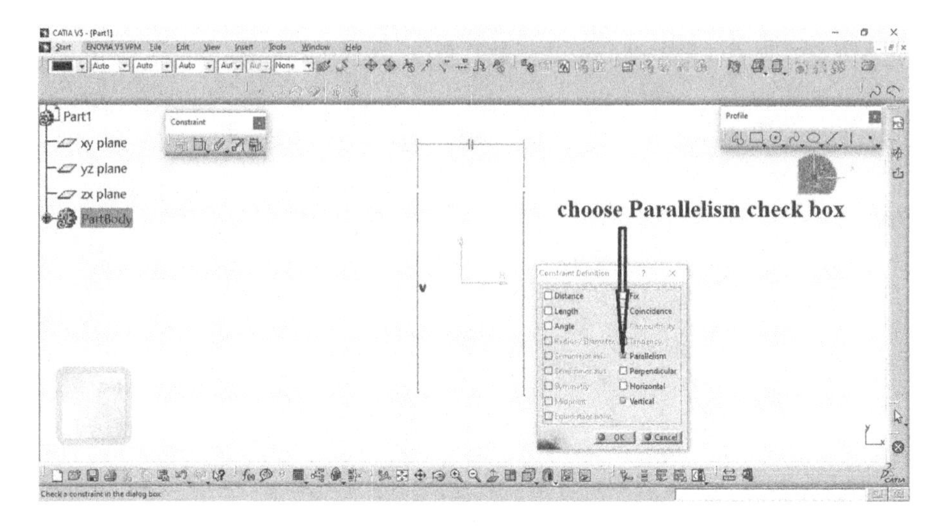

FIGURE 4.45 Selection of Parallelism Constraint in Constraint Definition dialog box.

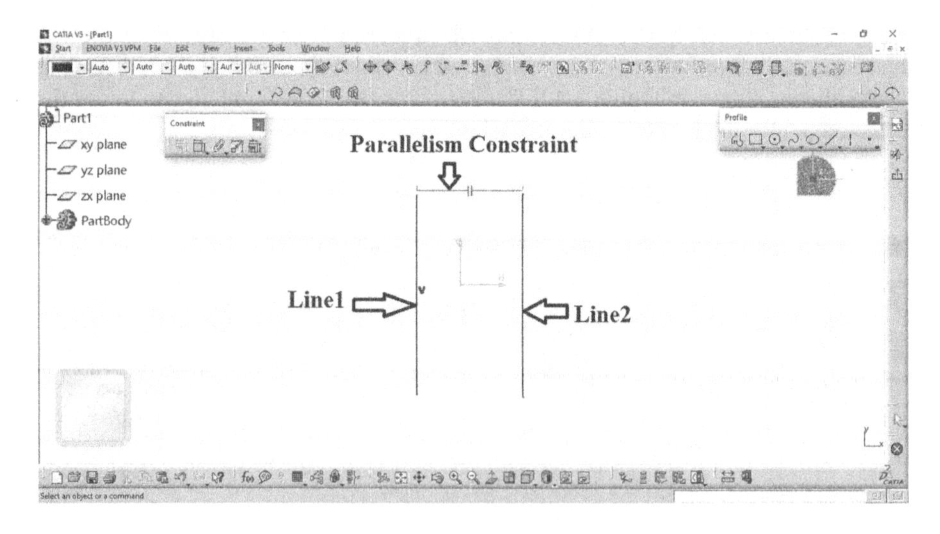

FIGURE 4.46 Constraining elements using Parallelism Constraint.

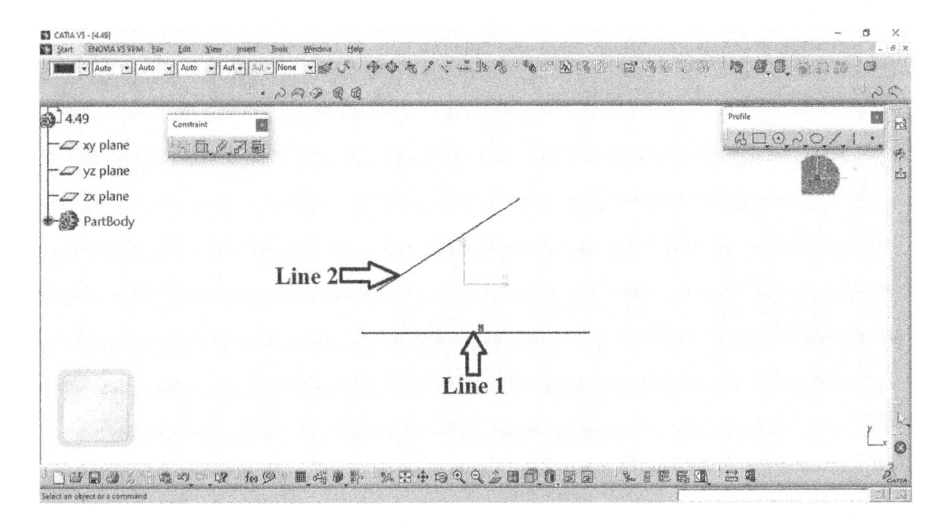

FIGURE 4.47 Sketched element for Perpendicular Constraint.

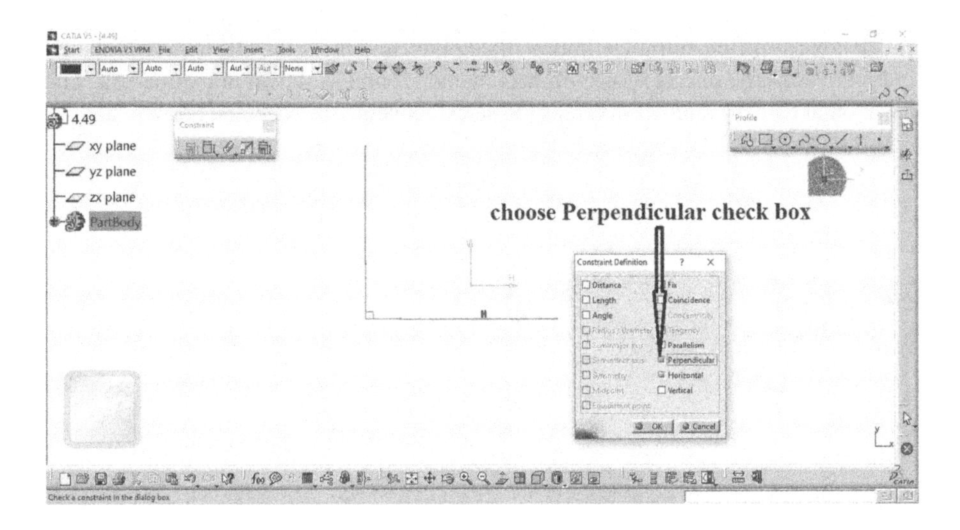

FIGURE 4.48 Selection of Perpendicular Constraint in Constraint Definition dialog box.

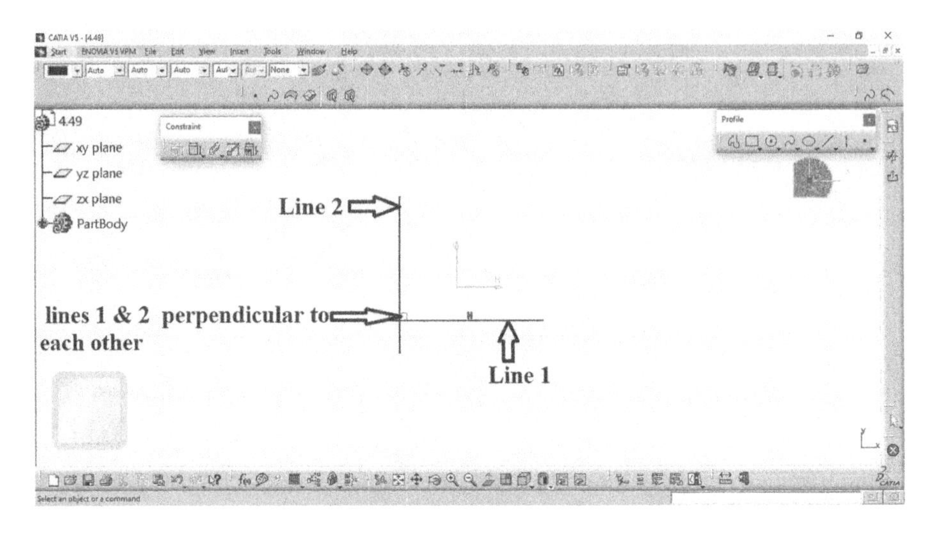

FIGURE 4.49 Constraining elements using Perpendicular Constraint.

After selecting sketched line as shown in Figure 4.50, if you choose Horizontal check box from Constraint Definition dialog box, and press the OK button as shown in Figure 4.51, the selected line will become horizontal as shown in Figure 4.52. The Horizontal Constraint always forces the selected element to become horizontal.

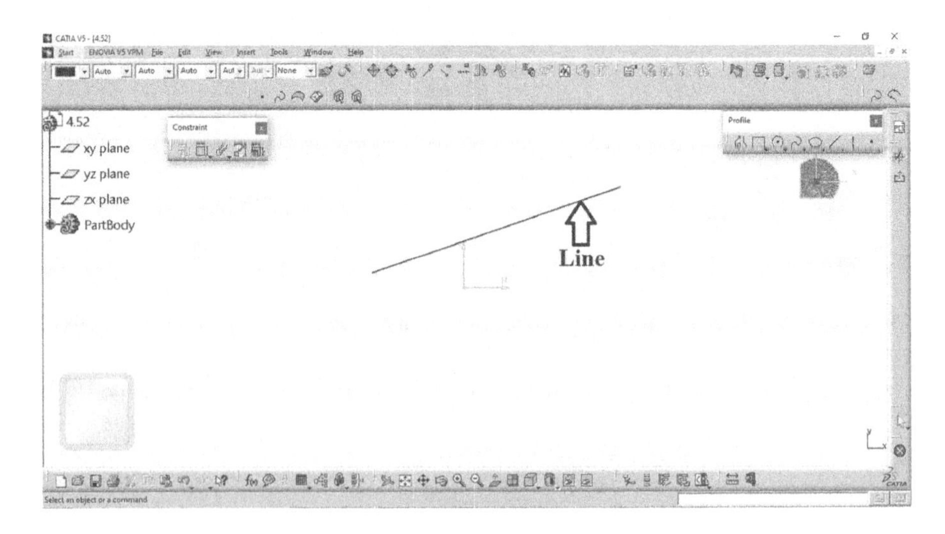

FIGURE 4.50 Sketched element for Horizontal Constraint.

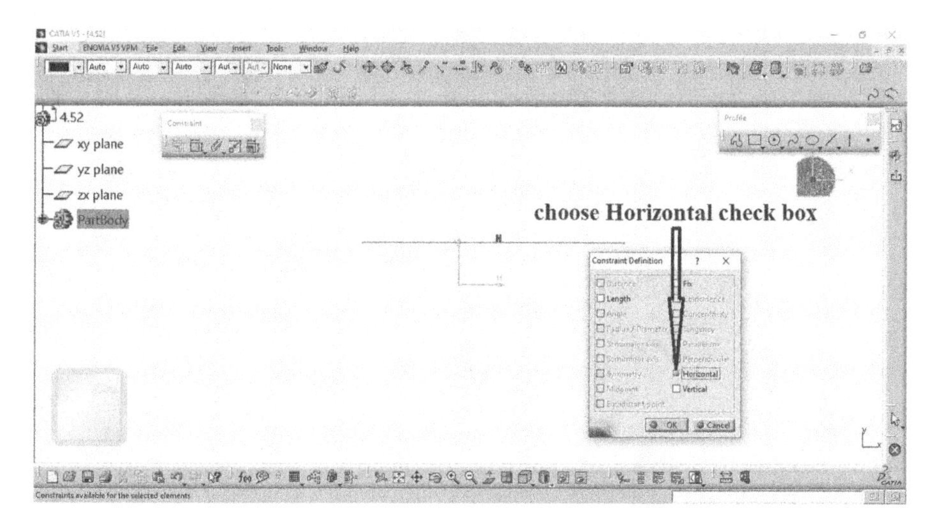

FIGURE 4.51 Selection of Horizontal Constraint in Constraint Definition dialog box.

After selecting sketched line as shown in Figure 4.53, if you choose Vertical check box from Constraint Definition dialog box, and press the OK button as shown in Figure 4.54, the selected line will become vertical as shown in Figure 4.55. The Vertical Constraint always forces the selected element to become vertical.

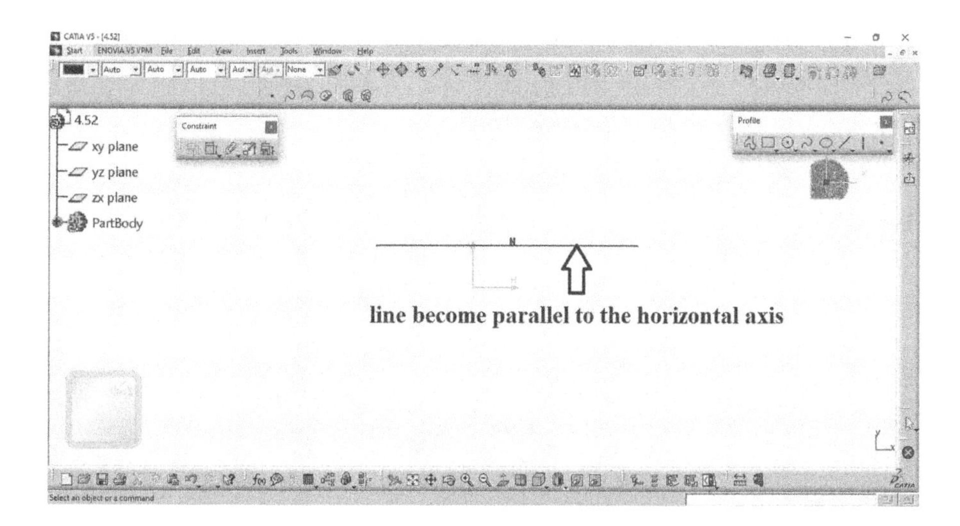

FIGURE 4.52 Constraining elements using Horizontal Constrain.

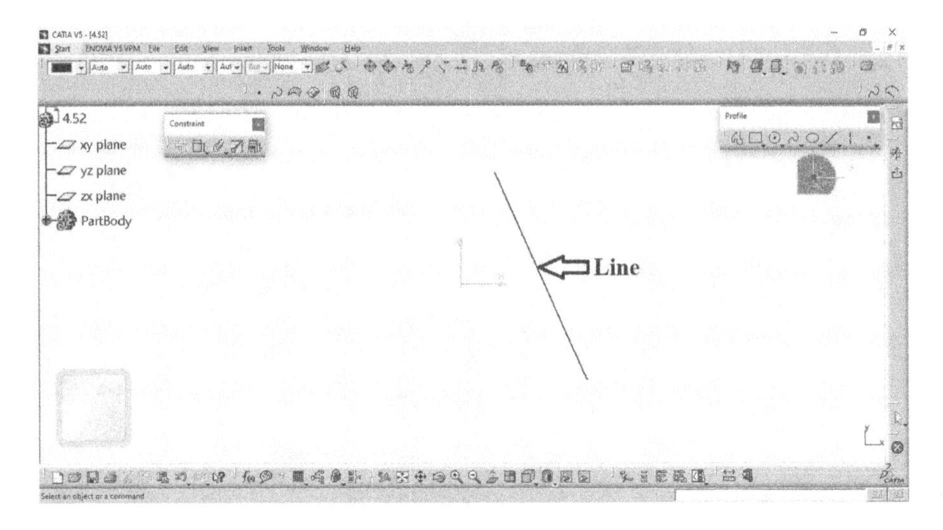

FIGURE 4.53 Sketched element for Vertical Constraint.

4.6 DIMENSIONAL CONSTRAINTS

After providing the geometric constraints, dimensional constraints are required to be defined sketch. To invoke Constraint Creation toolbar, choose the down arrow on the right of the Constraint button, as shown in Figure 4.3.

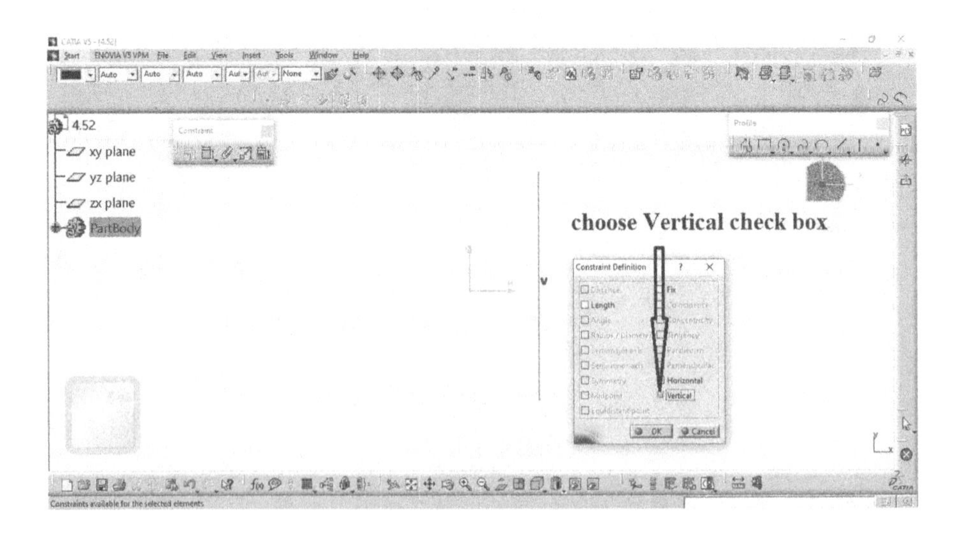

FIGURE 4.54 Selection of Vertical Constraint in Constraint Definition dialog box.

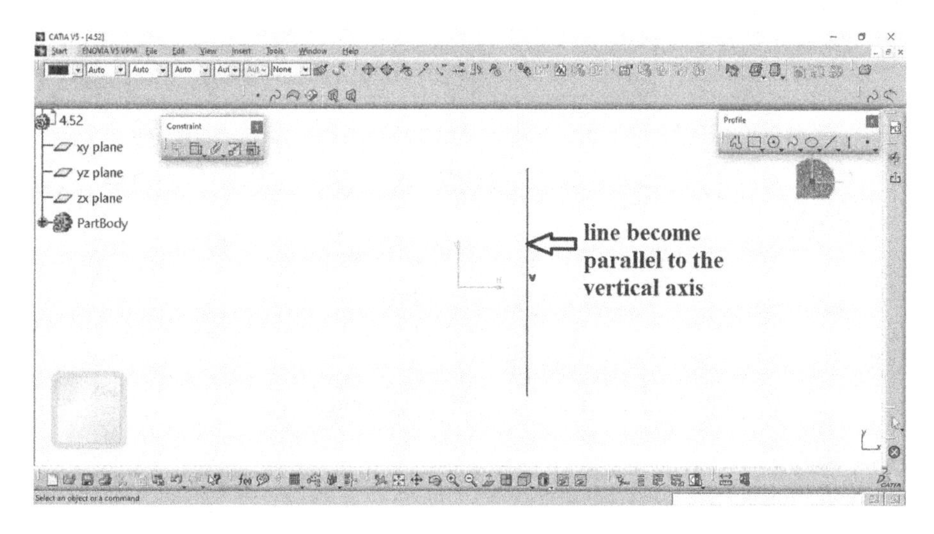

FIGURE 4.55 Constraining elements using Vertical Constraint.

4.6.1 Dimensioning of Straight Line

To apply dimension on straight line, invoke the Constraint tool from the Constraints toolbar, and select line from the geometry area as shown in Figure 4.56. A linear dimension is attached to the cursor. Click in geometry area to place the dimension, as shown in Figure 4.57.

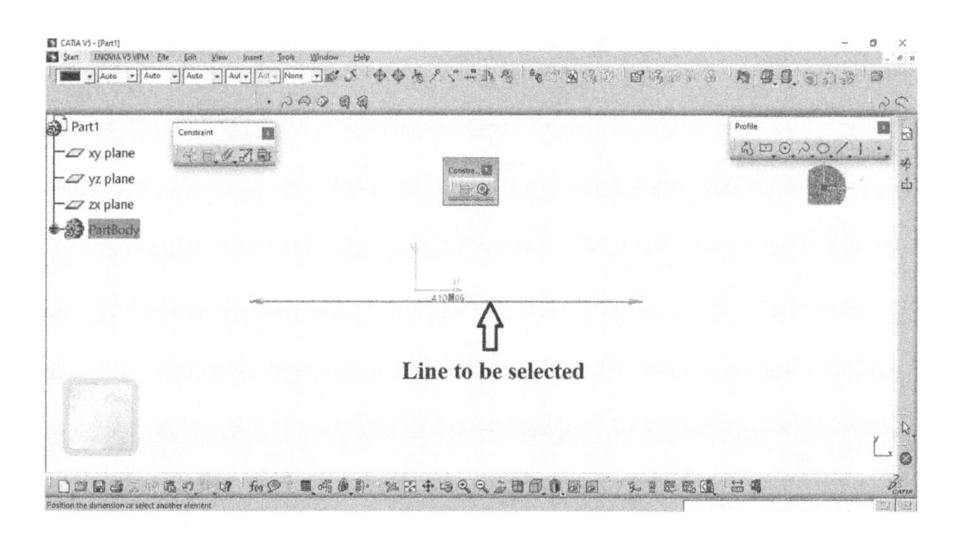

FIGURE 4.56 Selecting line for dimension.

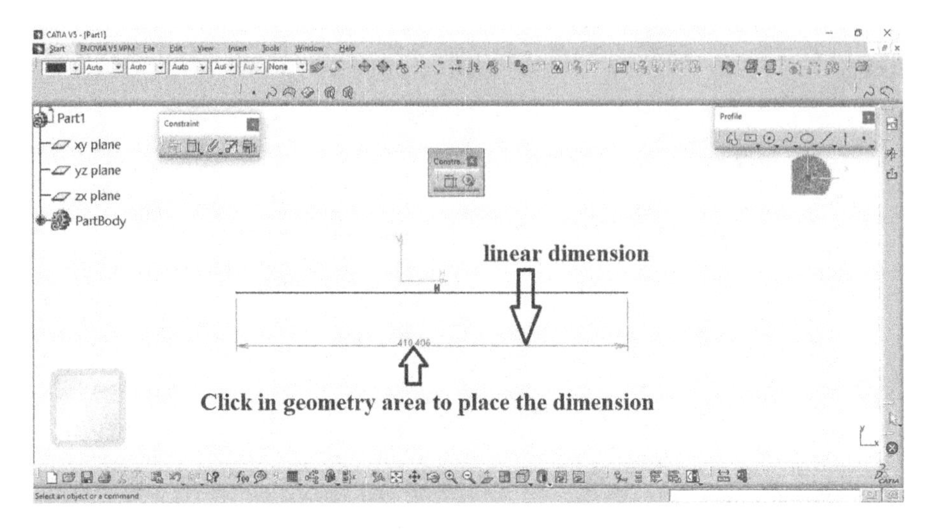

FIGURE 4.57 Dimension placement for selected line.

4.6.2 Dimensioning of Inclined Line

To apply dimension on inclined lines, invoke the constraint tool from the constraints toolbar, and select line from the geometry area as shown in Figure 4.58. An aligned dimension is attached to the cursor. Click in geometry area to place the dimension, as shown in Figure 4.59.

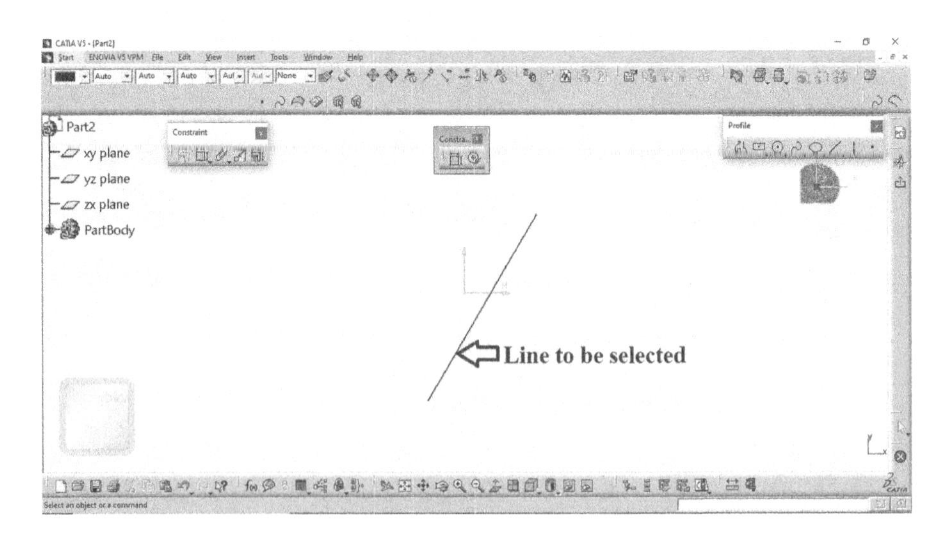

FIGURE 4.58 Selecting line for dimension.

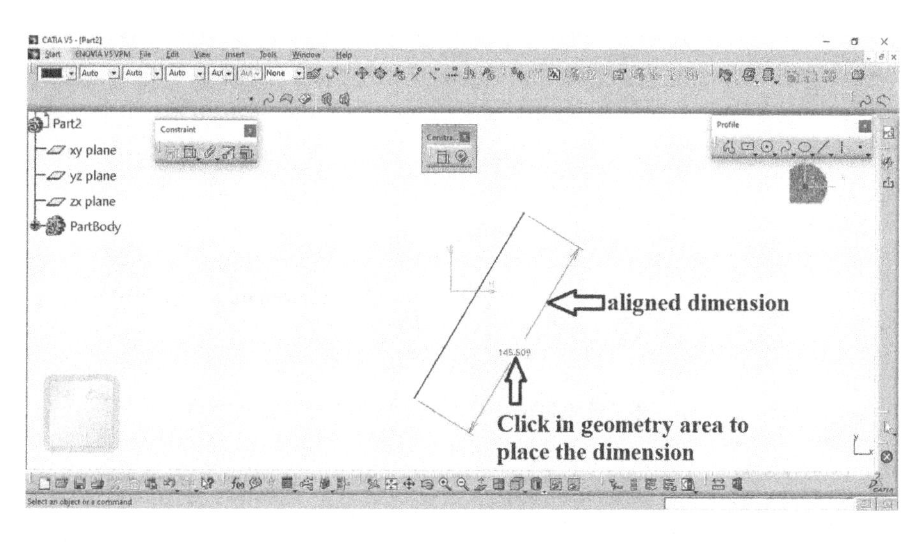

FIGURE 4.59 Dimension placement for selected line.

To apply a horizontal dimension on an inclined line, invoke the Constraint tool from the Constraints toolbar, and select inclined line from the geometry area as shown in Figure 4.58. An aligned dimension is attached to the cursor. Next, right-click to display the contextual menu on which choose the horizontal measure direction, as shown in Figure 4.60. Click in geometry area to place the dimension, as shown in Figure 4.61.

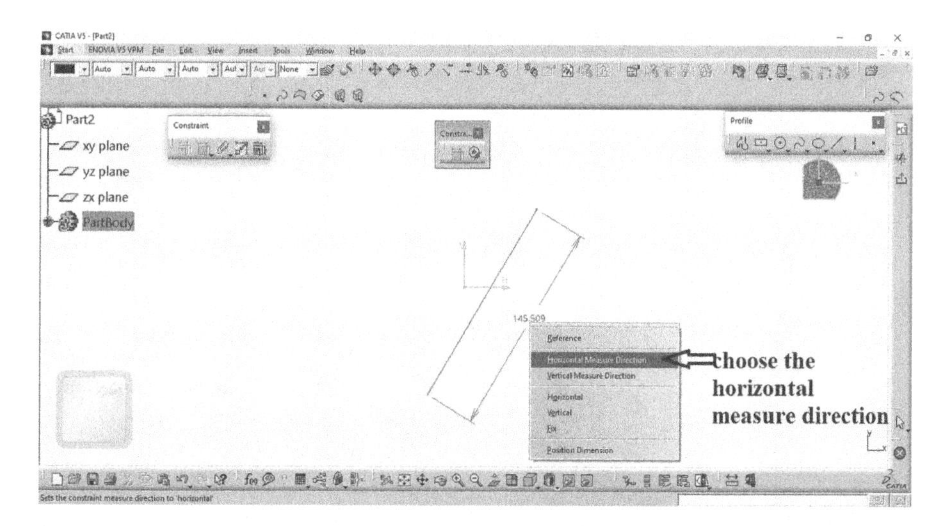

FIGURE 4.60 Choose the horizontal measure direction from contextual menu.

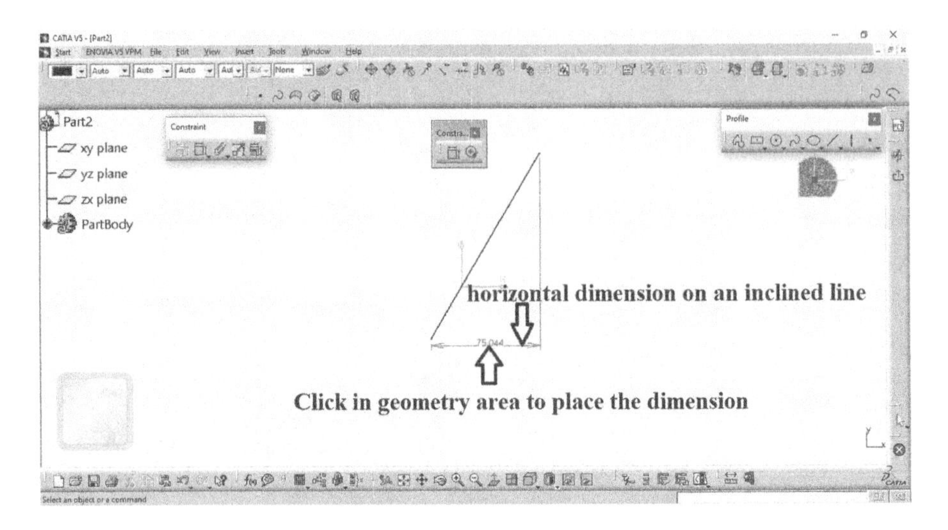

FIGURE 4.61 Dimension placement for inclined line in horizontal direction.

To apply a vertical dimension on an inclined line, invoke the Constraint tool from the Constraints toolbar, and select inclined line from the geometry area as shown in Figure 4.58. An aligned dimension is attached to the cursor. Next, right-click to display the contextual menu on which choose the vertical measure direction, as shown in Figure 4.62. Click in geometry area to place the dimension, as shown in Figure 4.63.

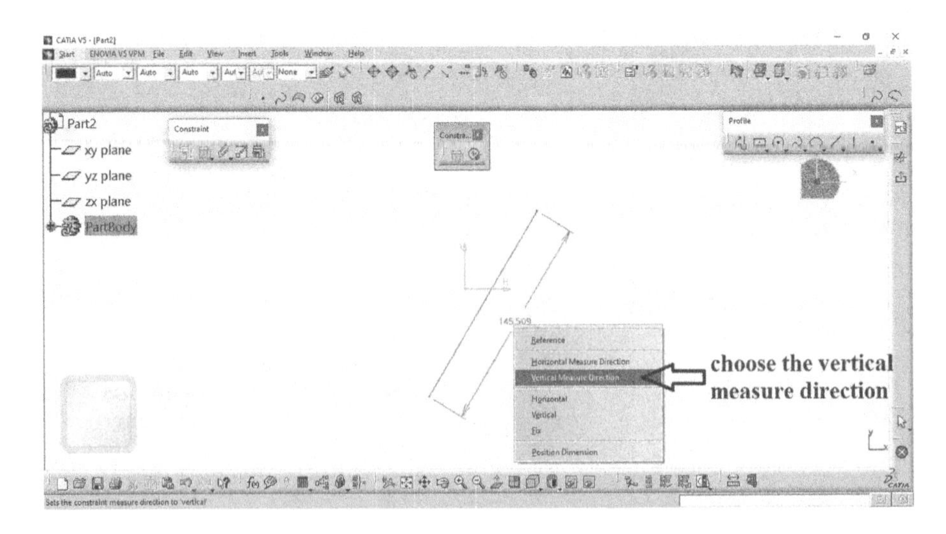

FIGURE 4.62 Choose the vertical measure direction from contextual menu.

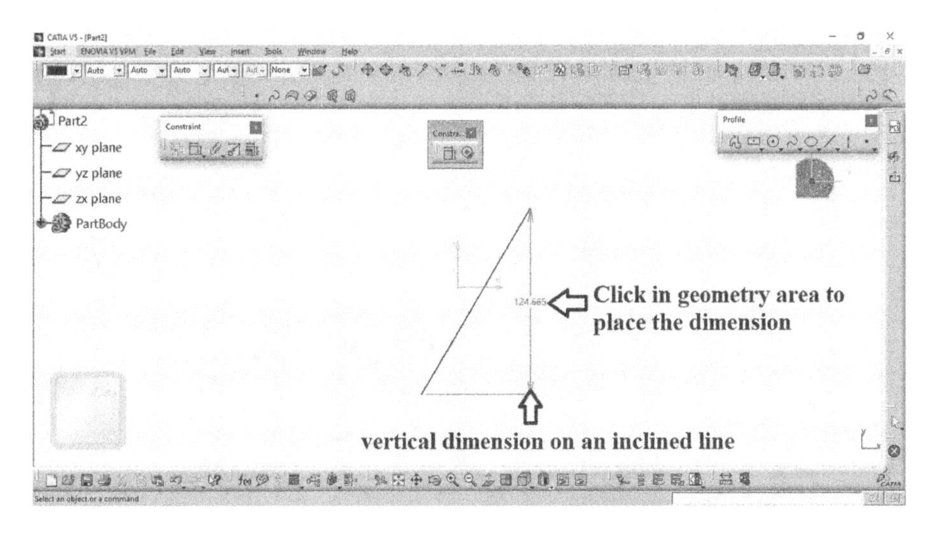

FIGURE 4.63 Dimension placement for inclined line in vertical direction.

4.6.3 Dimensioning between Two Points

To apply a horizontal dimension between two points, invoke the Constraint tool from the Constraints toolbar, and select point 1 and 2 from the geometry area as shown in Figure 4.64. A dimension is attached to the cursor. Next, right-click to display the contextual menu on which choose the horizontal measure direction, as shown in Figure 4.65. Click in geometry area to place the dimension, as shown in Figure 4.66.

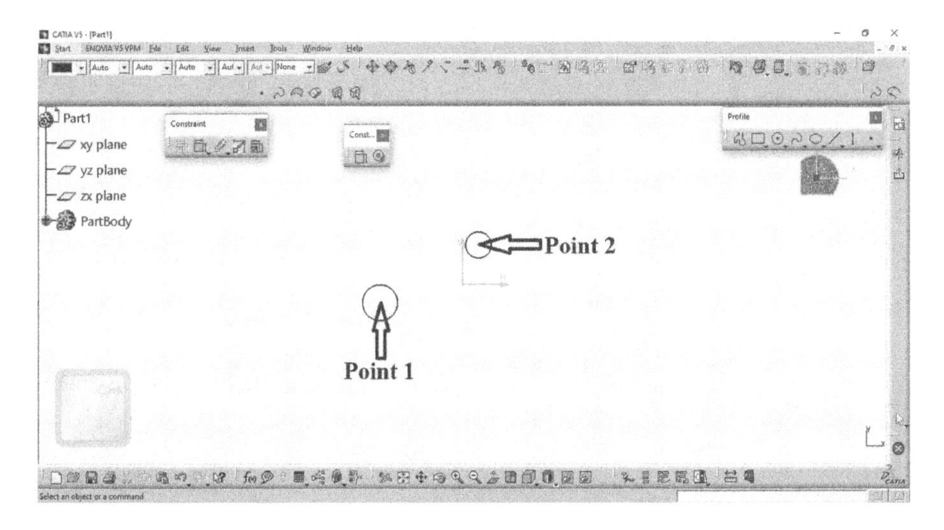

FIGURE 4.64 Selecting points 1 and 2 for dimension.

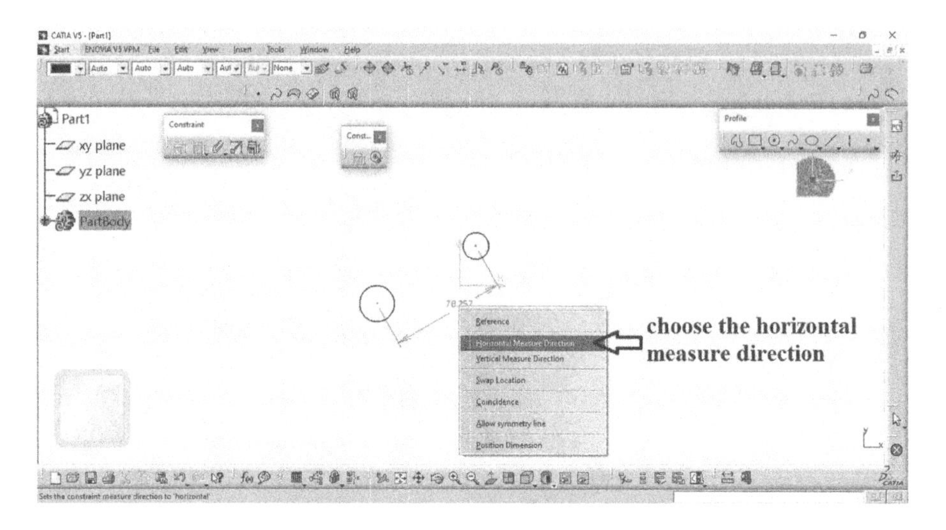

FIGURE 4.65 Choose the horizontal measure direction from contextual menu.

To apply a vertical dimension between two points, invoke the Constraint tool from the Constraints toolbar, and select points 1 and 2 from the geometry area as shown in Figure 4.64. A dimension is attached to the cursor. Next, right-click to display the contextual menu on which choose the vertical measure direction, as shown in Figure 4.67. Click in geometry area to place the dimension, as shown in Figure 4.68.

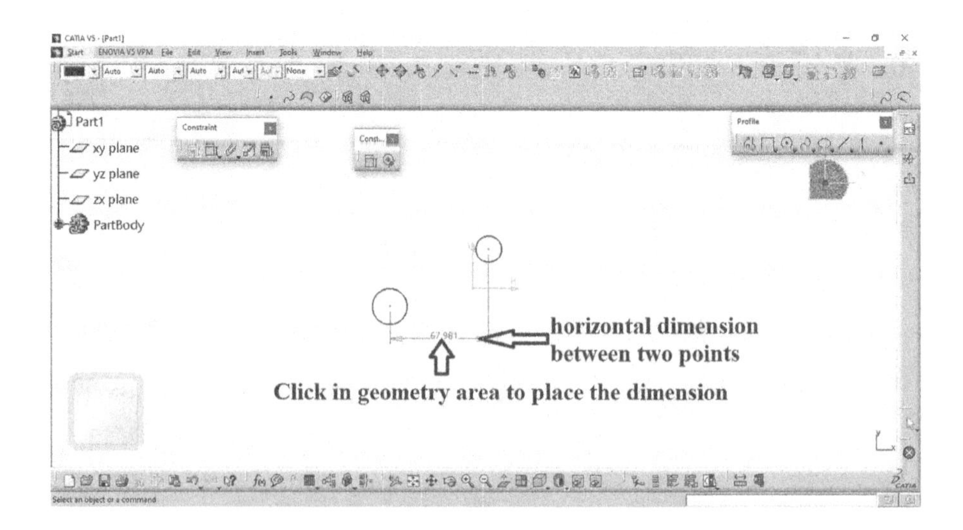

FIGURE 4.66 Dimension placement for horizontal dimension between two points.

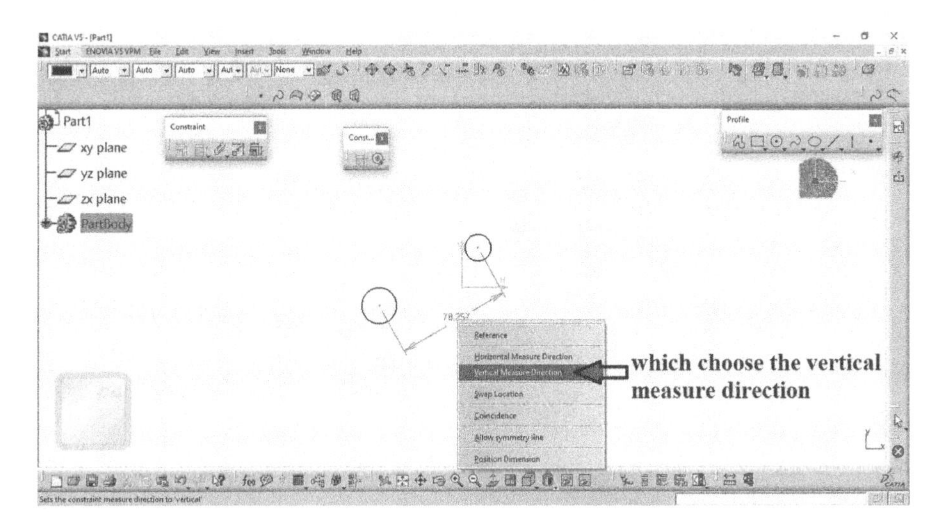

FIGURE 4.67 Choose the vertical measure direction from contextual menu.

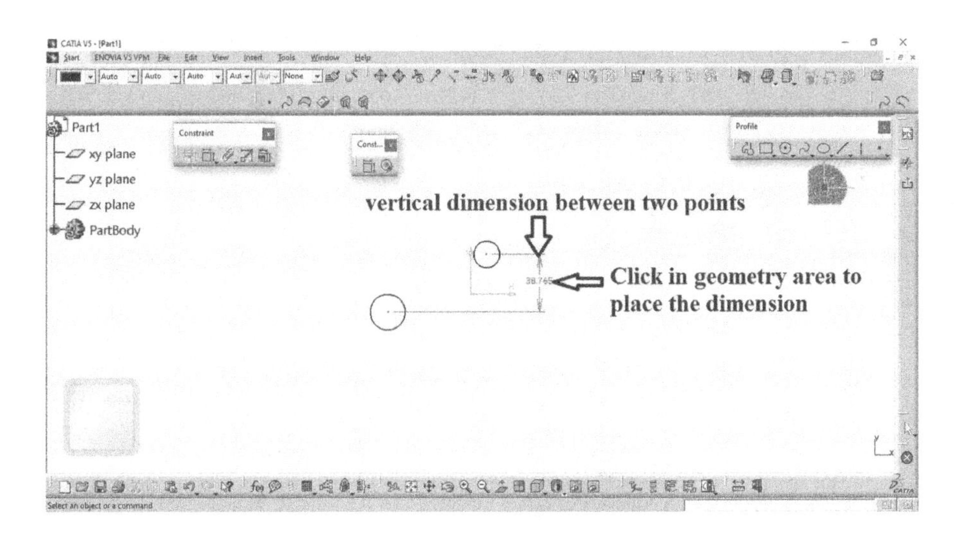

FIGURE 4.68 Dimension placement for vertical dimension between two points.

4.6.4 Dimensioning between Two Angular Lines

To apply a dimension between two angular lines, invoke the Constraint tool from the Constraints toolbar, and select lines 1 and 2 from the geometry area as shown in Figure 4.69. An angular dimension is attached to the cursor. Next, click in geometry area to place the angular dimension. Remember that the type of angular dimensions depends on its placement point. The angular dimensions are placed at different locations, as shown in Figures 4.70–4.73.

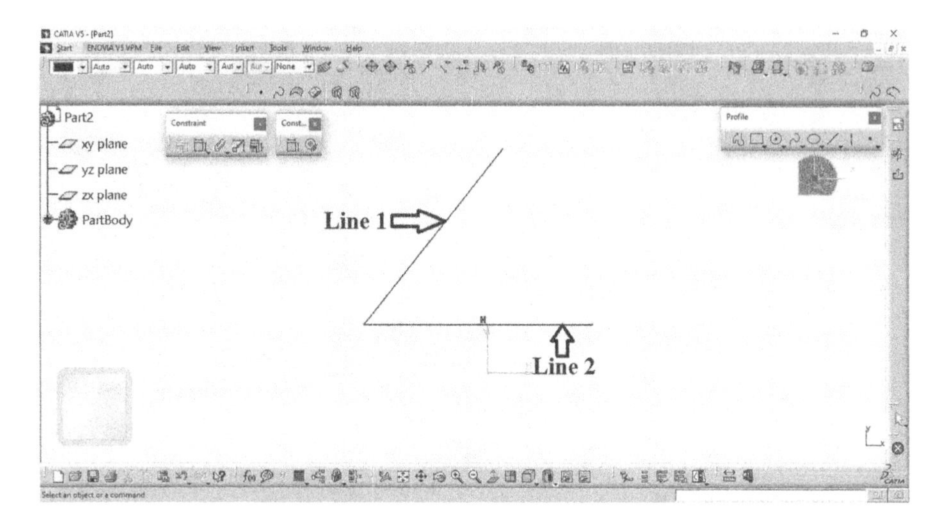

FIGURE 4.69 Selecting lines 1 and 2 for angular dimension.

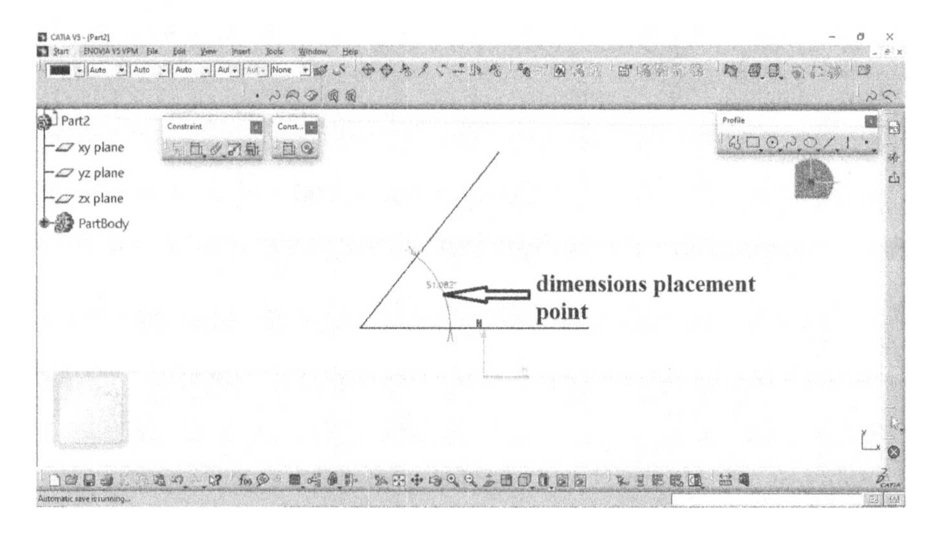

FIGURE 4.70 Placement of angular dimension point.

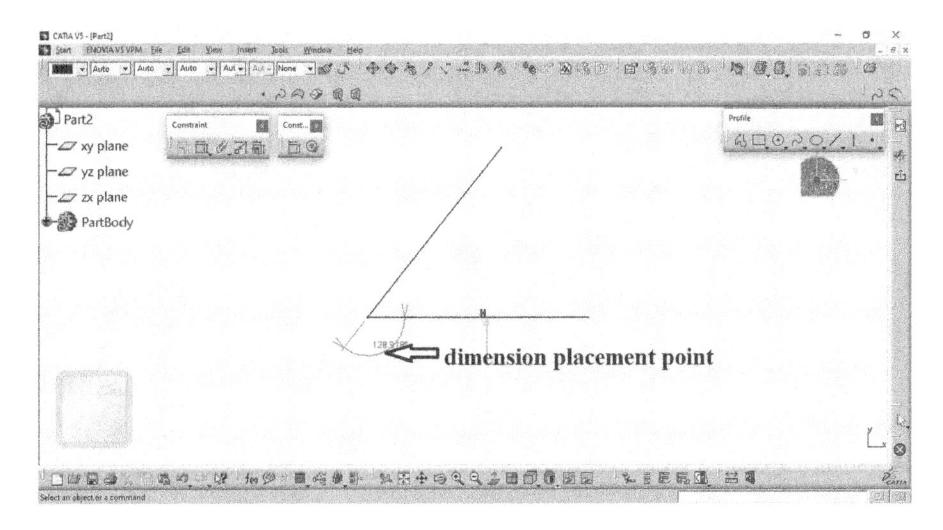

FIGURE 4.71 Placement of angular dimension point.

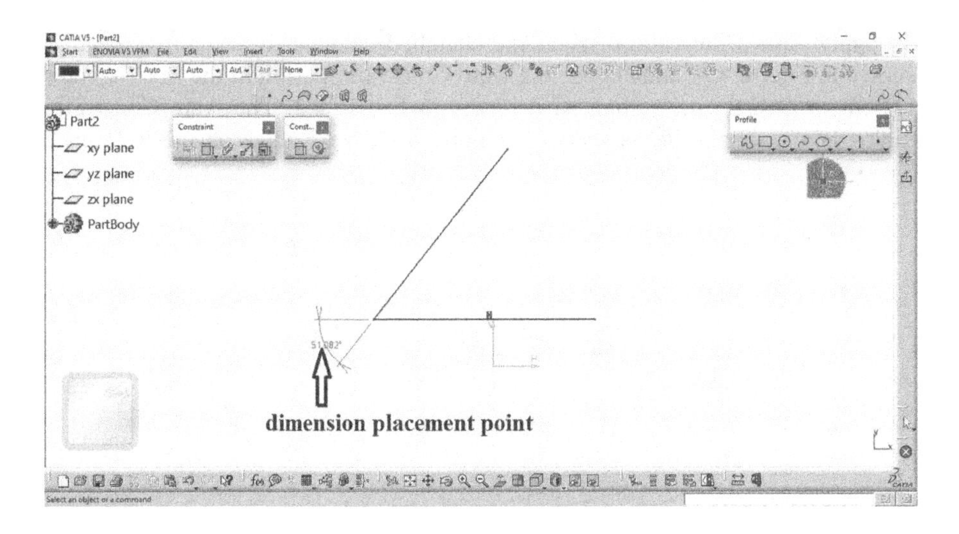

FIGURE 4.72 Placement of angular dimension point.

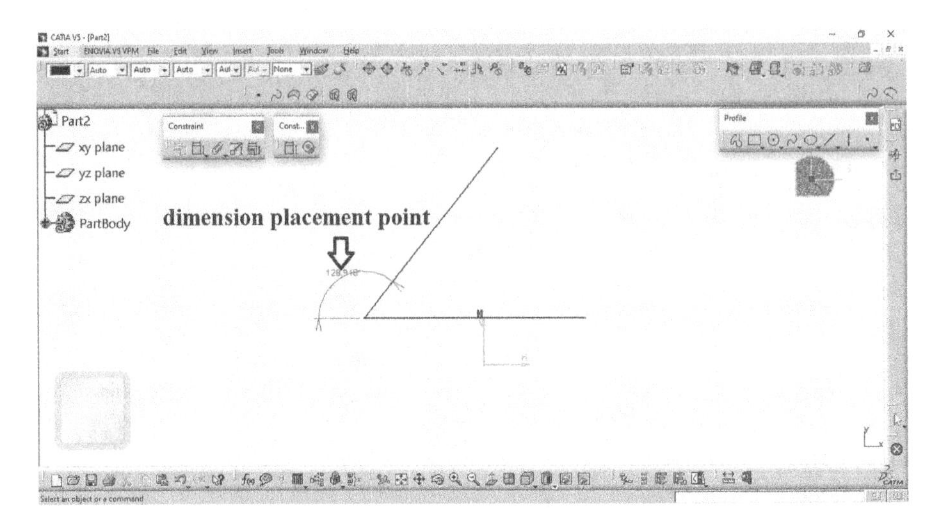

FIGURE 4.73 Placement of angular dimension point.

4.7 AUTO CONSTRAINT

To apply Auto Constraints, select the sketched element to be constrained, as shown in Figure 4.74 and choose the Auto Constraint button in the Fix Together fly out in Constraint toolbar, as depicted in Figure 4.3. The Auto Constraint dialog box is displayed, as shown in Figure 4.75. Next, click on the Reference

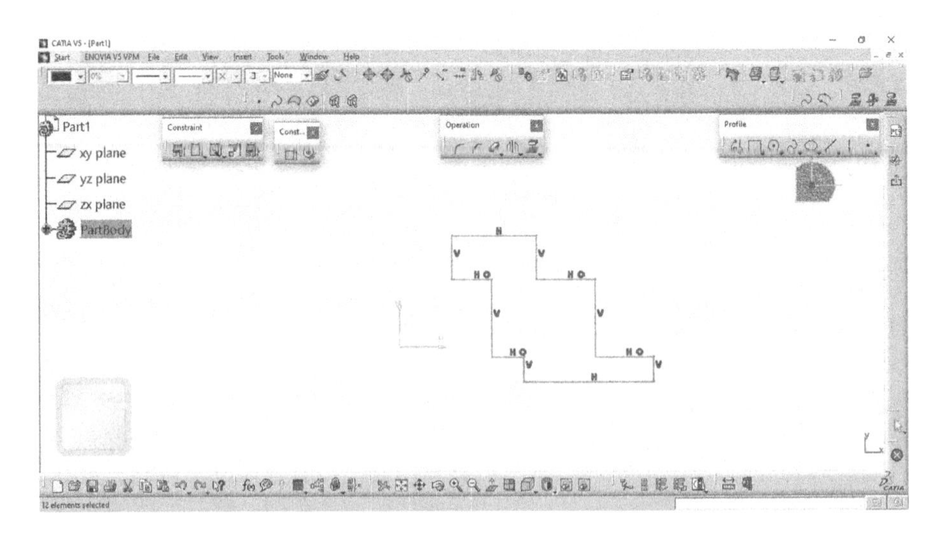

FIGURE 4.74 Selecting line for Auto Constraints.

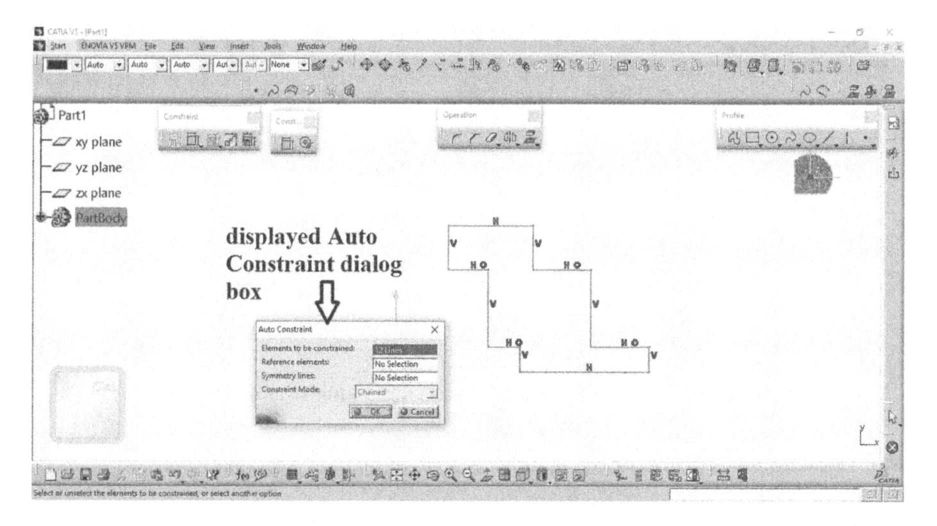

FIGURE 4.75 Displayed Auto Constraints dialog box.

Elements display box to select the reference elements. You can select the horizontal and vertical axis as the reference, as in Figure 4.76. Select type of dimensioning from the Constraint Mode drop-down menu. The options in this drop down menu are chained or the staked form. If you choose chained option and click the OK button from the Auto Constraint dialog box as shown in Figure 4.77.

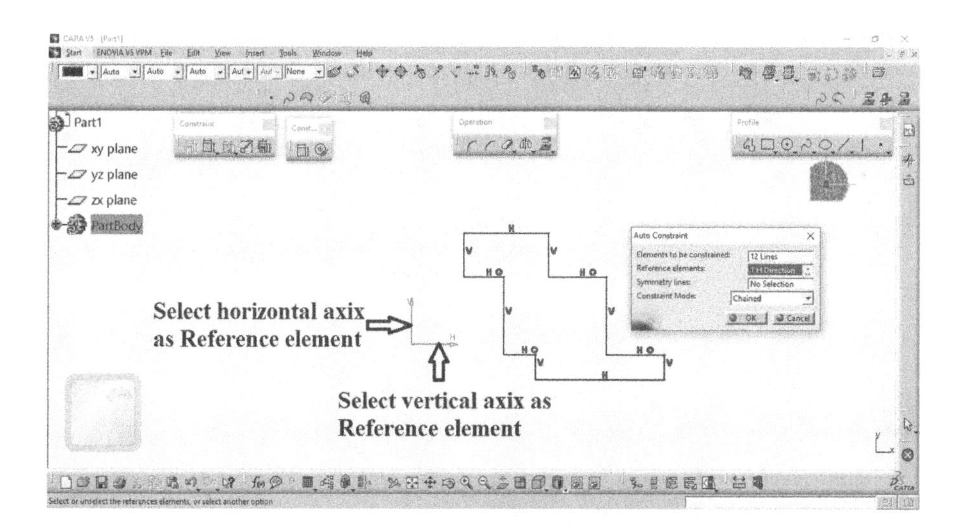

FIGURE 4.76 Selecting the horizontal and vertical axis as Reference.

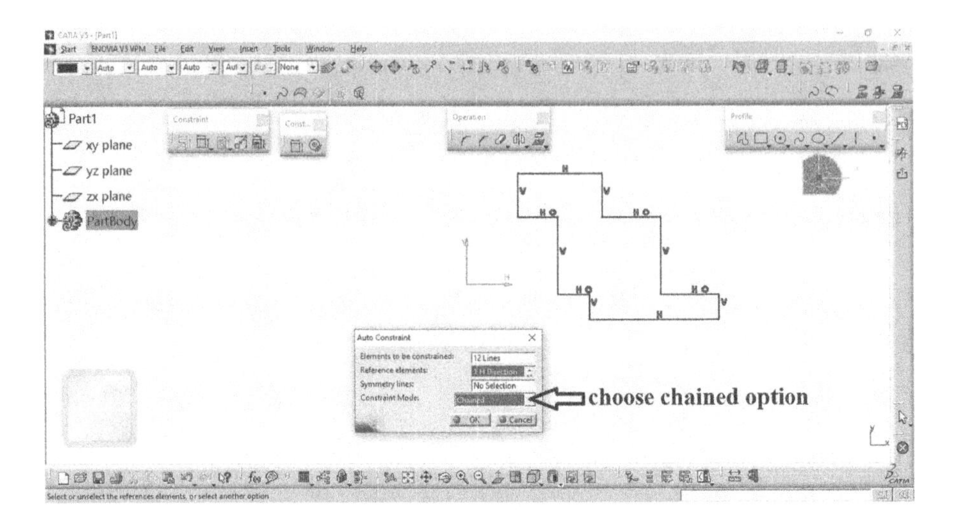

FIGURE 4.77 Selecting chained option from the Auto Constraint dialog box.

All possible constraints applied to the selected sketched elements as shown in Figure 4.78.

If you choose staked option and click the OK button from the Auto Constraint dialog box as shown in Figure 4.79, all possible constraints will applied to the selected sketched elements as shown in Figure 4.80.

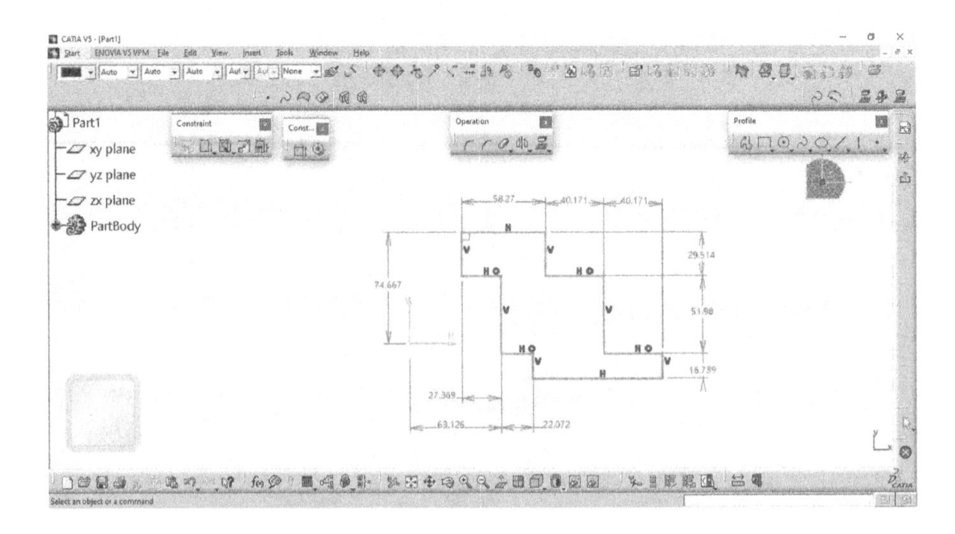

FIGURE 4.78 Auto constraints dimension using the chained option.

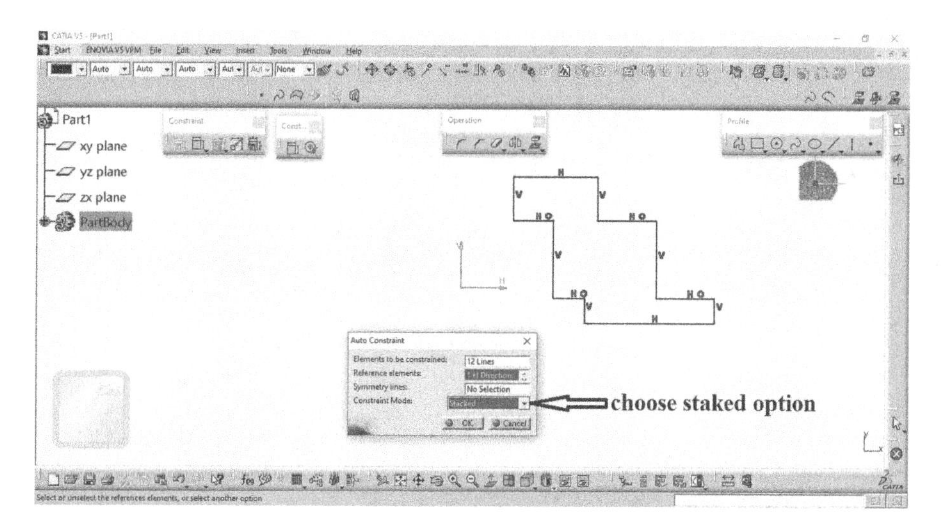

FIGURE 4.79 Selecting Stacked option from the Auto Constraint dialog box.

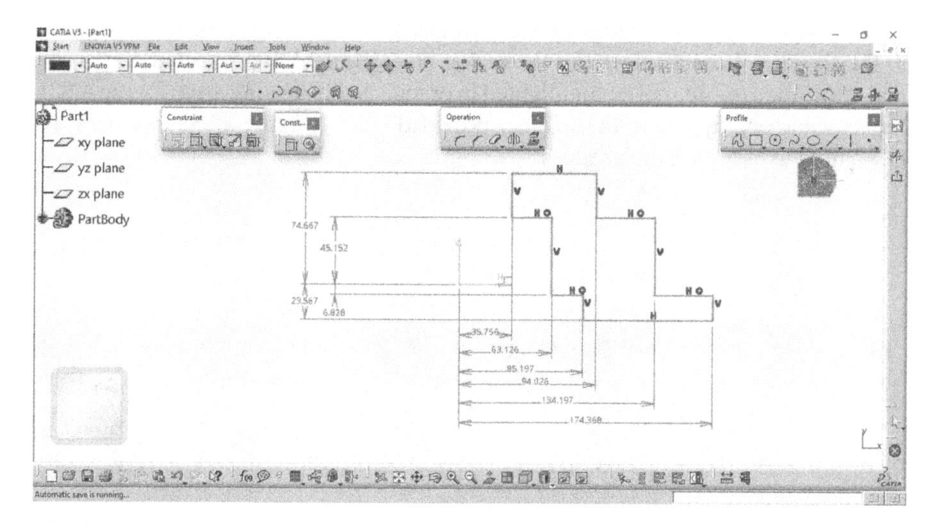

FIGURE 4.80 Auto constraints dimension using the stacked option.

4.8 EDIT MULTI-CONSTRAINT

This tool allows you to modify multiple dimensions using a single dialog box. To use this tool, you need to apply dimensions to the sketch as shown in Figure 4.80. Choose the Edit Multi-Constraint button from the Constraint toolbar the Edit Multi-Constraint dialog box will appear as provided in Figure 4.81. All dimensions applied

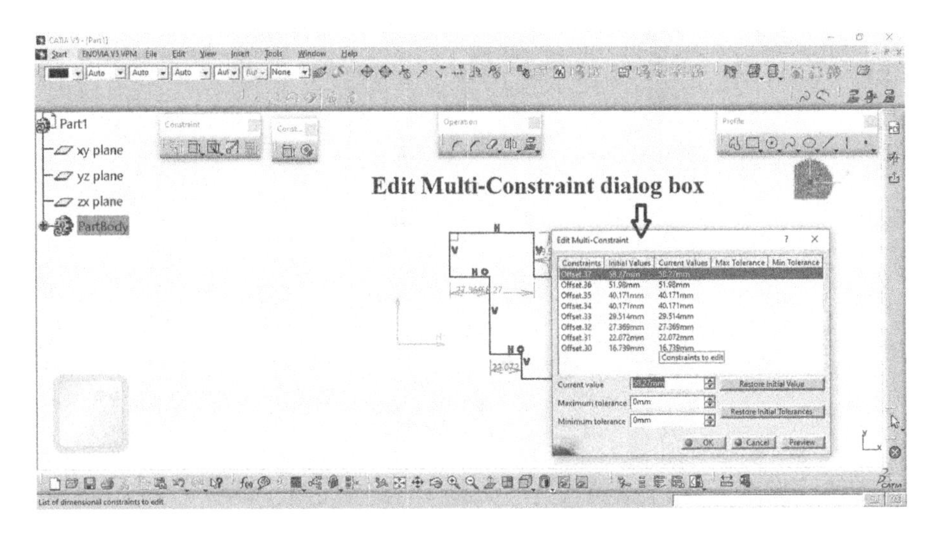

FIGURE 4.81 Displayed Edit Multi-Constraint dialog box.

to the sketch will be displayed in the dialog box (Figure 4.81). Selected dimension that are needed to edit will be highlighted in orange in the drawing area (Figure 4.82). Set its value one by one in the spanner provided below (Figure 4.83) and press the OK button to ascertain the changes.

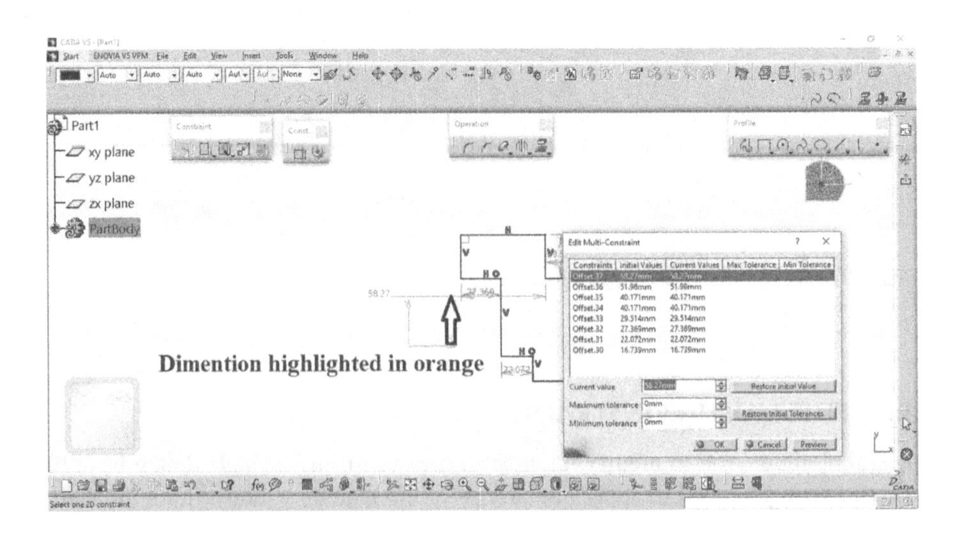

FIGURE 4.82 Selected dimension highlighted in orange.

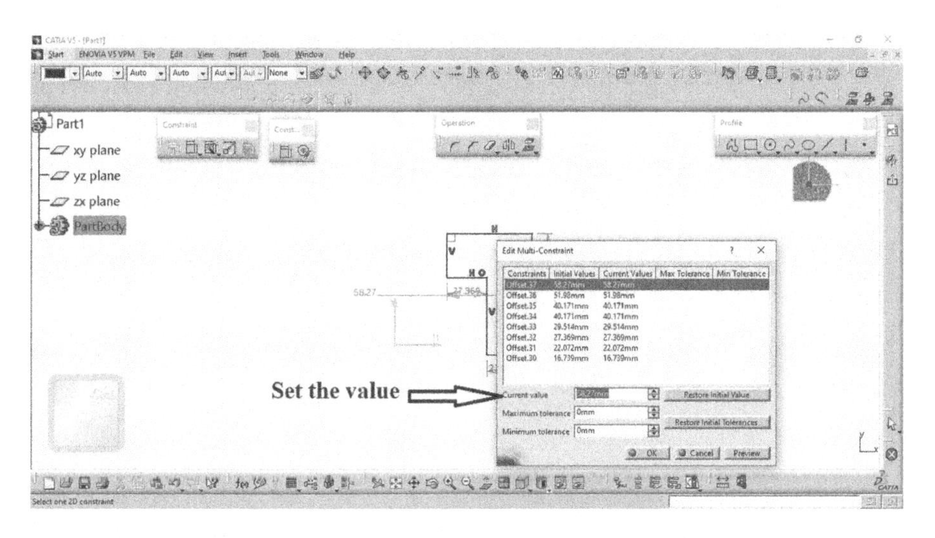

FIGURE 4.83 Set the value of Selected dimension.

4.9 MODIFYING DIMENSIONS AFTER PLACEMENT

As discussed in Chapter 1, CATIA software is parametric software. Therefore, you can change the dimensions of your design in any number of times at any stage. In this chapter, you learn to place various type of dimensions. While dimensioning the entity, the default value is placed that may not be your required value. To modify the dimension value, double-click on the value; the Constrain Definition dialog box appears (Figure 4.84). Set the value of the dimension in the value spanner and choose the OK button from the Constrain Definition dialog box.

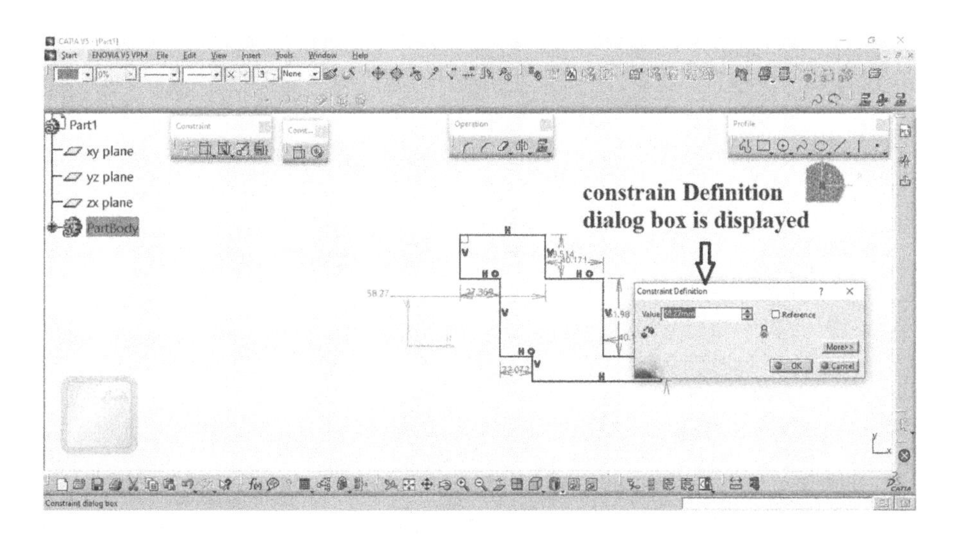

FIGURE 4.84 Displayed Constrain Definition dialog box.

Section III

Part Design

5 Part Design

5.1 INTRODUCTION

This chapter covers drawing differing profiles using different tools in CATIA software. The part design is used to create and modify solids. A solid is first created using an initial sketch in Sketcher workbench. Once the main solid is created, it may be modified using Sketch-Based Features commands or using commands that do not require a sketch. The general design process is to sketch the profile of the main pad, generate the main pad, create additional Sketch-Based Features, add Dress-Up Features, modify features as desired and insert new bodies for more complex parts.

5.2 ACCESSING THE PART DESIGN WORKBENCH

When you start CATIA software, a new Product file with the name Product1 appears as depicted in Figure 5.1

To use the Part Design workbench, go for

Select > Start > Mechanical Design > Part Design

from the Start drop-down menu (Figure 5.2).

The New Part dialog box will be displayed, as shown in Figure 5.3

Enter part name on New Part dialog box and click on OK. A new file in the Part Design workbench is displayed on the screen, as shown in Figure 5.4.

FIGURE 5.1 Initial screen after starting CATIA software.

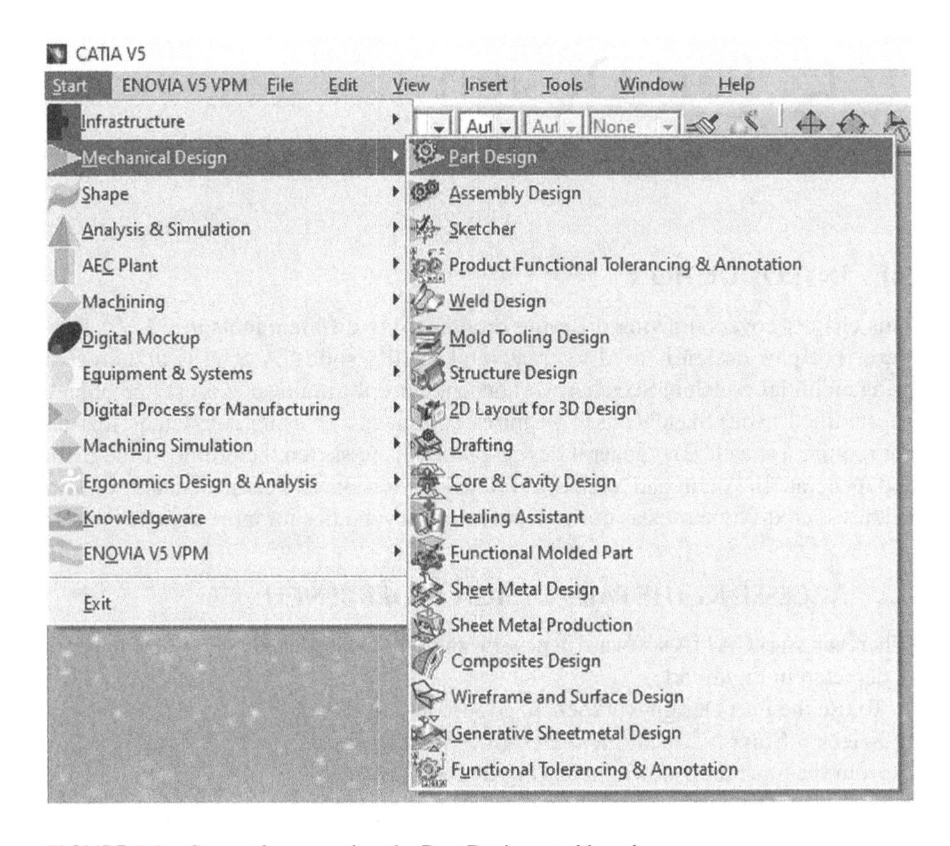

FIGURE 5.2 Screen for accessing the Part Design workbench.

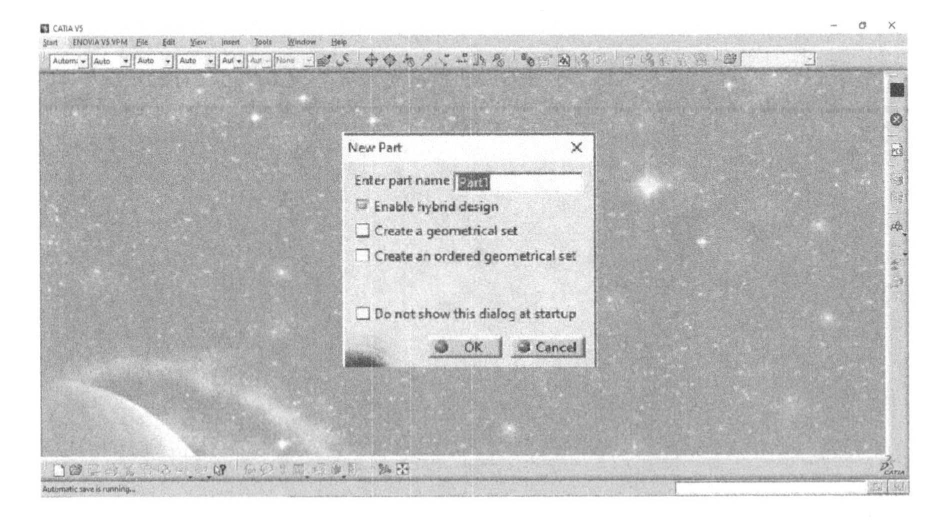

FIGURE 5.3 New Part dialog box.

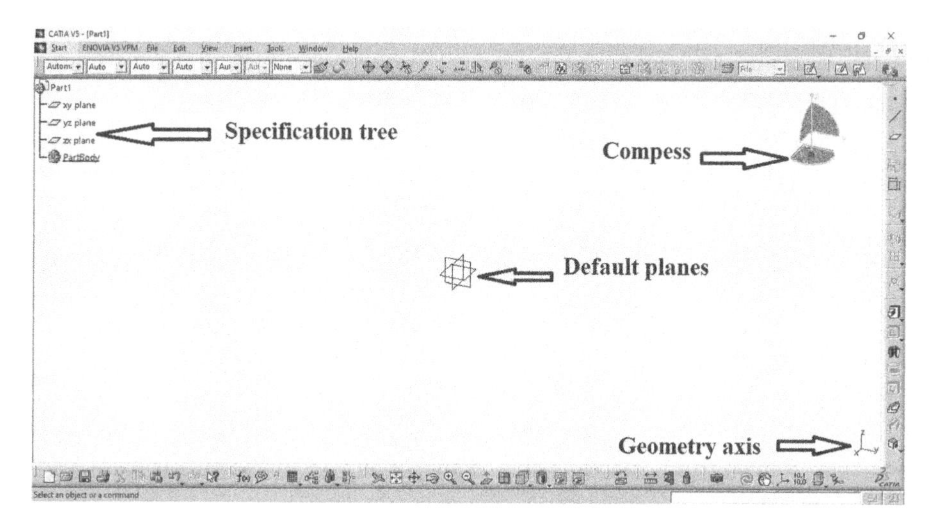

FIGURE 5.4 New Part dialog box.

Choose the Sketch button from Sketcher toolbar. Select the drawing plane from the available three default planes provided in the specification tree or can directly be accessed from the geometry area (Figure 5.5).

The Sketcher workbench that appears on selecting the xy plane, as shown in Figure 5.6

After drawing sketch in Sketcher workbench select Exit workbench button (Figure 5.7).

After selecting exit work, you will enter Part Design environment, as shown in Figure 5.8. We will be discussing with various types of Sketch-Based Features tools with the help of step-by-step tutorial using figures of the CATIA software.

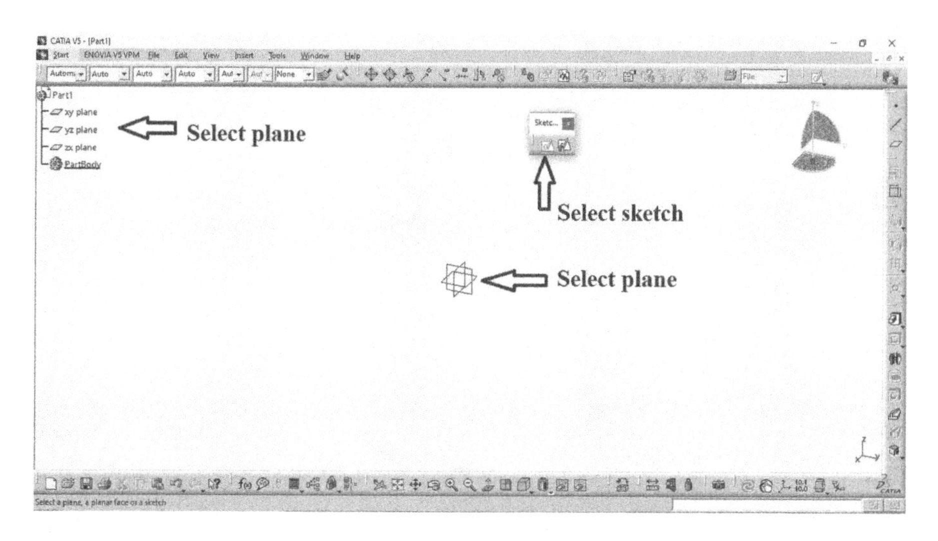

FIGURE 5.5 Sketcher workbench invoked Sketcher tool and plane.

FIGURE 5.6 Sketcher workbench invoked using the xy plane.

5.3 SKETCH-BASED FEATURES TOOLBAR

The Sketch-Based Features toolbar as shown in Figure 5.8 used to create and modify solids. The tools in this toolbar are pads, pocket, shaft, groove, hole, rib, slot, shell and thickness.

5.3.1 Pads

Pads are used to add or create solid material in a linear direction. The tools in this toolbar are Pad, Drafted Filleted Pad, and Multi-Pad. The Pad toolbar will come as shown in Figure 5.9.

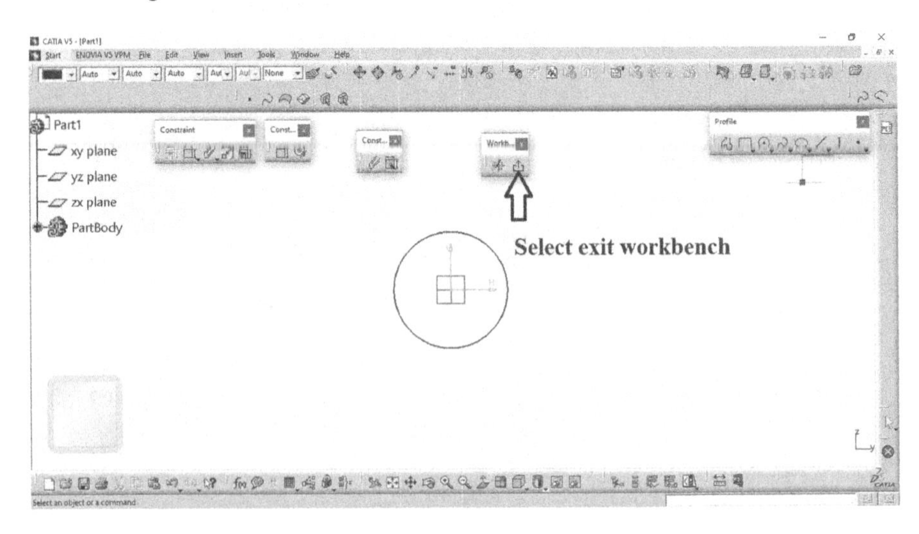

FIGURE 5.7 Selection of Exit workbench.

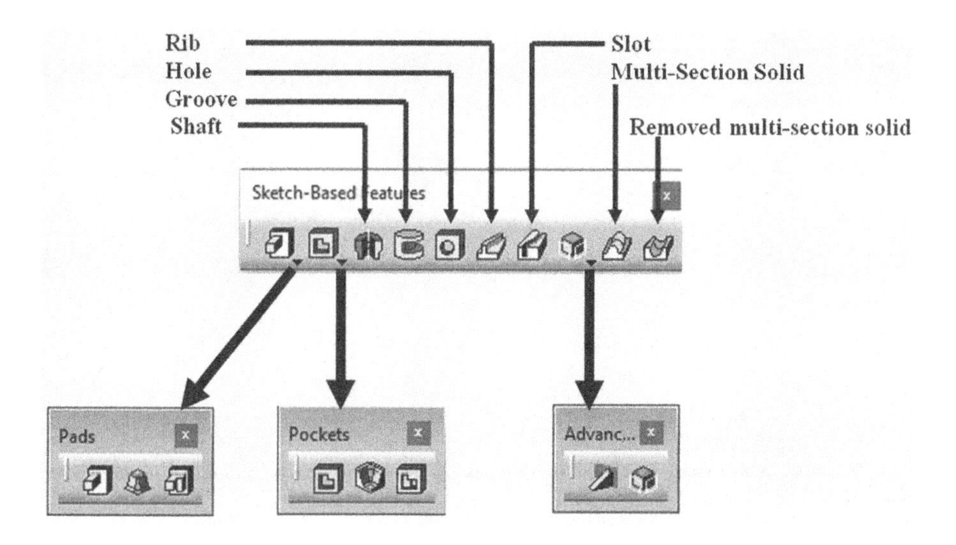

FIGURE 5.8 Sketch-Based Features toolbar.

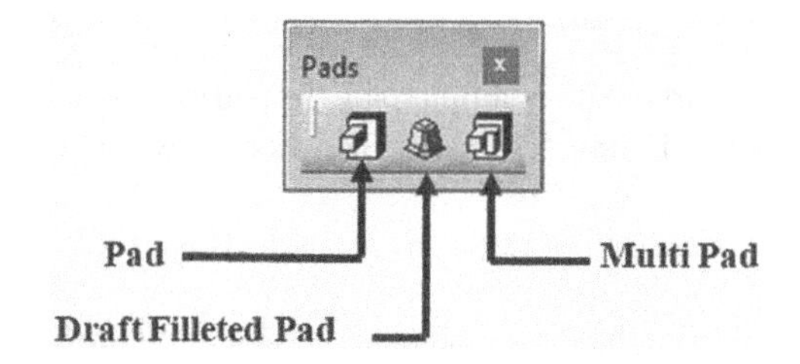

FIGURE 5.9 Pads toolbar.

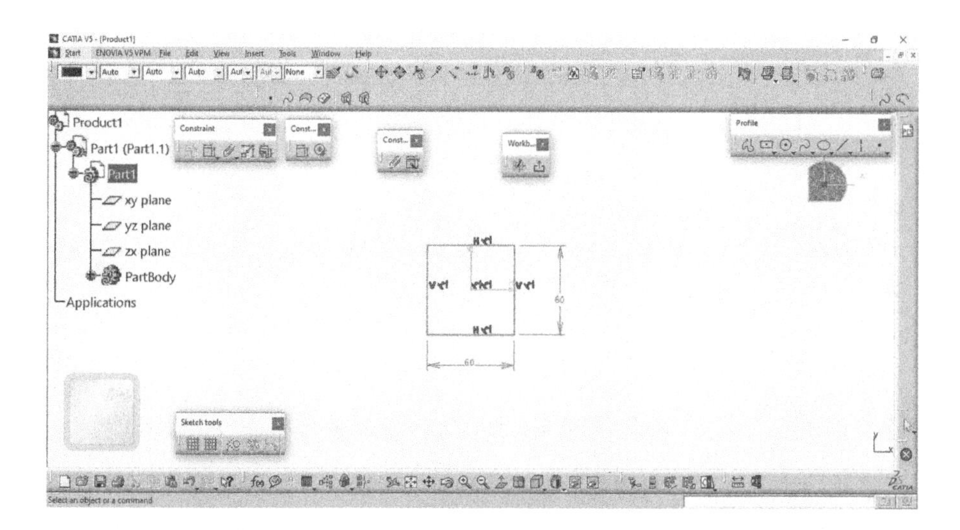

FIGURE 5.10 Constriction of sketch for Pad.

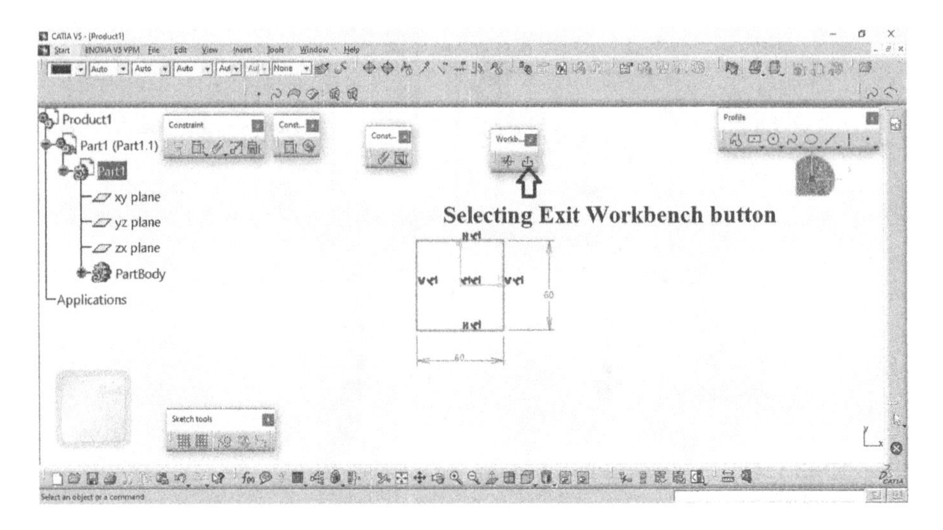

FIGURE 5.11 Selection of Exit workbench button.

5.3.1.1 Pad

This command is used to create a simple pad. To pad the object, create a sketch as per dimension shown in Figure 5.10 using Sketcher workbench as explained in the previous chapter.

Now, exit the Sketcher workbench by selecting Exit workbench button (Figure 5.11).

Select the Pad tool from the Pads toolbar, as shown in Figure 5.9. A Pad Definition window appears (Figure 5.12).

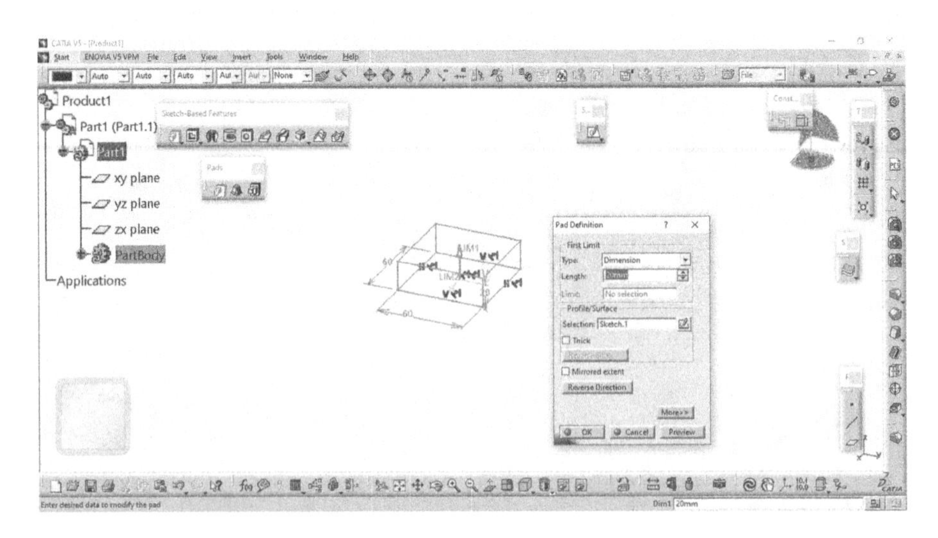

FIGURE 5.12 Selection of Pad tool.

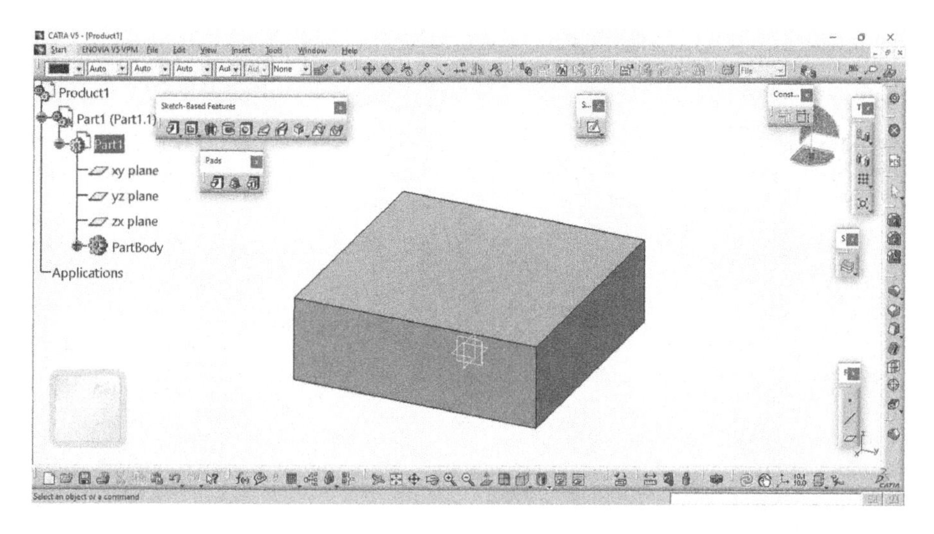

FIGURE 5.13 The sketch after padding.

Enter the value of length as 20mm in Pad Definition window and click on OK button. After clicking on the OK button, you will get the sketch as shown in Figure 5.13.

5.3.1.2 Drafted Filleted Pad

This command is used to create a pad that has angled sides and rounded edges. Select the face from sketch drawn in Figure 5.13 and click the Sketch tool, as shown in Figure 5.14.

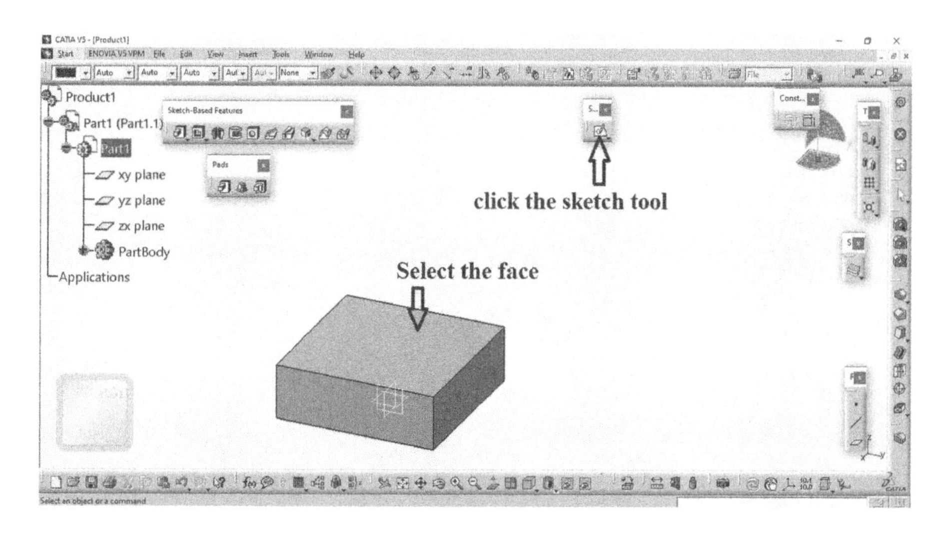

FIGURE 5.14 Selection of face and Sketch tool for Drafted Filleted Pad.

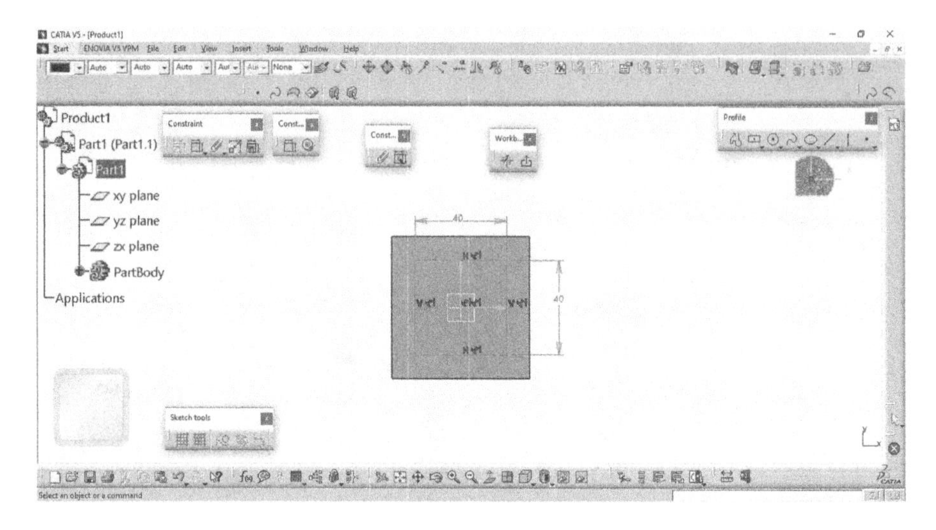

FIGURE 5.15 Constriction of sketch for Drafted Filleted Pad.

Draw the sketch as per dimension shown in Figure 5.15 using Sketcher workbench and then exit the Sketcher workbench.

Select the Drafted Filleted Pad tool from the Pads toolbar, as shown in Figure 5.9. A Drafted Filleted Pad Definition window appears as shown in Figure 5.16.

Next, to continue a reference is required to be provided for the second limit. For this, sketching plane being used for the sketch is selected. The name of the selected plane is displayed in the Limit selection area in the Second Limit area. Provide value 50mm of the depth of extrusion in the space provided for Length. Specify

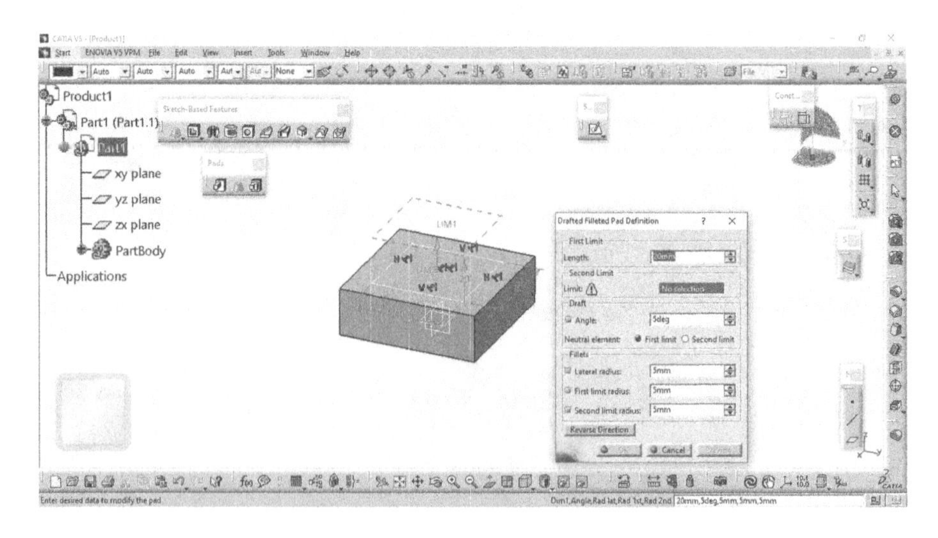

FIGURE 5.16 Selection of Drafted Filleted Pad tool.

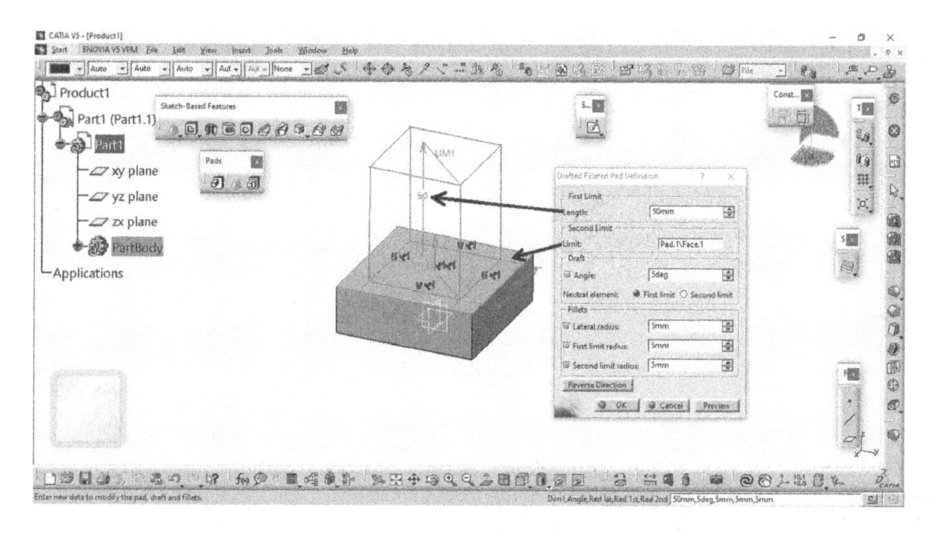

FIGURE 5.17 Value setting in Drafted Filleted Pad Definition window.

Draft Angle as 5deg in the space for angle value. By default, the upper surface of the extruded feature will be considered as a neutral plane for adding a draft. If you select the Second limit radio button, the second reference will be considered as a neutral plane. You can modify the default value of the fillet in the spinners in the Fillets area (Figure 5.17).

After setting the parameters, choose the OK button in the Drafted Filleted pad Definition dialog box and you will get the figure, as shown in Figure 5.18.

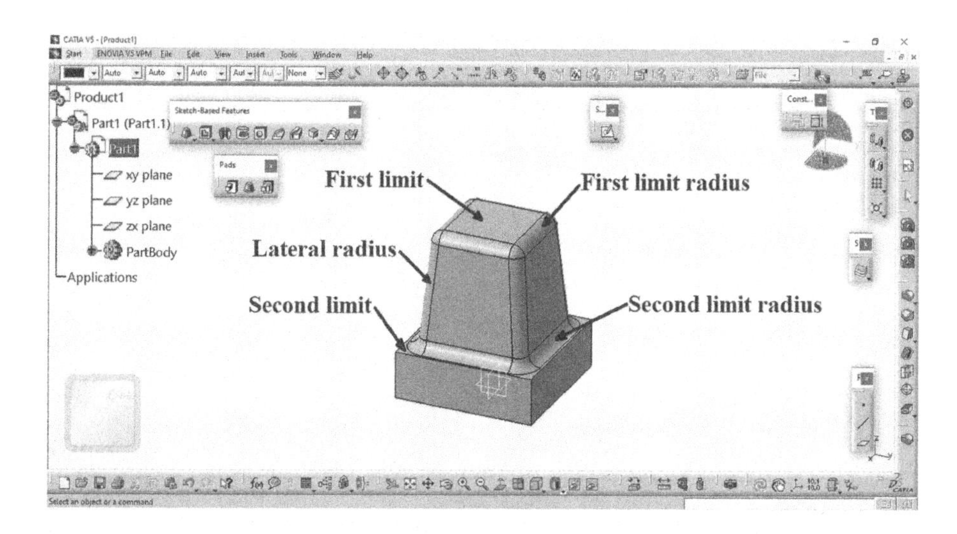

FIGURE 5.18 Sketch after Draft Filleted Padding.

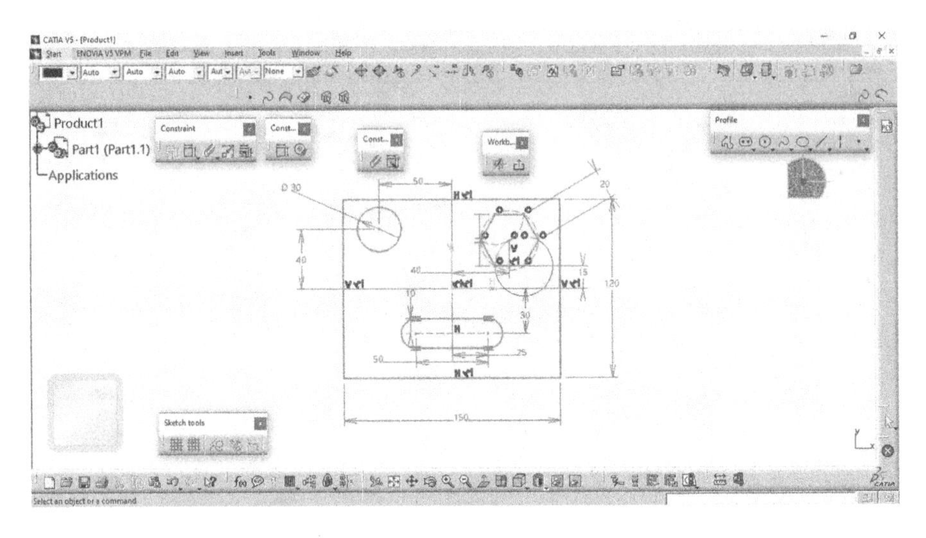

FIGURE 5.19 Constriction of sketch for Multi-Pad.

5.3.1.3 Multi-Pad

This command is used to create pad of varying depth all within a single sketch. To Multi-Pad the object creates a sketch as per dimension shown in Figure 5.19 using Sketcher workbench as explained in the previous chapter.

Now, exit the Sketcher workbench by selecting Exit workbench button, as shown in Figure 5.20.

Select the Multi-Pad tool from the Pads toolbar, as shown in Figure 5.9. A Multi-Pad Definition window appears as shown in Figure 5.21.

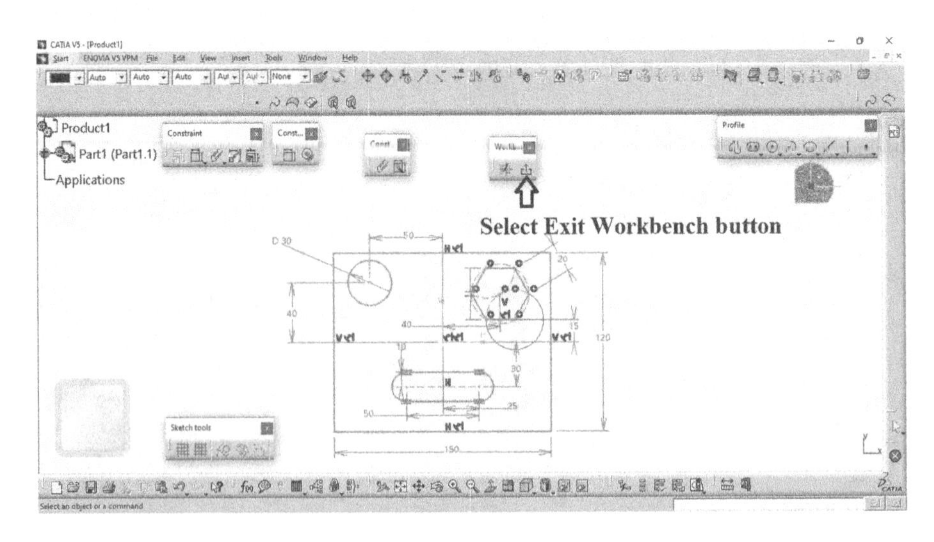

FIGURE 5.20 Selection of Exit workbench button.

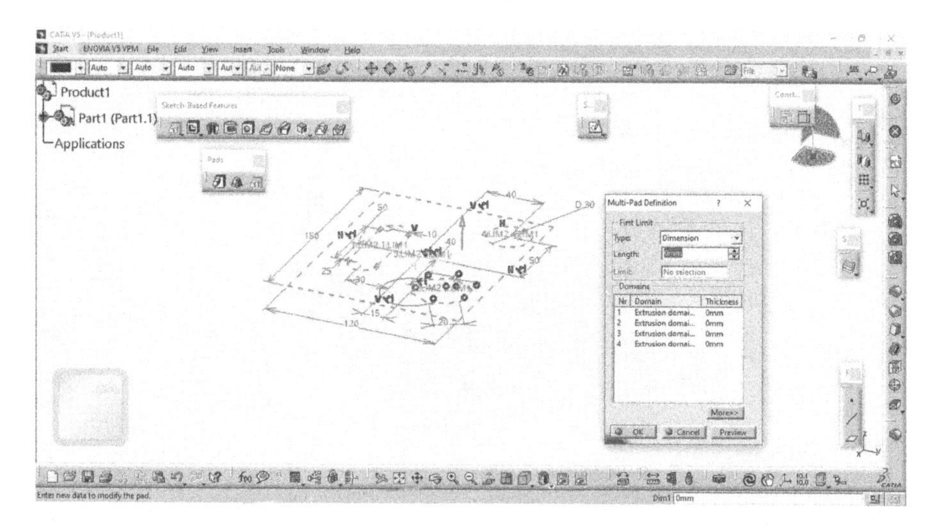

FIGURE 5.21 Selection of Multi-Pad tool.

The entire sketches are displayed in the list box provided in the Domains area of the Multi-Pad Definition dialog box (Figure 5.21). Select the name of the sketch from the list box. It is highlighted in the geometry area. Provide the value of depth of the selected loop using the Length spinner. Similarly, define the extrusion depth of all the sketches one by one in Multi-Pad Definition dialog box (Figure 5.22).

After setting all values, click the OK button and you will get the figure, as shown in Figure 5.23.

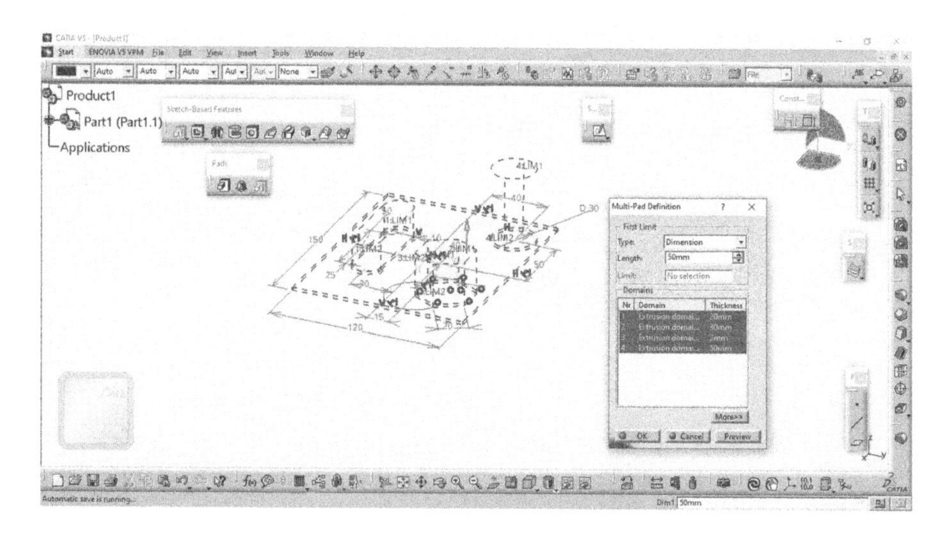

FIGURE 5.22 Value setting in Multi-Pad Definition dialog box.

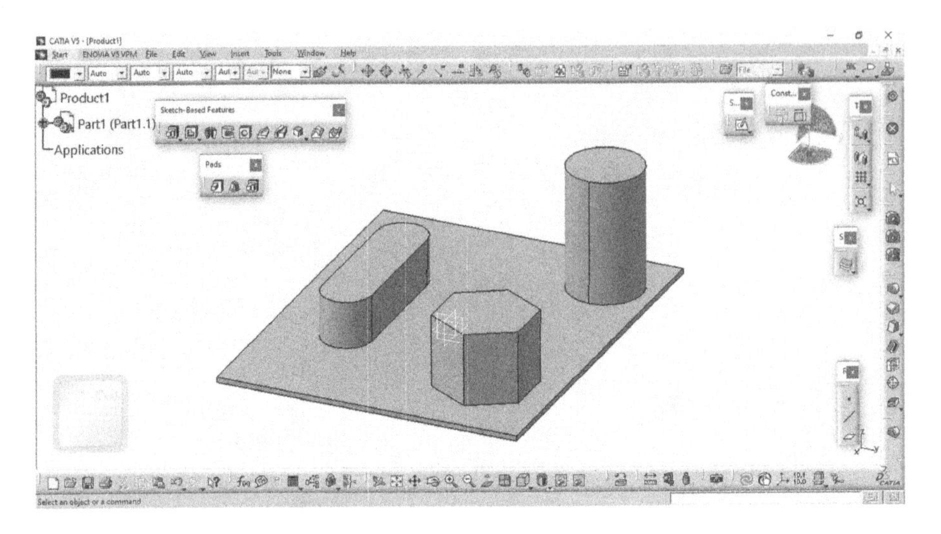

FIGURE 5.23 Sketch after Multi-Padding.

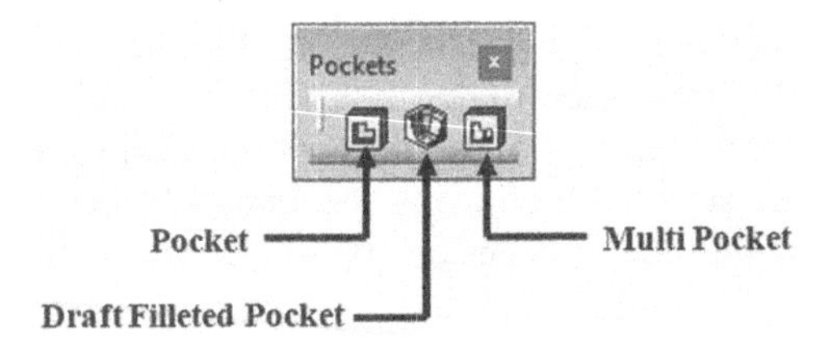

FIGURE 5.24 Pocket toolbar.

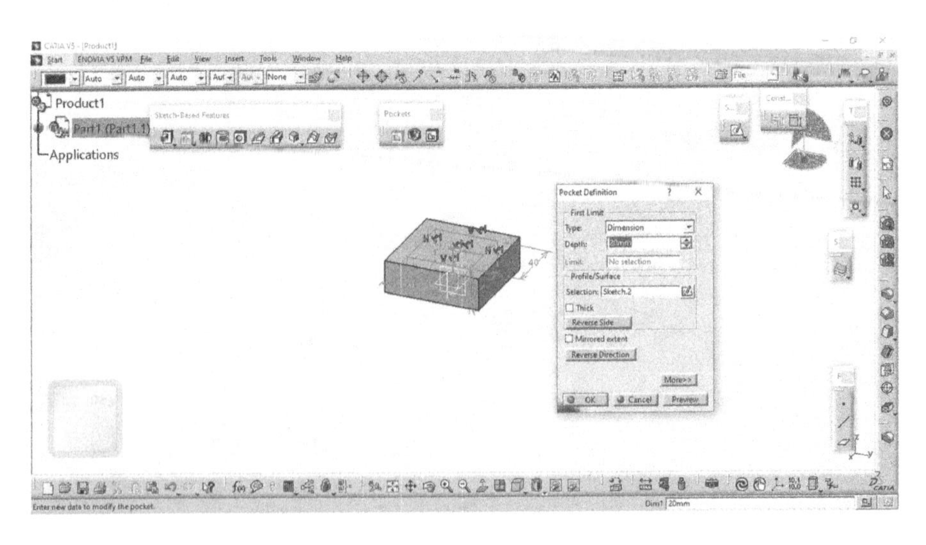

FIGURE 5.25 Selection of Pocket tool.

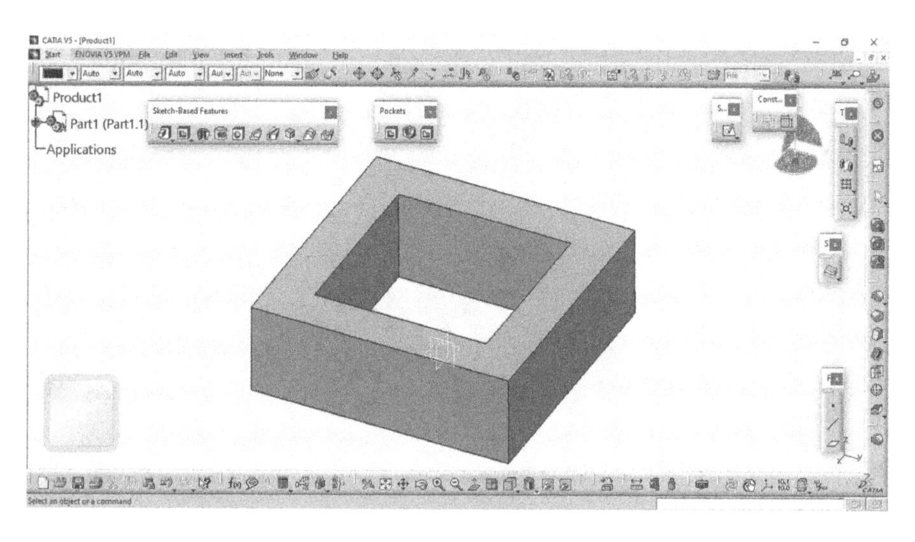

FIGURE 5.26 Sketch after Pocketing.

5.3.2 Pocket

Pockets are used to remove material in a linear direction. The tools in this toolbar are Pocket, Drafted Filleted Pocket and Multi-Pocket. The Pad toolbar will appear as shown in Figure 5.24.

5.3.2.1 Pocket

This command is used to create a simple pocket. It also defined as material removal tools that remove the material from an existing feature by existing feature by extruding the sketch to the given depth or feature termination condition. Pad the object and select the face from Padded object as shown in Figure 5.13 and click the Sketch tool as shown in Figure 5.14. Draw the sketch as per dimension shown in Figure 5.15 using Sketcher workbench and then exit the Sketcher workbench. Choose the Pocket tool from Pocket toolbar (Figure 5.24). The Pocket Definition dialog box will appear as shown in Figure 5.25.

Enter the value of length as 20mm in Pocket Definition window and click the OK button. After clicking the OK button, you will get the figure, as shown in Figure 5.26.

5.3.2.2 Drafted Filleted Pocket

This command is used to create a pocket that has angled sides and rounded edges. Pad the object and select the face from Padded object (Figure 5.13) and click the Sketch tool, as shown in Figure 5.14. Draw the sketch as per dimension (Figure 5.15) using Sketcher workbench and then exit the Sketcher workbench. Then choose the Drafted Filleted Pocket tool from Pocket toolbar, as shown in Figure 5.24. The Pocket Definition dialog box will be displayed, as shown in Figure 5.27.

Next, to continue a reference is required to be provided for the second limit. For this, sketching plane being used for the sketch is selected. The name of the selected

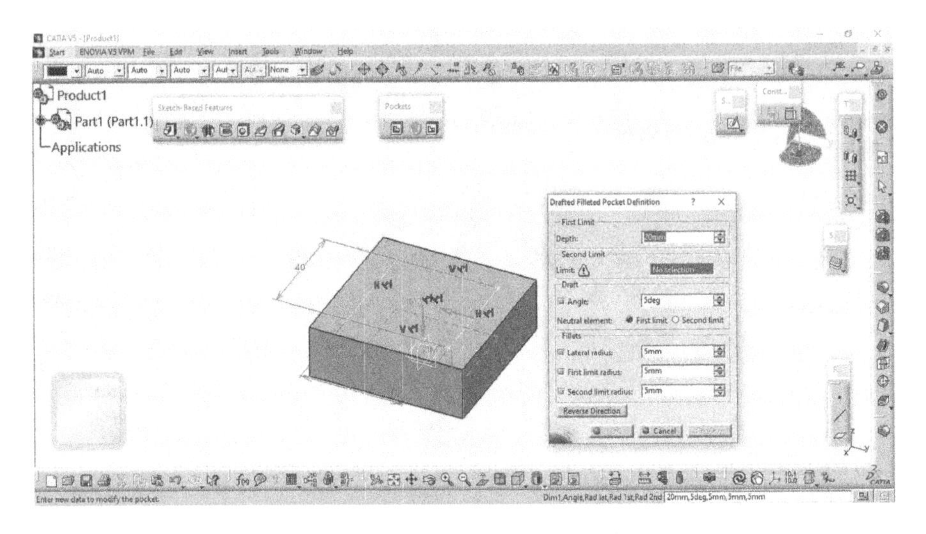

FIGURE 5.27 Selection of Drafted Filleted Pocket tool.

plane is displayed in the Limit selection area in the Second Limit area. Provide value 18mm of the depth of pocket extrusion in the space provided for length. Specify Draft Angle as 5deg in the space for angle value. By default, the upper surface of the extruded feature will be considered as a neutral plane for adding a draft. If you select the second limit radio button, then the second reference will be considered as a neutral plane. You can modify the default value of the fillet in the spinners in the Fillets area (Figure 5.28).

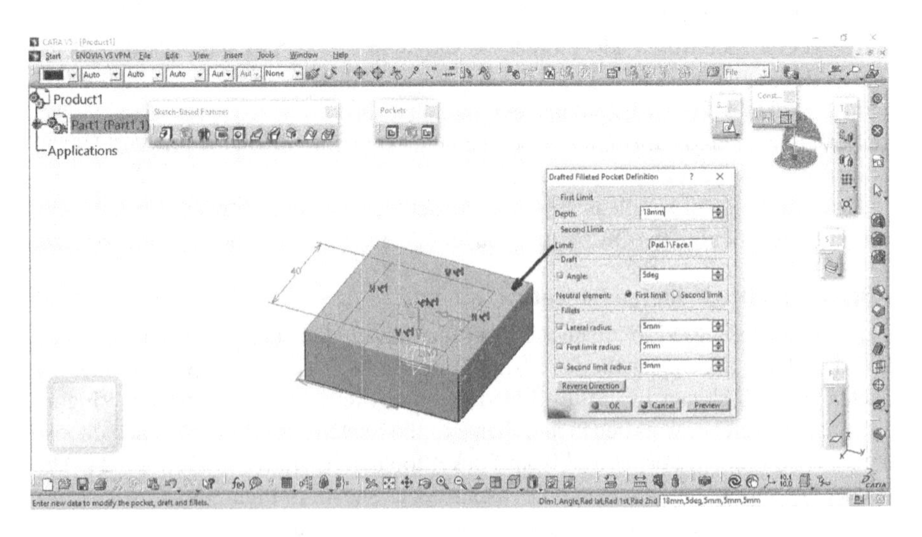

FIGURE 5.28 Selection of Drafted Filleted Pocket tool.

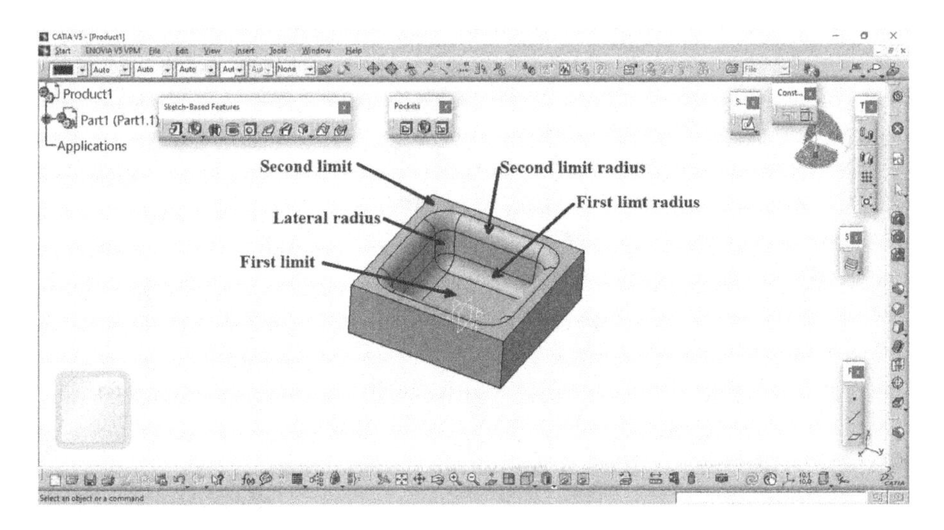

FIGURE 5.29 Sketch after Drafted Filleted Pocketing.

After setting the parameters, choose the OK button in the Drafted Filleted Pocket Definition dialog box and you will get the figure, as shown in Figure 5.29.

5.3.2.3 Multi-Pocket

This command is used to create pockets of varying depth all within a single sketch. To Multi-Pocket the object creates a sketch and pad, as shown in Figure 5.30.

Select upper face of the object and draw sketch for multi-pocket, as shown in Figure 5.31.

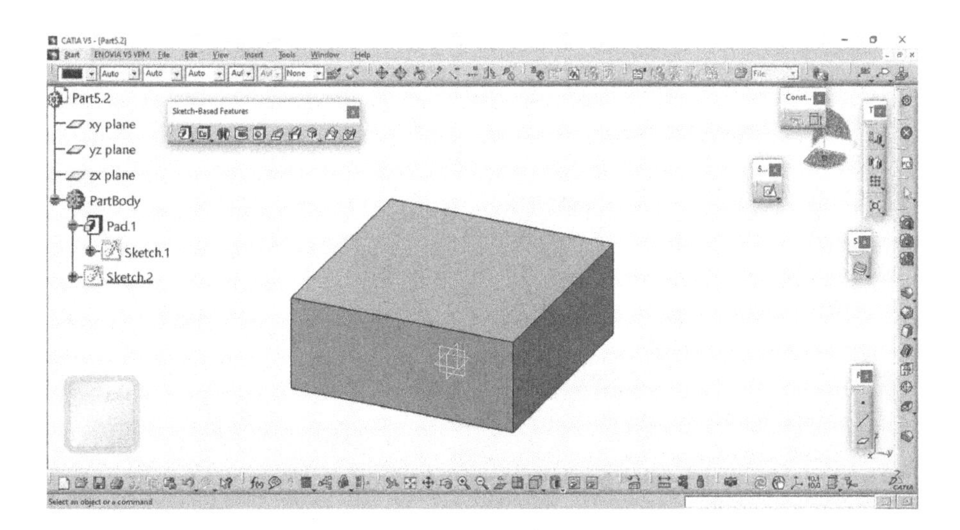

FIGURE 5.30 Padded base object.

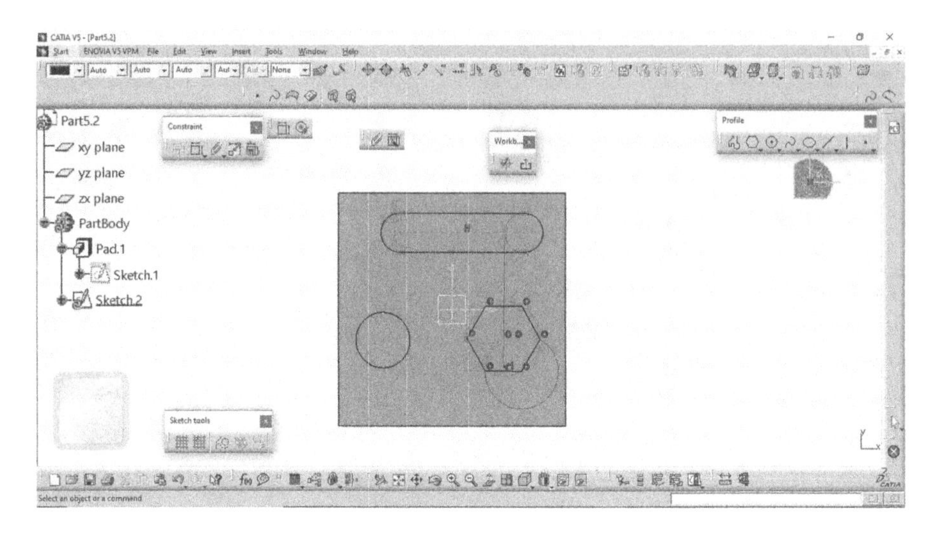

FIGURE 5.31 Sketch for Multi-Pocket.

Select the Multi-Pocket tool from the Pocket toolbar, as shown in Figure 5.24. A Multi-Pocket Definition window appears, as shown in Figure 5.32.

The entire sketches are displayed in the list box provided in the Domains area of the Multi-Pocket Definition dialog box (Figure 5.32). Select the name of the sketch from the list box. It is highlighted in the geometry area. Provide value of depth of the selected loop using the Length spinner. Similarly, define the extrusion depth of all the sketches one by one in Multi-Pocket Definition dialog box (Figure 5.33).

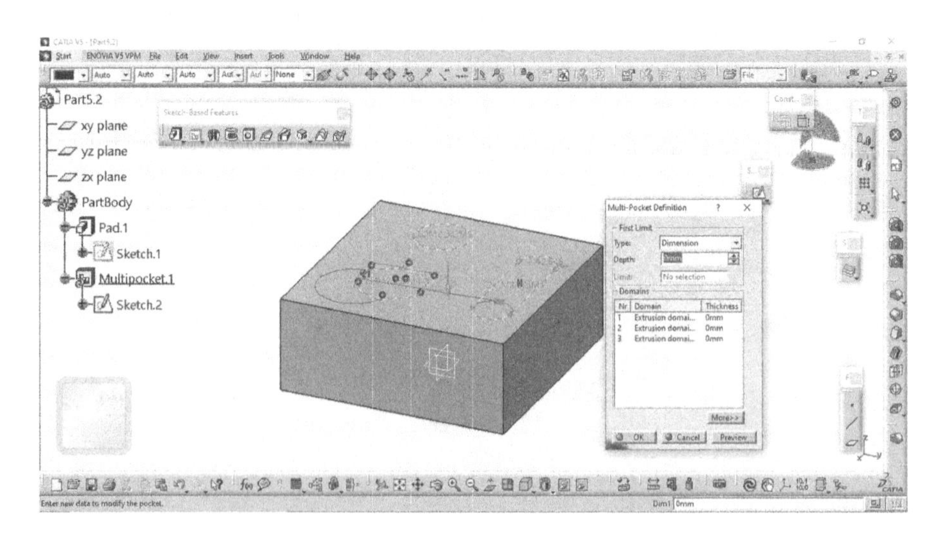

FIGURE 5.32 Selection of Multi-Pocket tool.

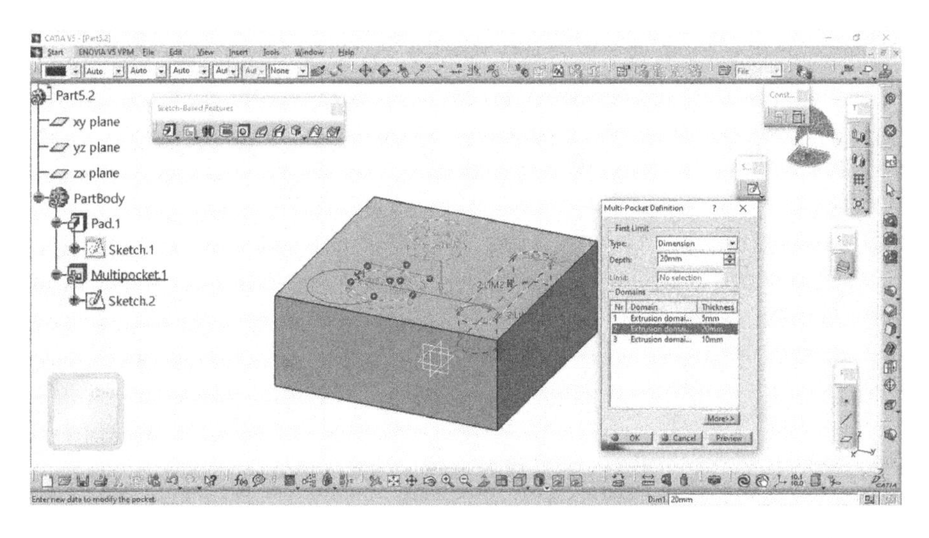

FIGURE 5.33 Value setting in Multi-Pocket Definition dialog box.

After setting all values, click on the OK button and you will get the figure, as shown in Figure 5.34.

5.3.3 Shaft

The Shaft tool is used to generate rotating elements. It required a sketch having an axis of revolution. Draw the sketch as per dimension as shown in Figure 5.35 using Sketcher workbench and then exit the Sketcher workbench.

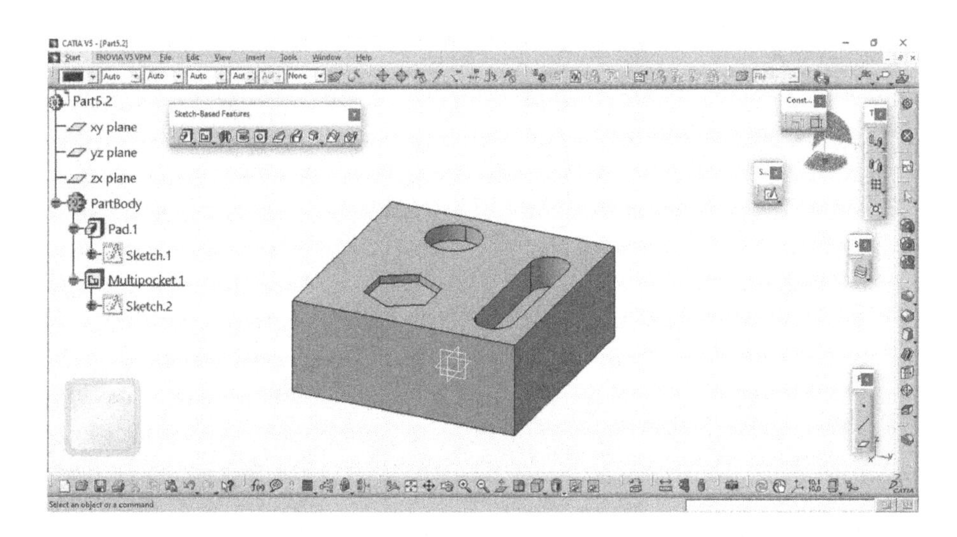

FIGURE 5.34 Sketch after Multi-Pocketing.

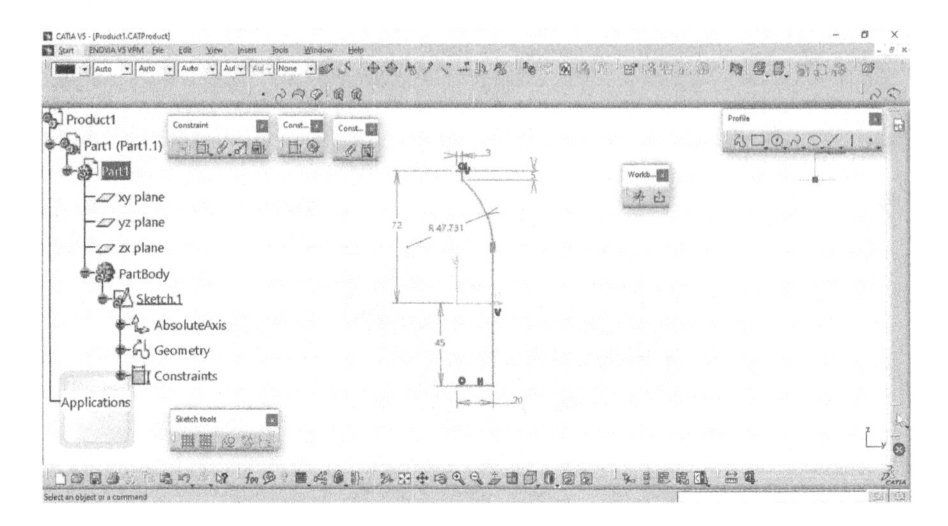

FIGURE 5.35 Sketch for using Shaft tool.

Choose the Shaft tool from Sketch-Based Features toolbar (Figure 5.8). The Shaft Definition dialog box will appear as shown in Figure 5.36.

Next, you need to specify First angle as 360deg and Second angle as 0deg in the Limits area. Select the Sketch in selection spanner. Select the vertical axis in selection spanner in Axis area. Select and fill all the value in Shaft Definition dialog box, as shown in Figure 5.37.

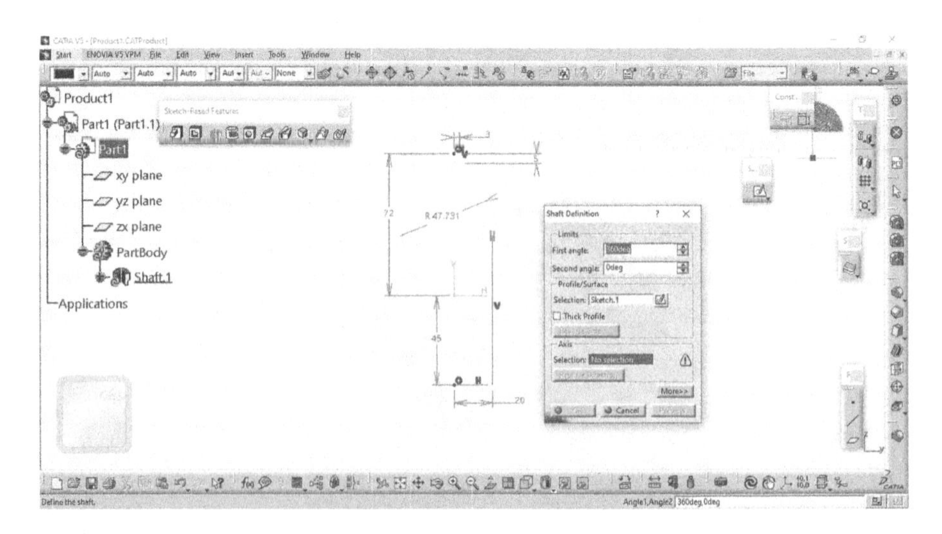

FIGURE 5.36 Selection of Shaft tool.

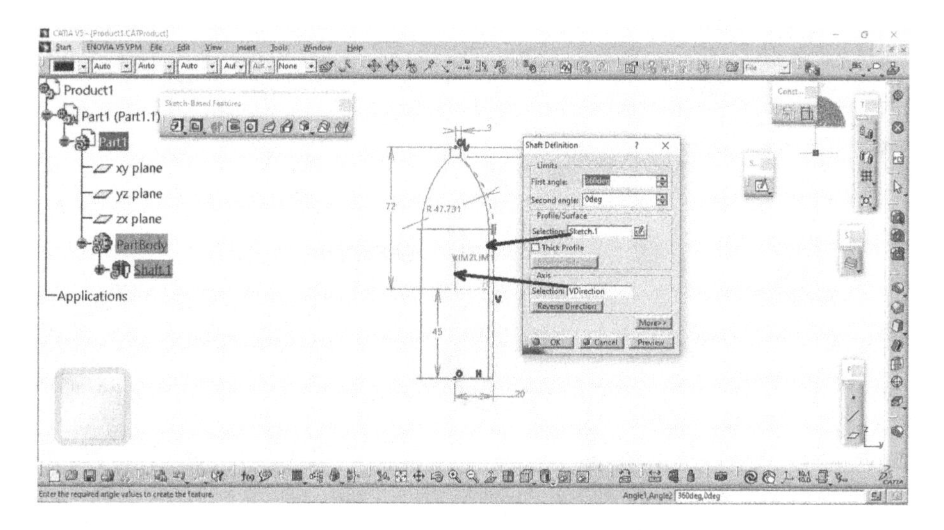

FIGURE 5.37 Value setting in Shaft Definition dialog box.

After setting all values, click the OK button and then you will get the figure, as shown in Figure 5.38.

5.3.4 Groove

The Groove tool is used to remove the material by revolving the sketch around the axis of revolution. The working of this tool is like that of the Shaft tool. The only difference is that this is a material removal operation. To create a groove feature, draw

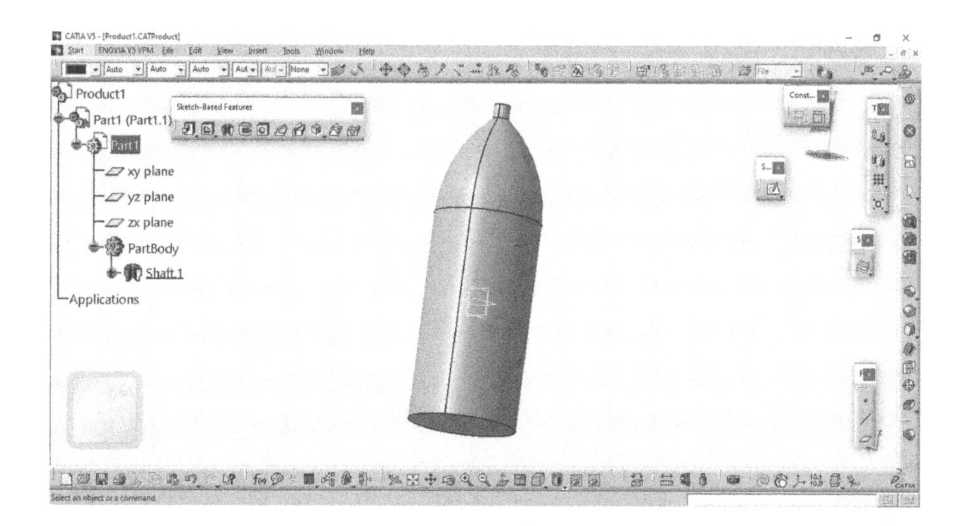

FIGURE 5.38 Sketch after the use of Shaft.

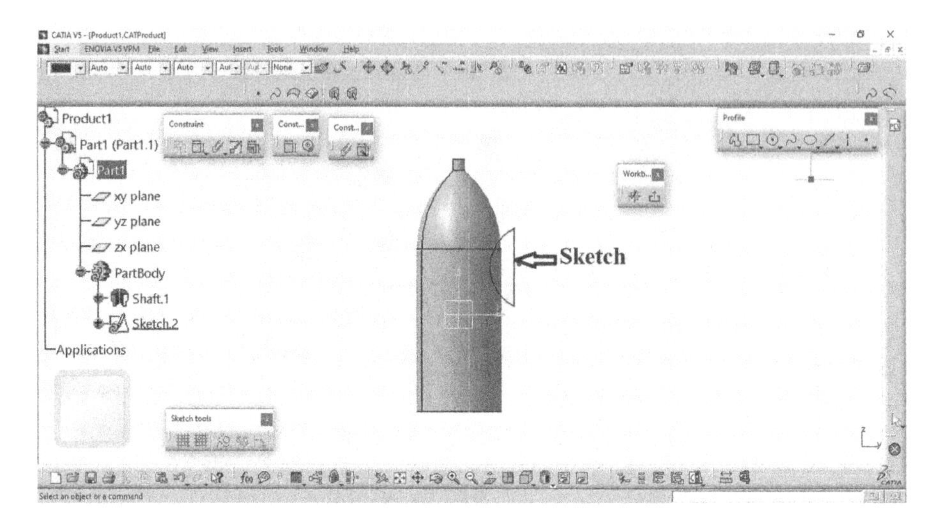

FIGURE 5.39 Sketch for using Groove tool.

the sketch as shown in Figure 5.39 by selecting yz plane in created object as shown in Figure 5.38 and then exit the Sketcher workbench.

Next, close the Groove button from Sketch-Based Features Toolbar, as shown in Figure 5.8. The Groove Definition dialog box is displayed as shown in Figure 5.40.

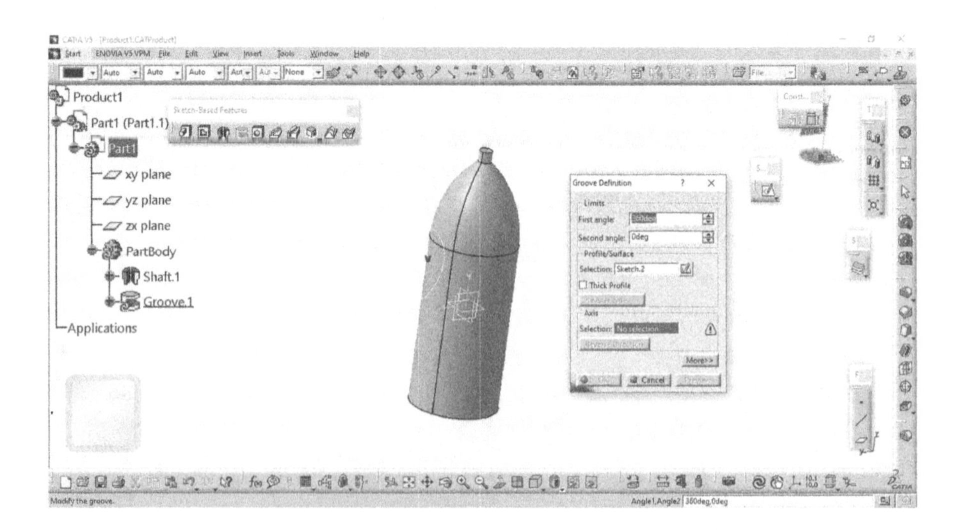

FIGURE 5.40 Selection of Groove tool.

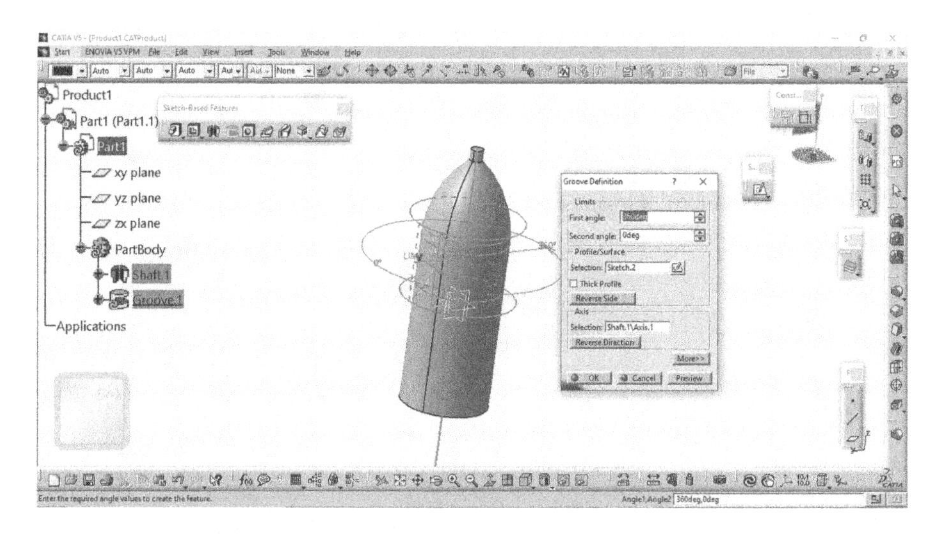

FIGURE 5.41 Value setting in Groove Definition dialog box.

Next, you need to specify First angle as 360deg and Second angle as 0deg in the Limits area. Select the Sketch in selection spanner. Select the vertical axis in selection spanner in Axis area. Select and fill all the value in Groove Definition dialog box, as shown in Figure 5.41.

After setting all values, click on the OK button and then you will get the figure, as shown in Figure 5.42.

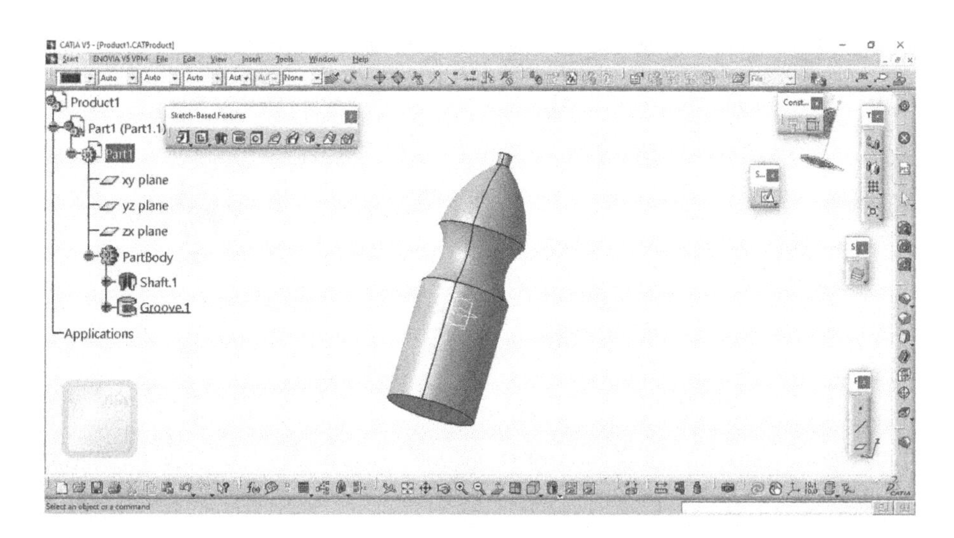

FIGURE 5.42 Sketch after the use of Groove.

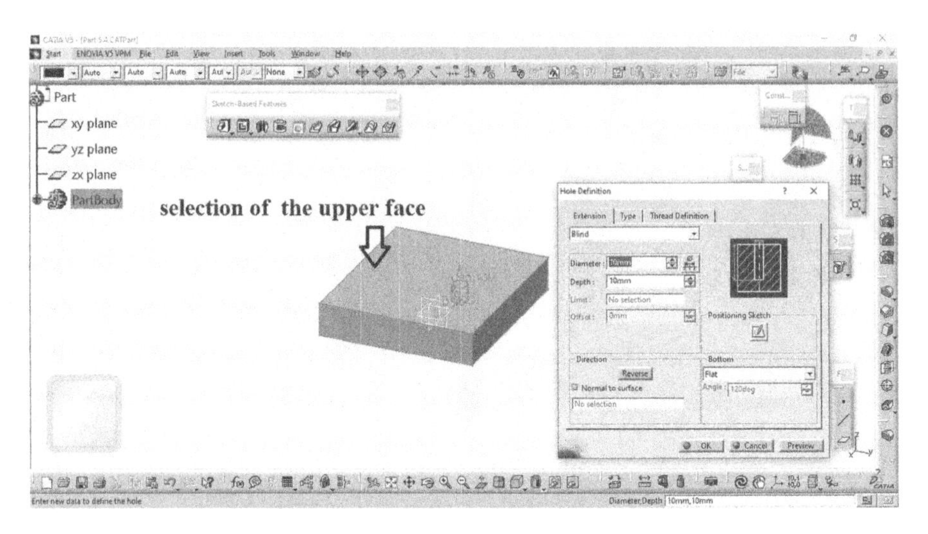

FIGURE 5.43 Selection of face for using Hole tool.

5.3.5 Hole

The Hole tool is used for creating hole and thread in existing body. Pad the object, as shown in Figure 5.13. Choose the Hole tool from Sketch-Based Features toolbar as shown in Figure 5.8 and select the upper face of the object. The Hole Definition dialog box will be displayed as shown in Figure 5.43.

Set the value of Diameter as 60mm in Diameter spanner and Depth as 8mm in Depth spanner. Choose Flat option in Bottom area. Select and fill all the value in Hole Definition dialog box, as shown in Figure 5.44.

After setting all values, click on the OK button, you will get the Hole, as shown in Figure 5.45.

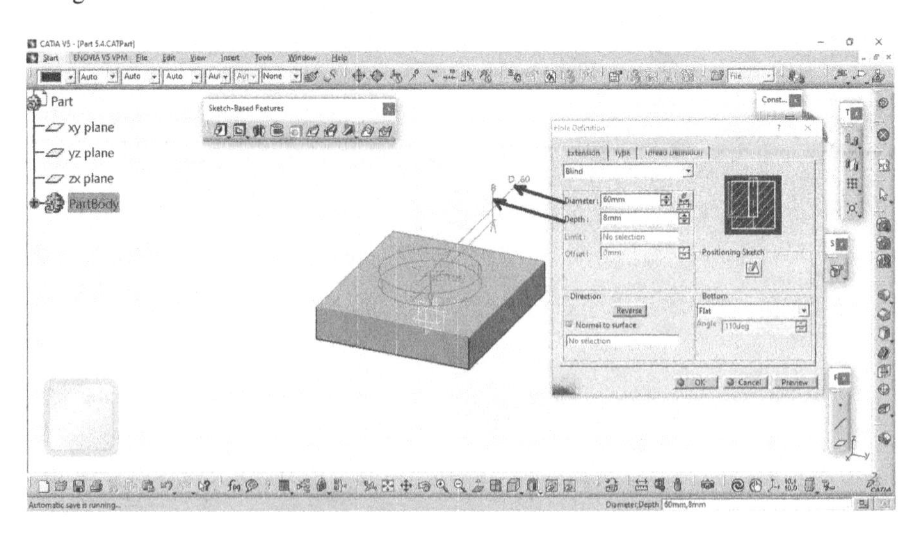

FIGURE 5.44 Value setting in Hole Definition dialog box.

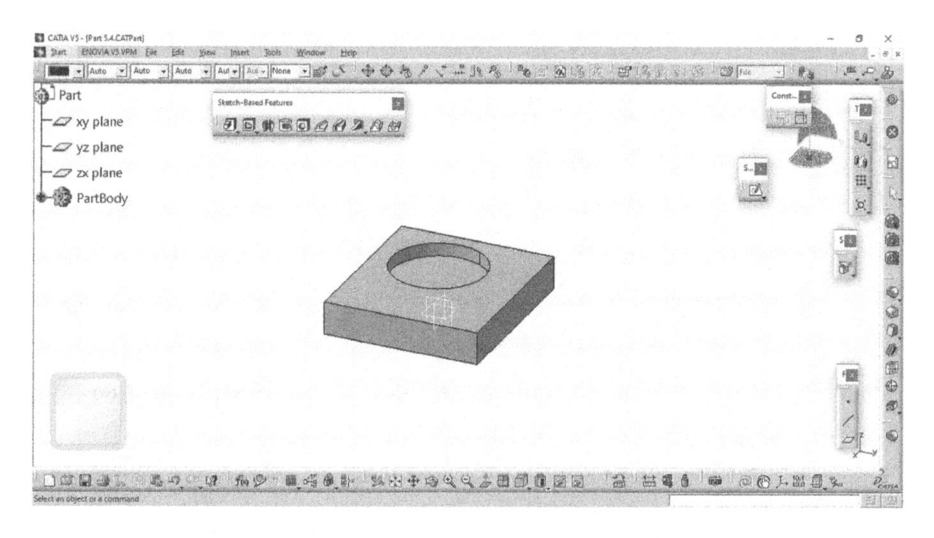

FIGURE 5.45 Sketch after the use of Hole tool.

5.3.6 Rib

One of the most important advanced modeling tools is Rib tool. This tool is used to sweep an open or closed profile along an open or closed center curve. A profile is the section for the rib feature and the center curve is the course taken by the profile while creating the rib feature. To create a rib feature, first, you have to draw profile sketch as shown in Figure 5.46 selecting yz plane using Sketcher workbench and then exit the Sketcher workbench.

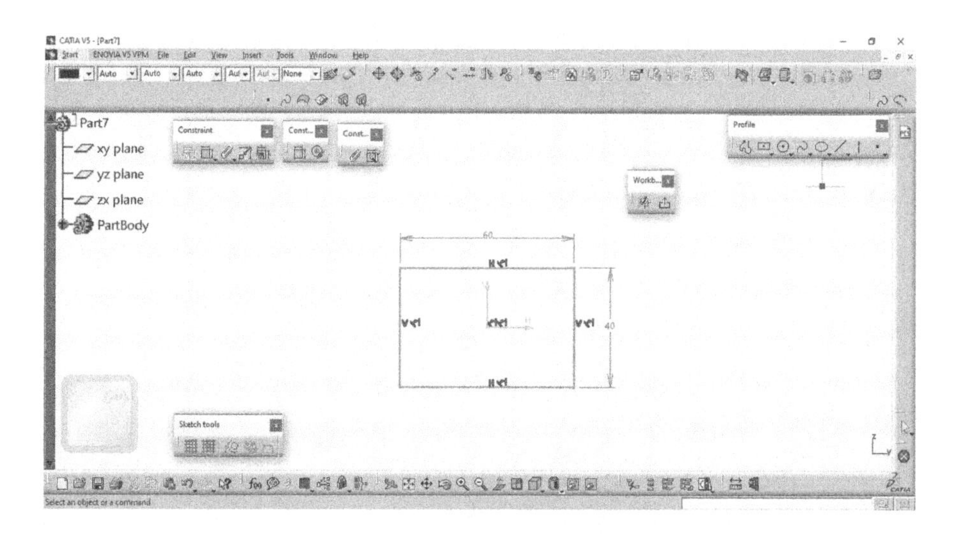

FIGURE 5.46 Profile Sketch for using Rib tool.

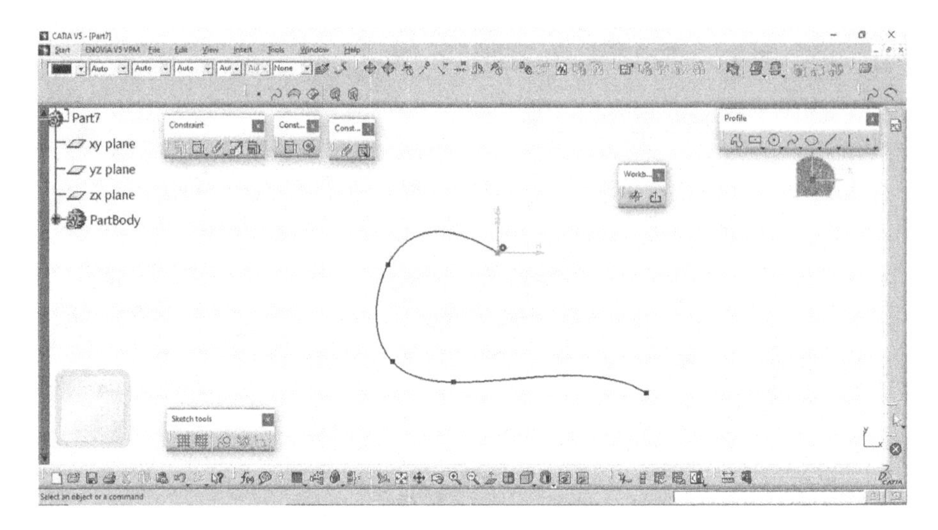

FIGURE 5.47 Center Curve Sketch for using Rib tool.

For making center curve choose xy plane and draw center curve sketch as shown in Figure 5.47 using Sketcher workbench and then exit the Sketcher workbench.

Next, Rib button from the Sketch-Based Features toolbar is to be chosen (Figure 5.8). The Rib Definition dialog box will appear as shown in Figure 5.48.

Select the Profile and Center curve in Rib Definition dialog box (Figure 5.49).

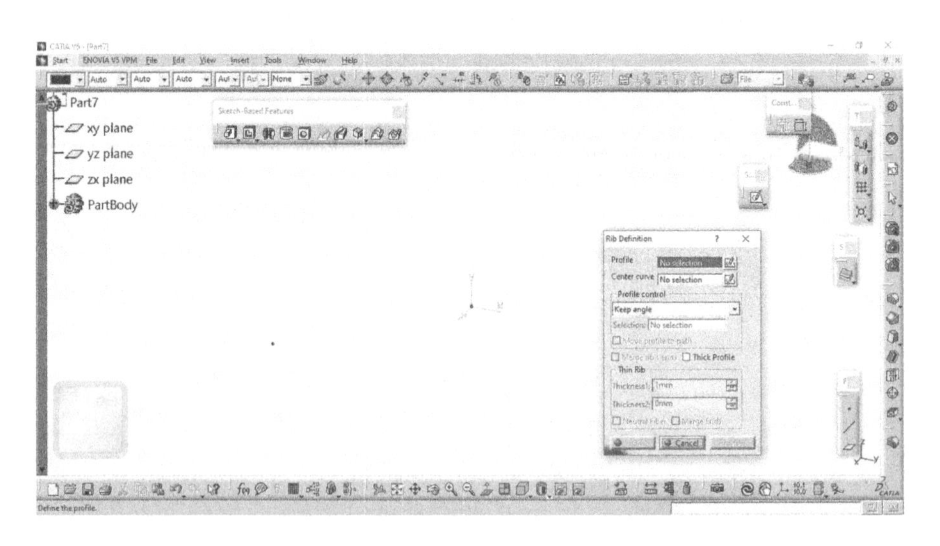

FIGURE 5.48 Selection of Rib tool.

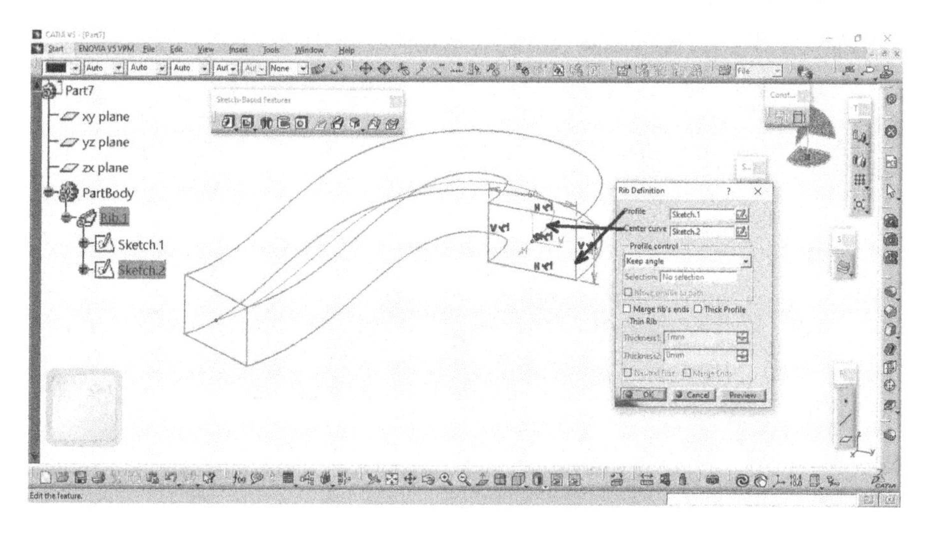

FIGURE 5.49 Value setting in Rib Definition dialog box.

After setting all values, click on the OK button and then you will get the Rib, as shown in Figure 5.50.

5.3.7 Slot

To create a slot feature, first you have to draw profile sketch as drawn in Figure 5.50. As shown in Figure 5.51, select yz plane using Sketcher workbench and then exit the Sketcher workbench.

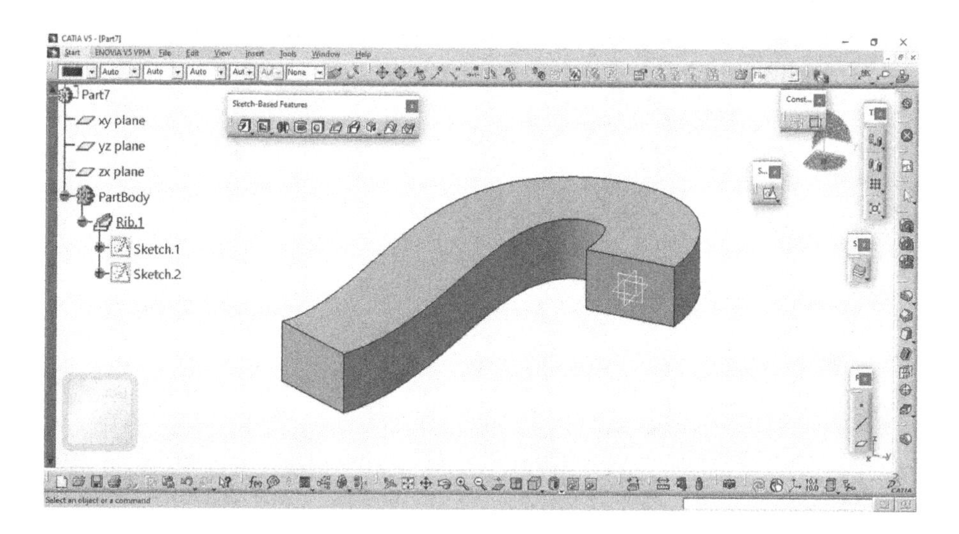

FIGURE 5.50 Sketch after the use of Rib tool.

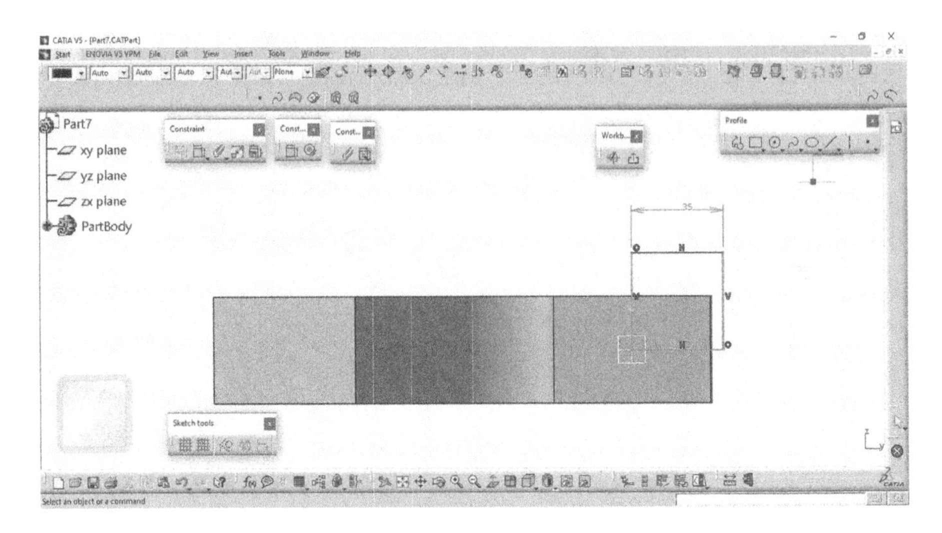

FIGURE 5.51 Profile Sketch for using Slot tool.

For making center curve, choose xy plane and draw center curve sketch as shown in Figure 5.52 using Sketcher workbench and then exit the Sketcher workbench.

Next, Slot button from the Sketch-Based Features toolbar is to be chosen (Figure 5.8). The Slot Definition dialog box will appear (Figure 5.53).

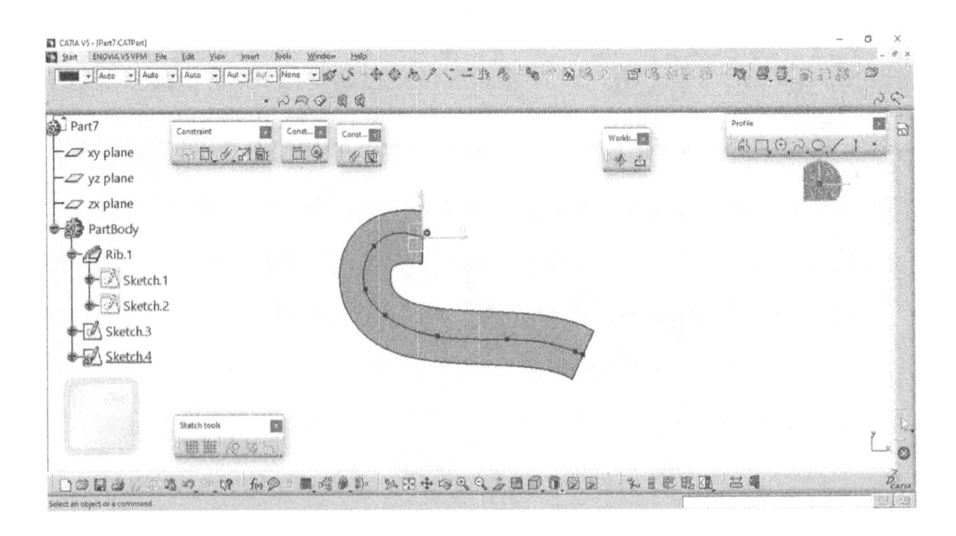

FIGURE 5.52 Center Curve Sketch for using Slot tool.

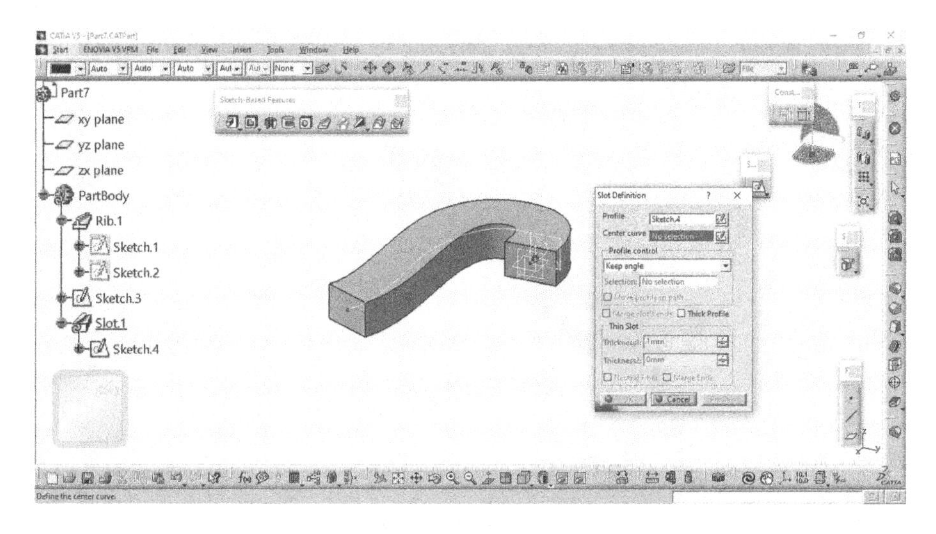

FIGURE 5.53 Selection of Slot tool.

Select the Profile and Center curve in Slot Definition dialog box as shown in Figure 5.54.

After setting all values, click on the OK button and then you will get the rib, as shown in Figure 5.55.

5.3.8 Solid Combine

The tools in this toolbar are Stiffener and Solid Combine. The Solid Combine toolbar will appear as shown in Figure 5.56.

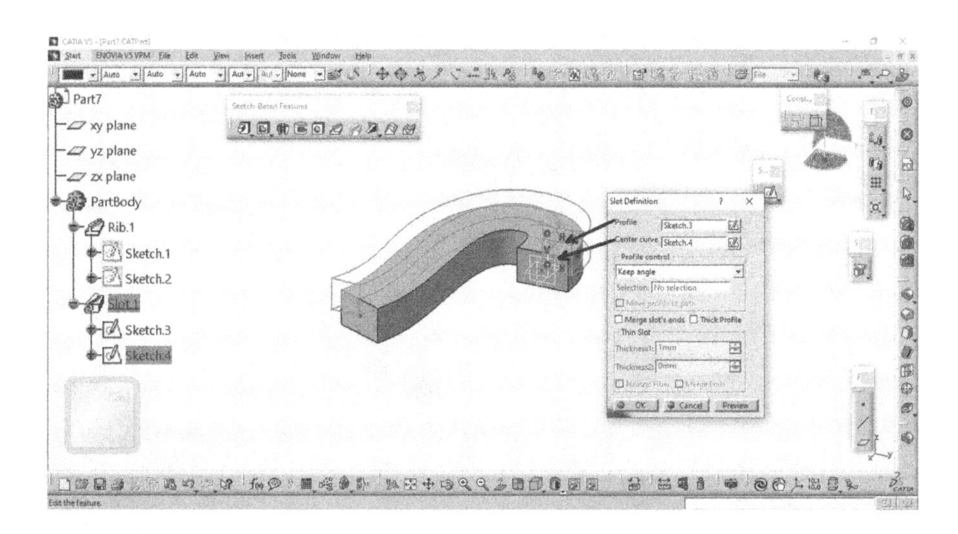

FIGURE 5.54 Value setting in Slot Definition dialog box.

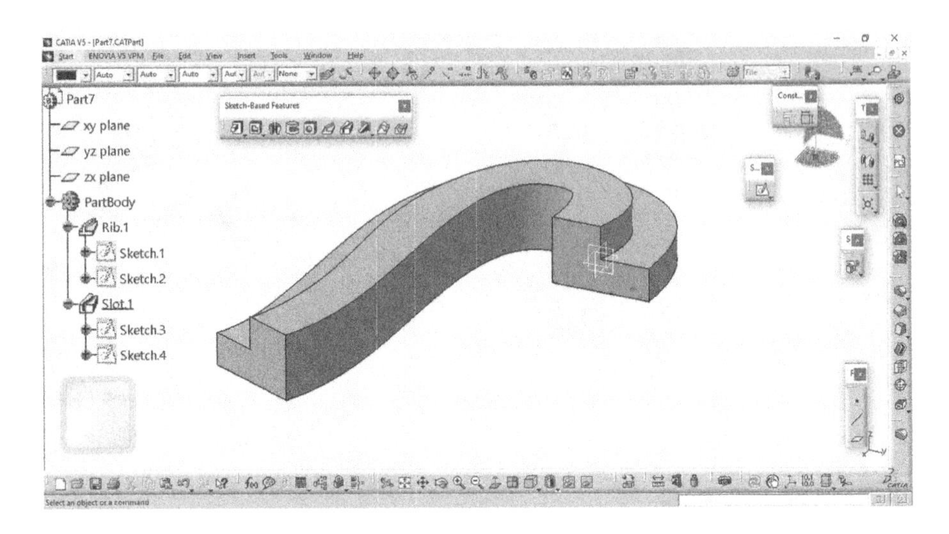

FIGURE 5.55 Sketch after the use of Rib tool.

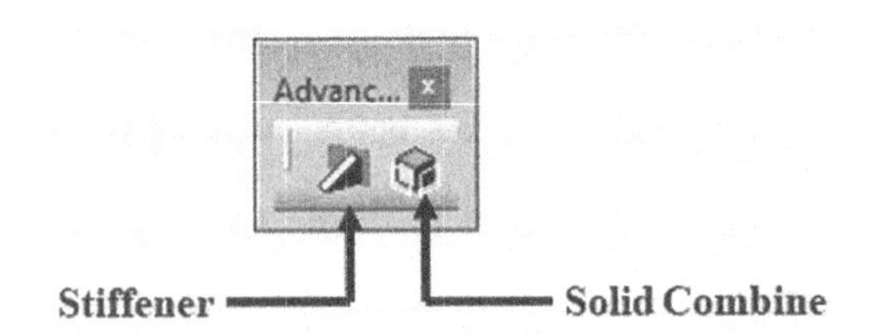

FIGURE 5.56 Solid Combine toolbar.

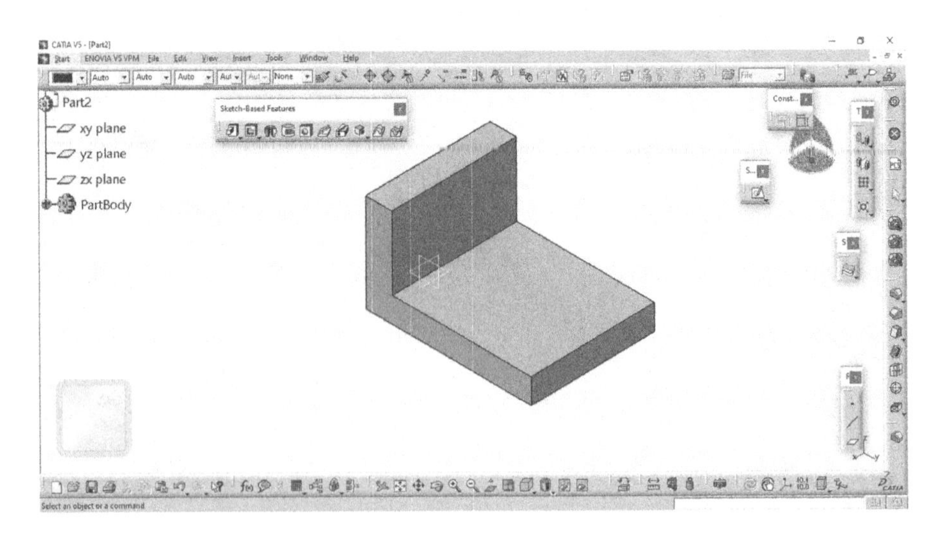

FIGURE 5.57 Padded object.

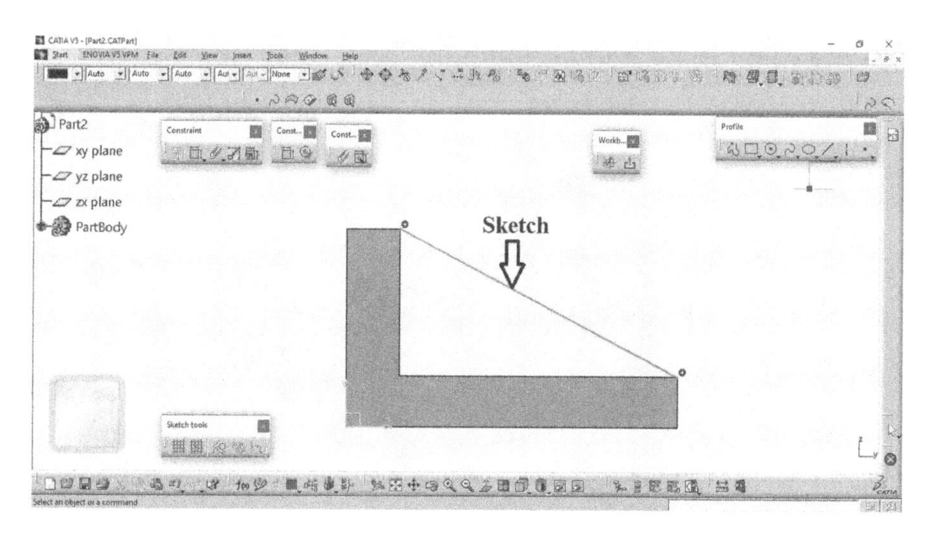

FIGURE 5.58 Sketch for the use of Stiffener tool.

5.3.8.1 Stiffener

Draw the padded object as shown in Figure 5.57.

Select the yz plane and draw the sketch as shown in Figure 5.58 using Sketcher workbench and then exit the Sketcher workbench.

Next Stiffener button from the Sketch-Based Features toolbar is to be chosen as shown in Figure 5.8. The Stiffener Definition dialog box will appear as shown in Figure 5.59.

Enter the Thickness1 value as 10mm in Thickness area in Stiffener Definition dialog box as shown in Figure 5.60.

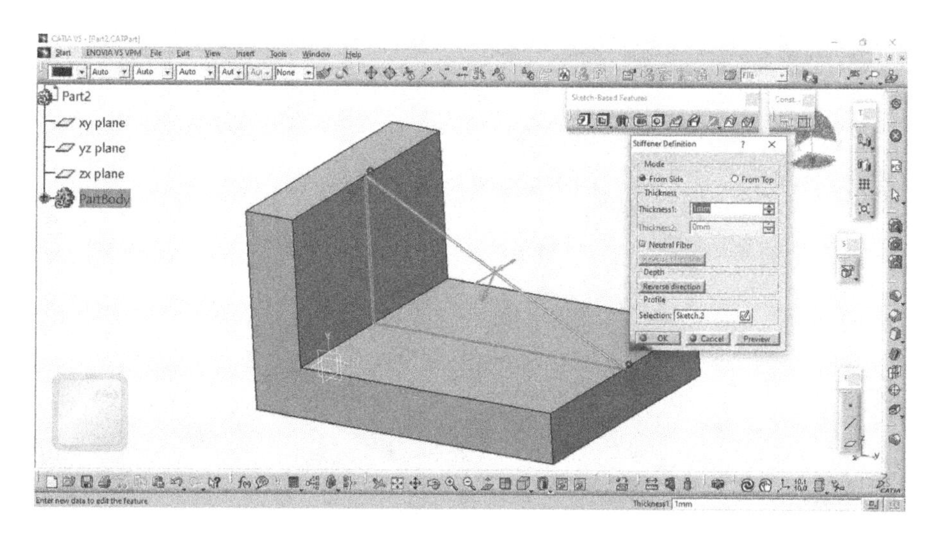

FIGURE 5.59 Selection of Stiffener tool.

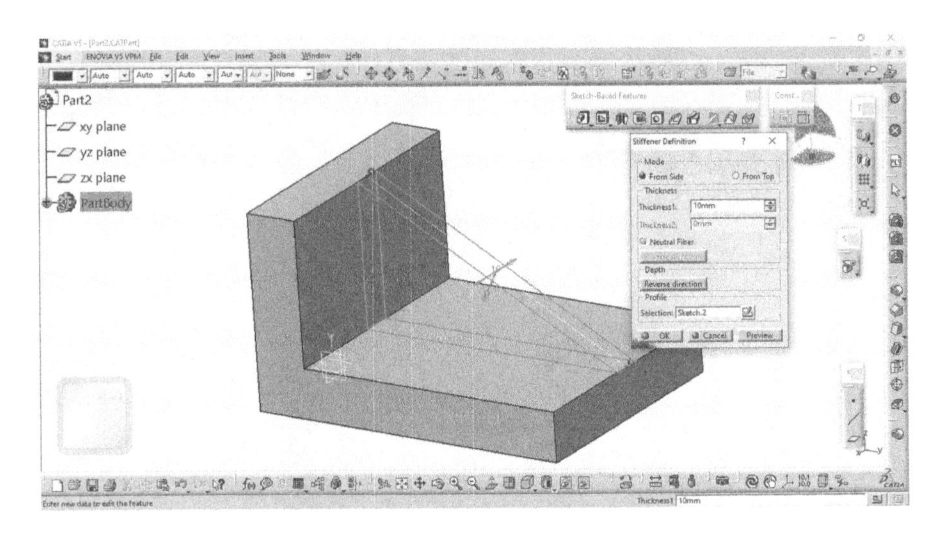

FIGURE 5.60 Value setting in Stiffener Definition dialog box.

After setting all values, click on the OK button and then you will get the Stiffener, as shown in Figure 5.61.

5.3.8.2 Solid Combine

Solid Combine tool is used for combining two sketched elements into single solid element. Select the xy plane and draw the Sketch.1 as shown in Figure 5.62 using Sketcher workbench and then exit the Sketcher workbench.

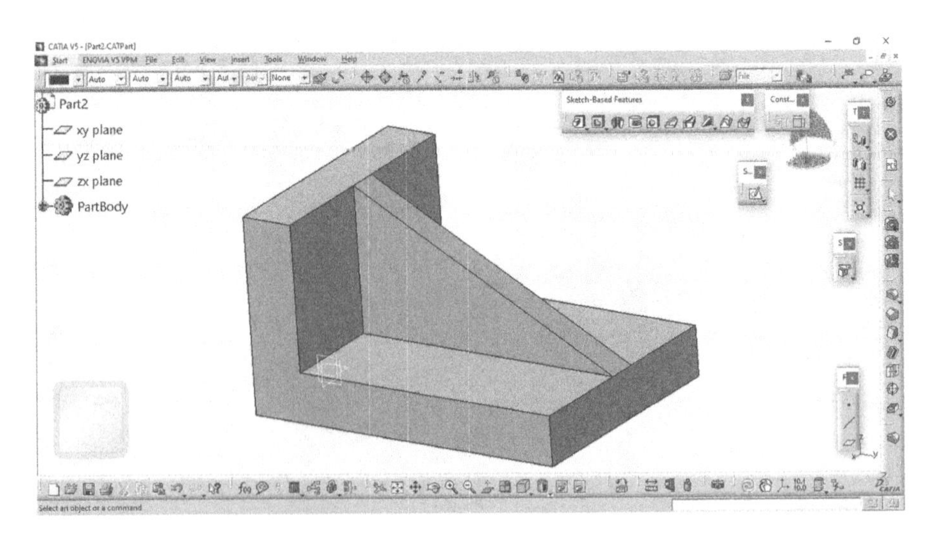

FIGURE 5.61 Sketch after the use of Stiffener tool.

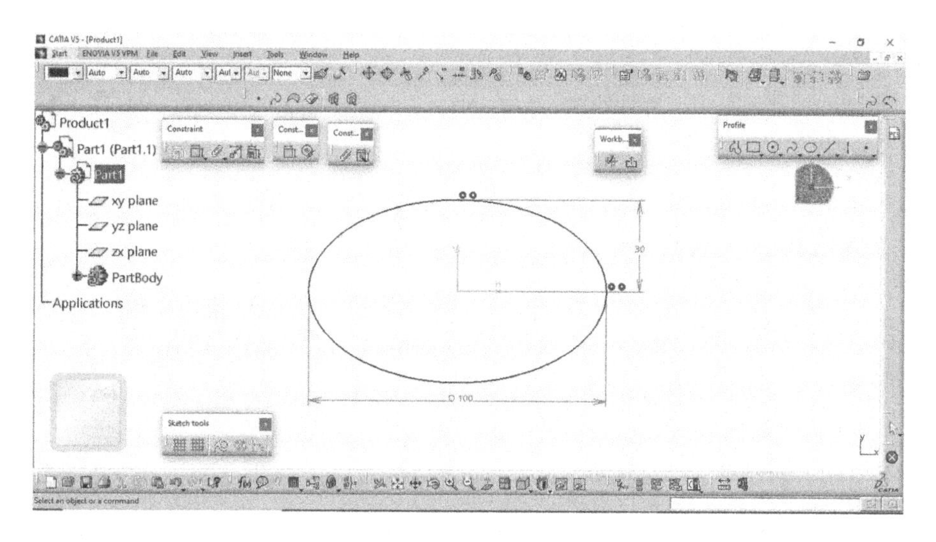

FIGURE 5.62 Sketch.1 for Solid Combine tool.

Next, select the xy plane and draw the Sketch.2 as shown in Figure 5.63 using Sketcher workbench and then exit the Sketcher workbench.

Next, choose the Solid Combine button from the Sketch-Based Features as shown in Figure 5.8. The Combine Definition dialog box will appear as shown in Figure 5.64.

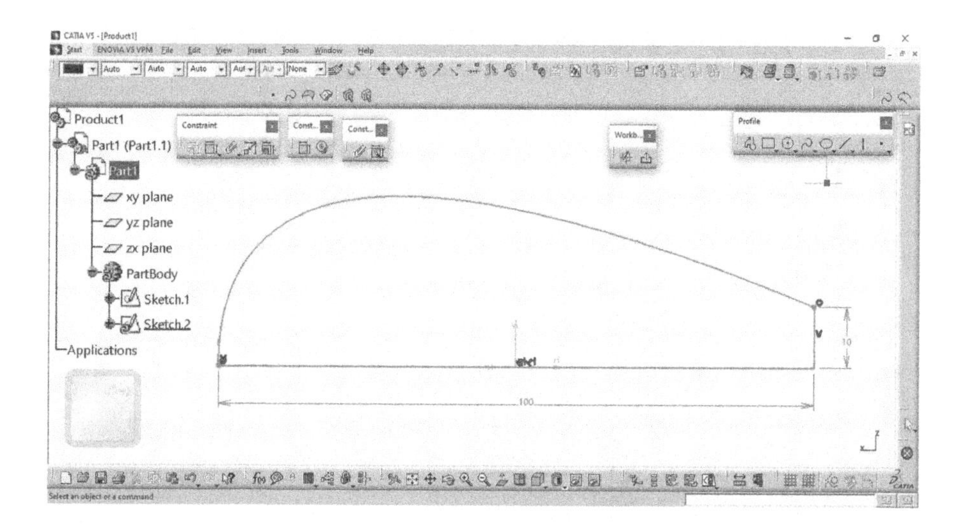

FIGURE 5.63 Sketch.2 for Solid Combine tool.

FIGURE 5.64 Selection of Solid Combine tool.

Select Sketch.2 as First component and Sketch.1 as Second component in Profile spanner in Combine Definition dialog box Figure 5.65.

After setting all values, click the OK button, you will get the combine sketches solid as shown in Figure 5.66.

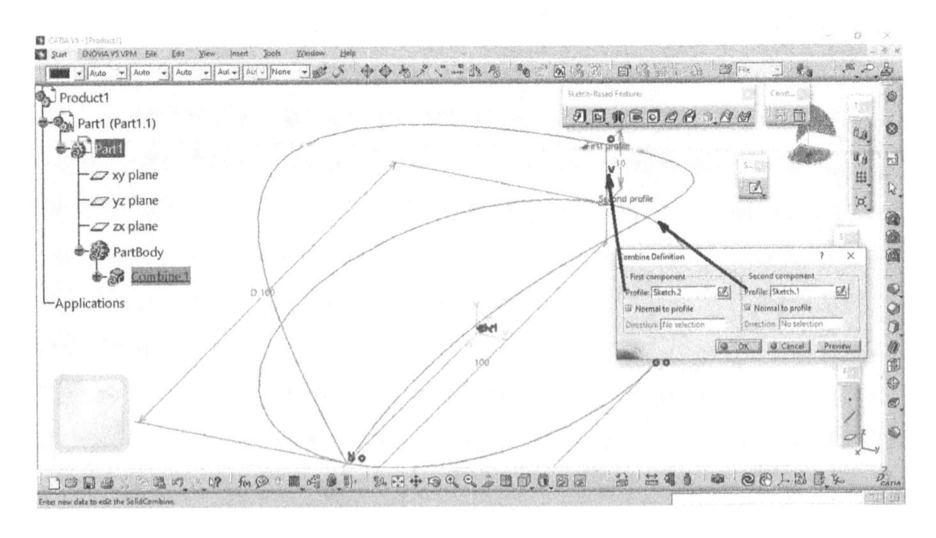

FIGURE 5.65 Selection of sketches in Combine Definition dialog box.

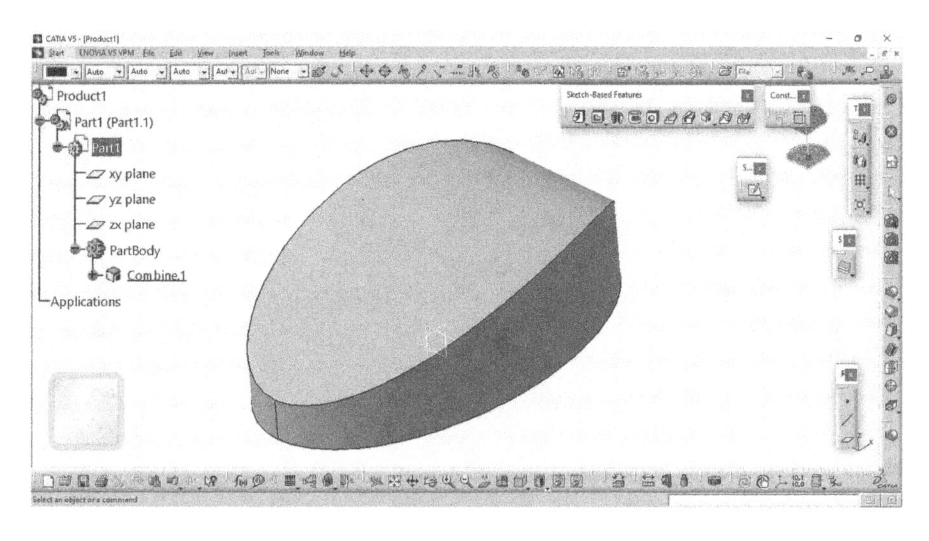

FIGURE 5.66 Sketch after the use of Solid Combine tool.

5.3.9 Multi-Section Solid

Multi-Section Solid is used for combining more than two sketched elements into single solid element. For creating sketch, you must create plane. Select xy plane and choose plane tool from reference element tool bar, as shown in Figure 5.67.

Plane Definition dialog box will appear as shown in Figure 5.68. Select reference plane as xy plane in Reference tab and set Offset value as 60mm in Offset tab Figure 5.68.

FIGURE 5.67 Selection of Plane tool.

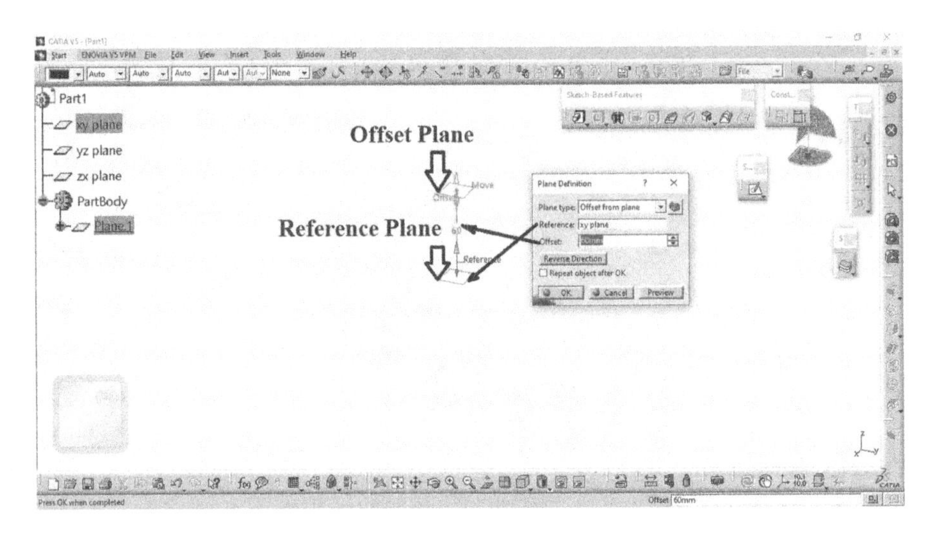

FIGURE 5.68 Setting value in Plane Definition dialog box.

Select the Plane.1 and draw the Sketch.1 as shown in Figure 5.69 using Sketcher workbench and then exit the Sketcher workbench.

Select the xy plane and draw the Sketch.2 as shown in Figure 5.70 using Sketcher workbench and then exit the Sketcher workbench.

Next, choose the Multi-Section Solid button from the Sketch-Based Features toolbar (Figure 5.8). The Multi-Section Solid dialog box will appear, as shown in Figure 5.71.

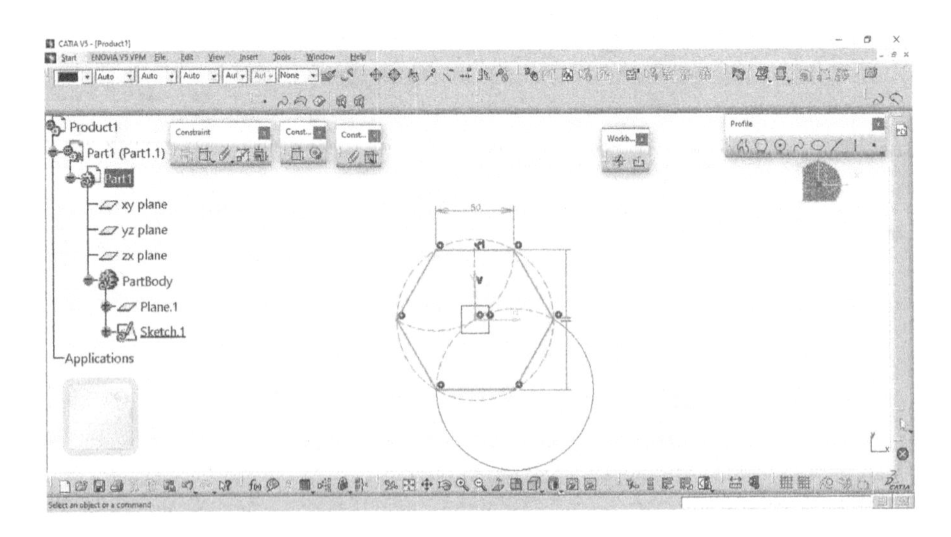

FIGURE 5.69 Sketch.1 for Multi-Section Solid tool.

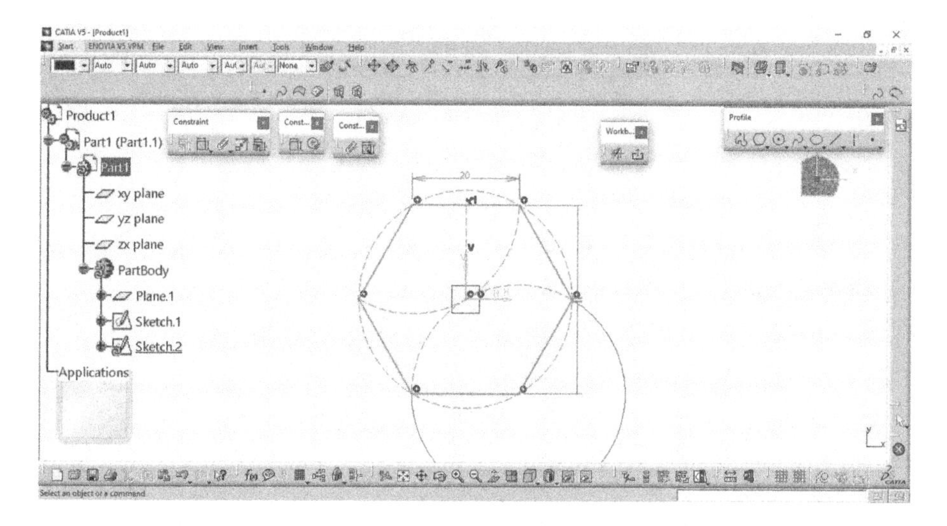

FIGURE 5.70 Sketch.2 for Multi-Section Solid tool.

Select Sketch.1 and Sketch.2 in Multi-Section Solid dialog box in Figure 5.72.

After setting all values, click the OK button and then you will get the Multi-Section Solid, as shown in Figure 5.73.

5.3.10 Removed Multi-Section Solid

Select the Plane.1 and draw the Sketch.1 as shown in Figure 5.74 using Sketcher workbench and then exit the Sketcher workbench.

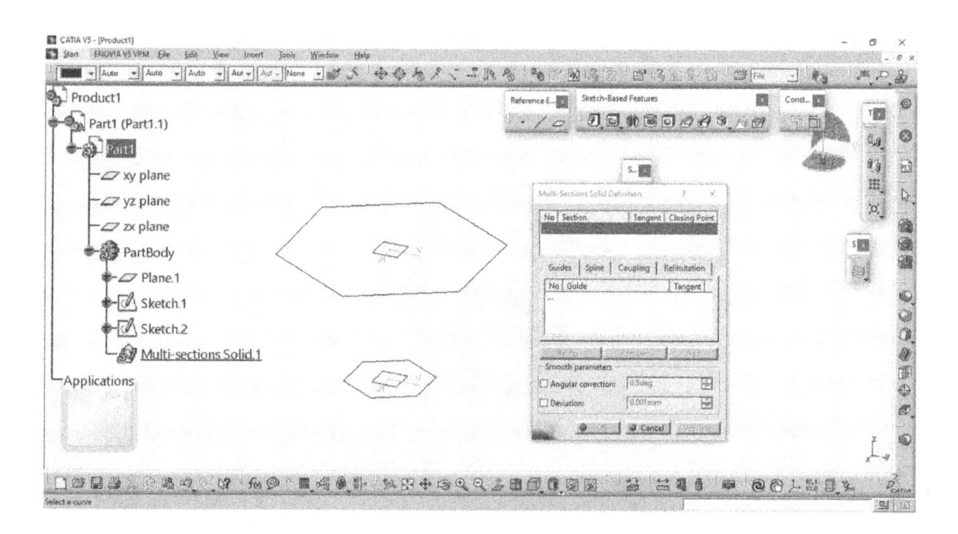

FIGURE 5.71 Selection of Multi-Section Solid tool.

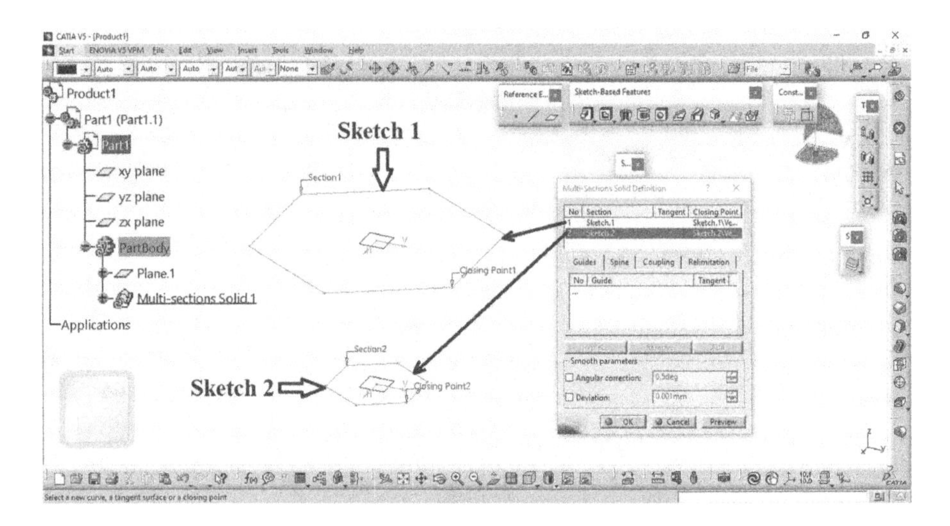

FIGURE 5.72 Selection of sketches in Multi-Section Solid dialog box.

Select the xy plane and draw the Sketch.2 as shown in Figure 5.75 using Sketcher workbench and then exit the Sketcher workbench.

Next, the Removed Multi-Section Solid button from the Sketch-Based Features toolbar is chosen as shown in Figure 5.8. The Multi-Section Solid dialog box will appear as shown in Figure 5.76.

Select Sketch.3 and Sketch.4 in Removed Multi-Section Solid dialog box in Figure 5.77.

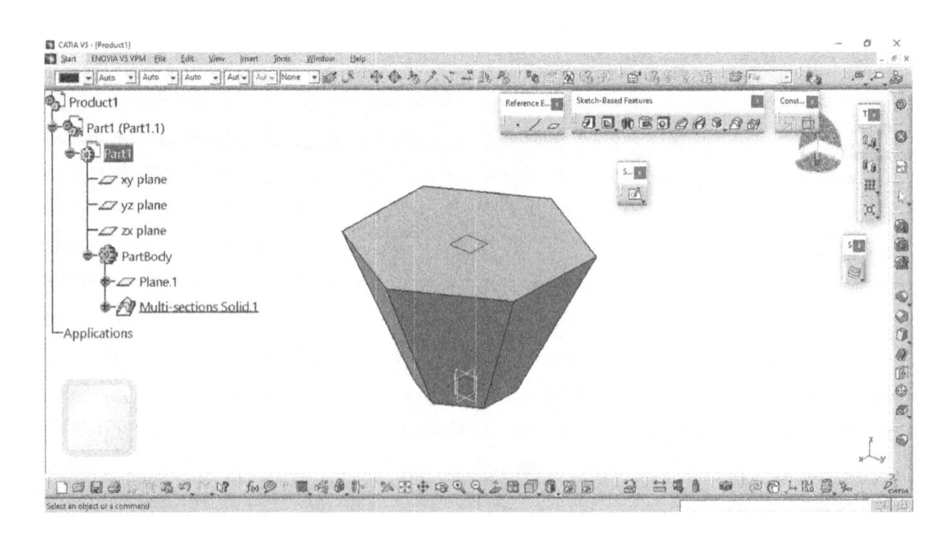

FIGURE 5.73 Sketch after the use of Multi-Section Solid tool.

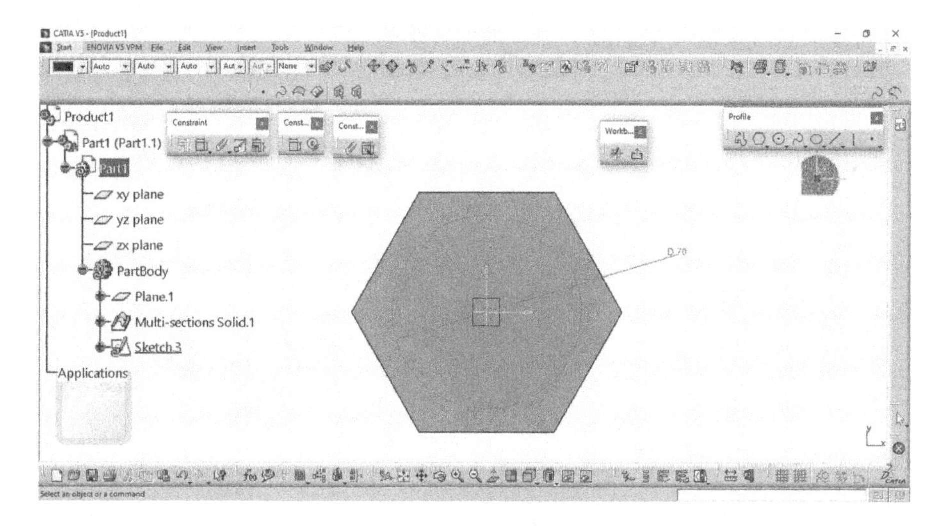

FIGURE 5.74 Sketch.1 for Removed Multi-Section Solid tool.

After selecting all sketches, click the OK button and then you will get the Removed Multi-Section Solid as shown in Figure 5.78.

5.4 DRESS-UP FEATURES TOOLBAR

Dress-Up Features tools are used to finish the created solid parts. The tools in this toolbar are Fillets, Chamfer, Drafts, Shell, Thickness, Thread and Remove Face. The Dress-Up Features toolbar will appear as shown in Figure 5.79.

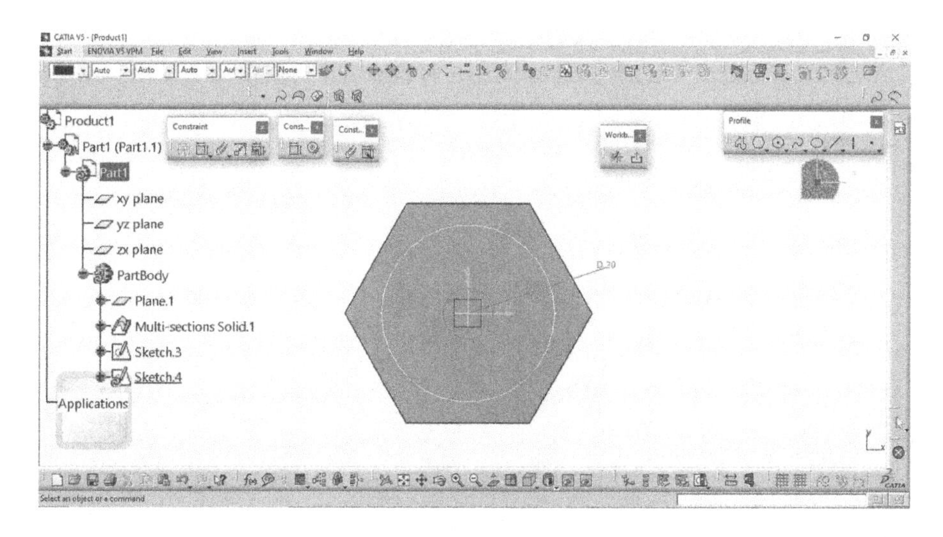

FIGURE 5.75 Sketch.2 for Removed Multi-Section Solid tool.

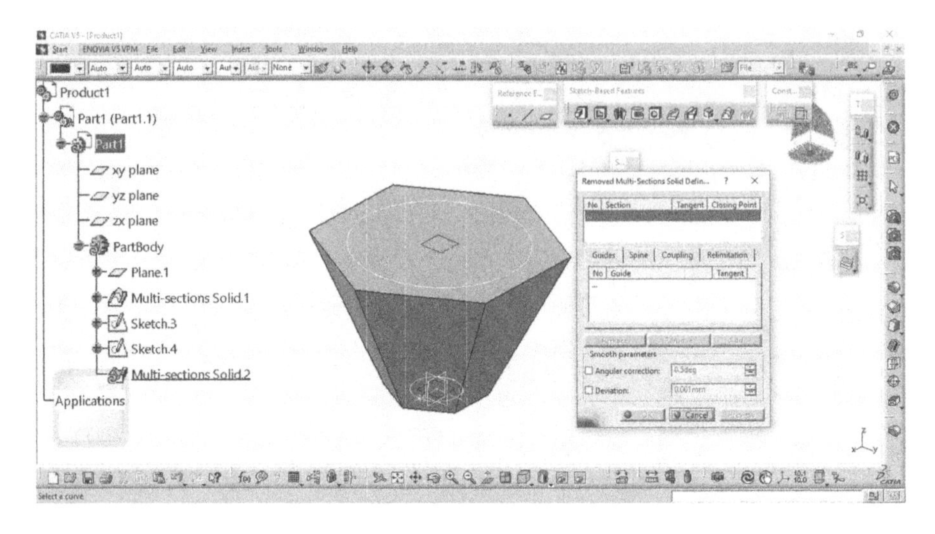

FIGURE 5.76 Selection of Removed Multi-Section Solid tool.

5.4.1 Fillets Toolbar

Fillets toolbar is used to round edge or corner of the same radius or variable radius on created solid parts. The tools in this toolbar are Edge fillet, Variable radius fillet, Chordal fillet, Face-face fillet and Tritangent Fillet. The Fillets toolbar will appear as shown in Figure 5.80.

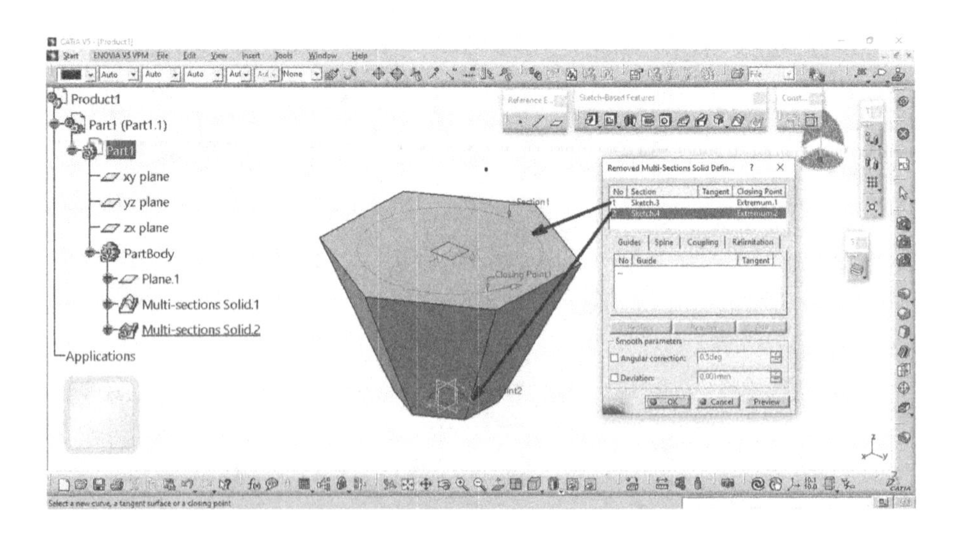

FIGURE 5.77 Selection of sketches in Removed Multi-Section Solid dialog box.

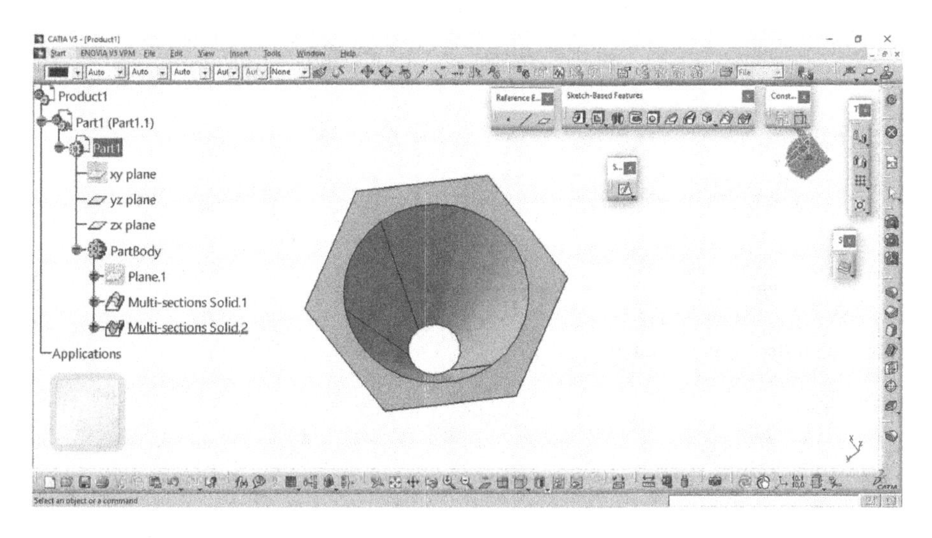

FIGURE 5.78 Sketch after the use of Removed Multi-Section Solid tool.

5.4.1.1 Edge Fillet

The Edge fillet tool creates smooth fillet on edge. Draw the padded object as shown in Figure 5.81.

Next, choose the Edge fillet tool from the Fillets toolbar as shown in Figure 5.80. The Edge Fillet Definition dialog box will be displayed as shown in Figure 5.82.

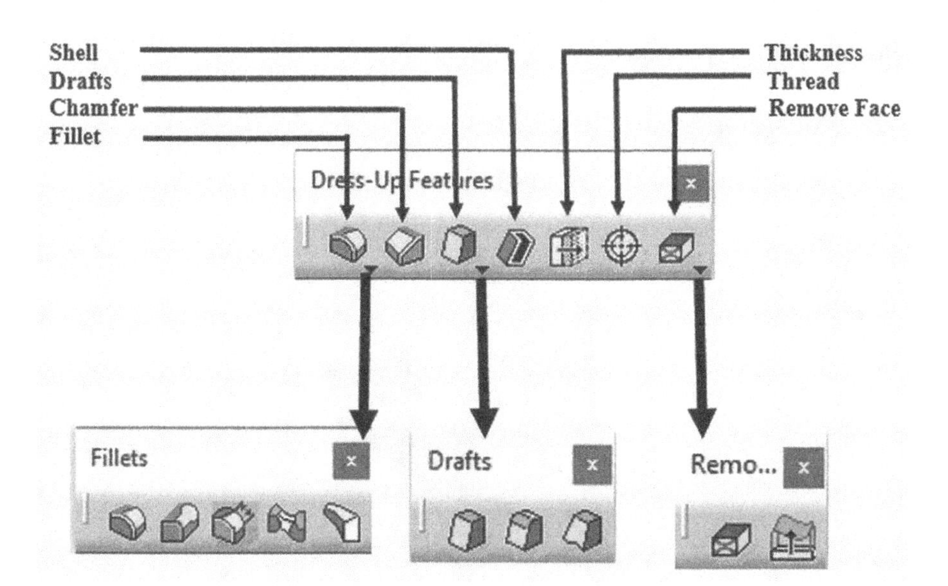

FIGURE 5.79 Dress-Up Features toolbar.

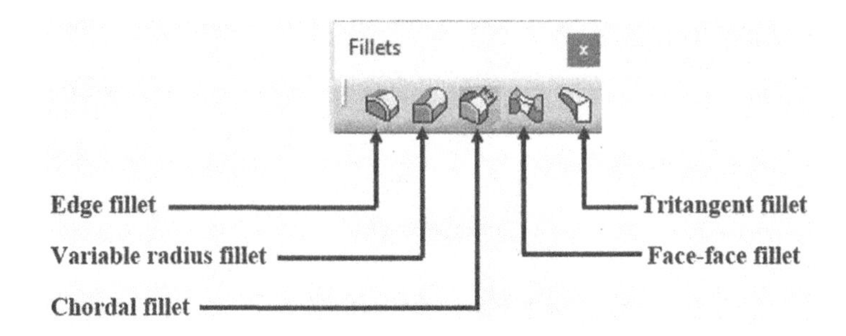

FIGURE 5.80 Fillets toolbar.

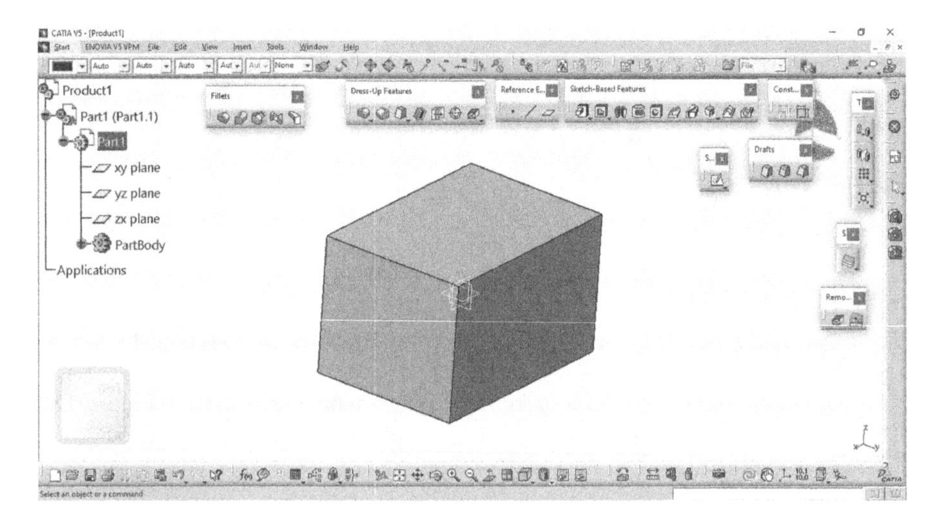

FIGURE 5.81 Padded object.

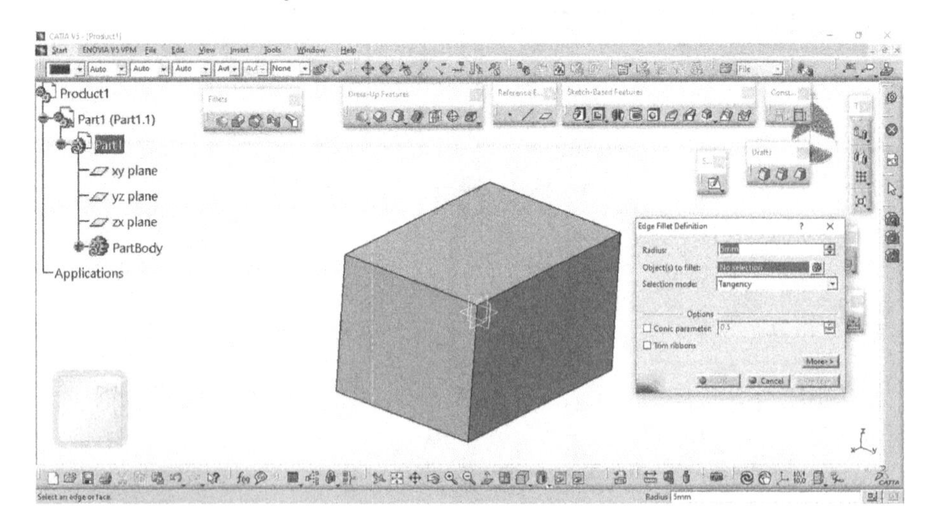

FIGURE 5.82 Selection of Edge fillet tool.

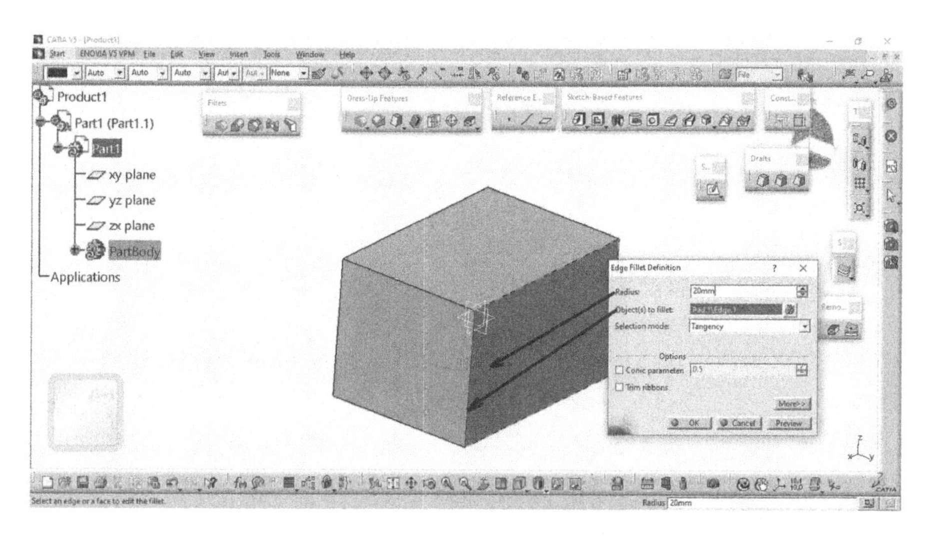

FIGURE 5.83 Selection of edge and radius value filling in Edge Fillet Definition dialog box.

Enter the Radius value as 20mm and select Edge under Radius that is Object(s) to fillet spanner in Edge Fillet Definition dialog box, as shown in Figure 5.83.

After selecting and filling, click on the OK button and then you will get the Edge fillet, as shown in Figure 5.84.

5.4.1.2 Variable Radius Fillet

This command allows the fillet to have more than one radius. Draw the padded object as shown in Figure 5.81. Choose the Variable Radius Fillet tool from the Fillets toolbar, as shown in Figure 5.80. The Variable Radius Fillet Definition dialog box will appear as shown in Figure 5.85.

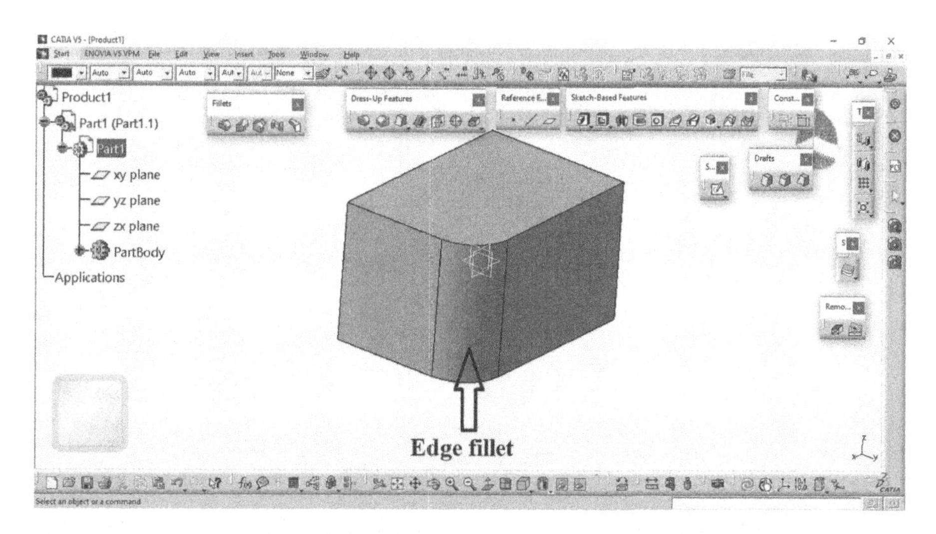

FIGURE 5.84 Solid Edge after the use of Edge fillet tool.

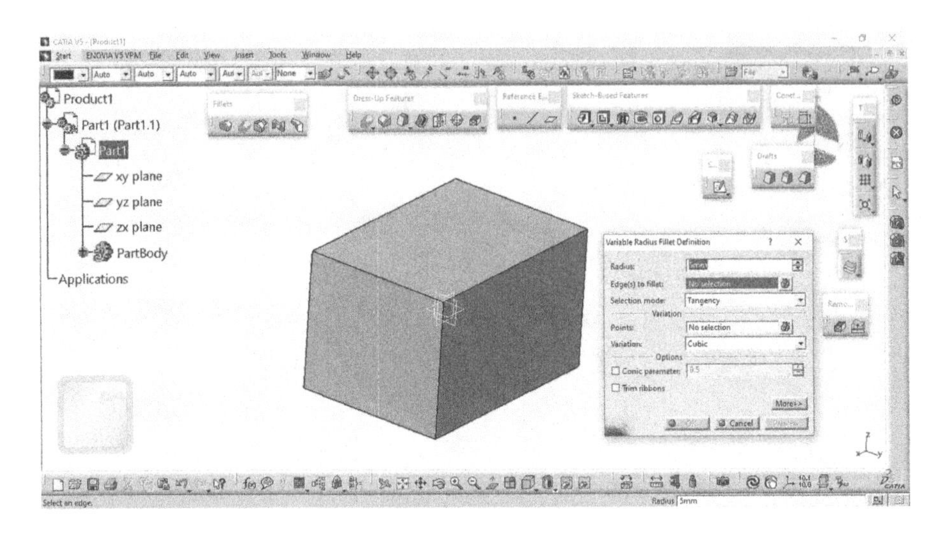

FIGURE 5.85 Selection of Variable Radius Fillet tool.

Next, select the edge and locate the point on the Edge. Enter the Radius value as 30, 20, and 5mm for points 1, 2, and 3, respectively, in Variable Radius Fillet dialog box (Figure 5.86).

After selecting and filling, click on the OK button and then you will get the Variable Radius Fillet, as shown in Figure 5.87.

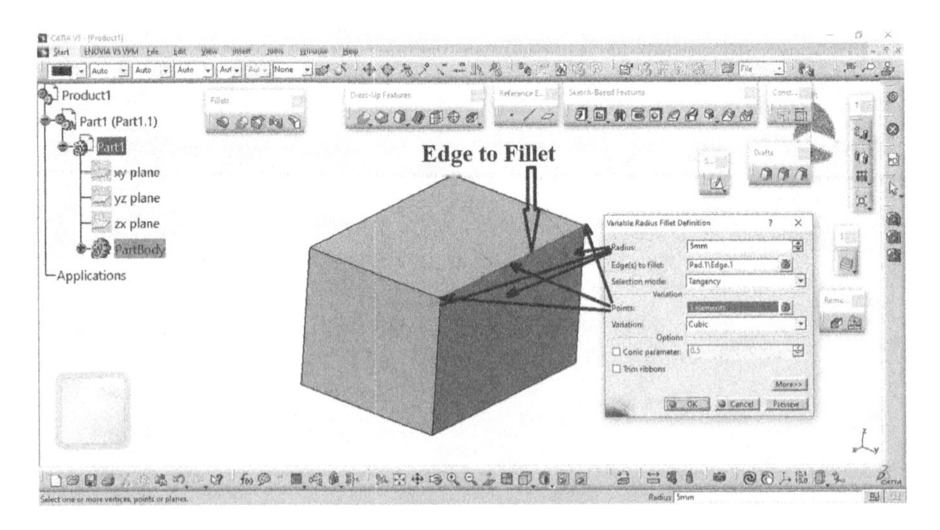

FIGURE 5.86 Selection of edge and radius value filling in Variable Radius Fillet Definition dialog box.

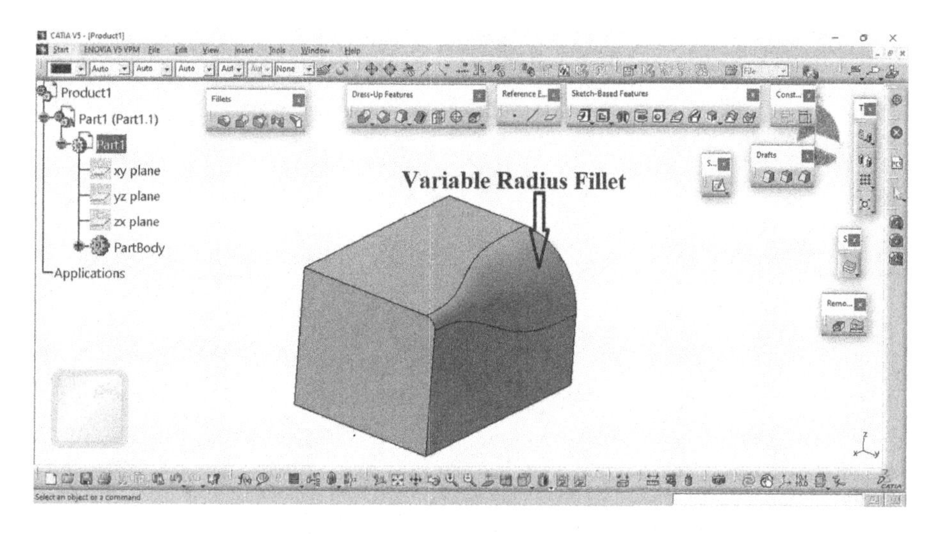

FIGURE 5.87 Solid Edge after the use of Variable Radius Fillet tool.

5.4.1.3 Chordal Fillet

This command allows the fillet to have more than one chordal length. Draw the padded object as shown in Figure 5.81. Choose the Chordal Fillet tool from the Fillets toolbar, as shown in Figure 5.80. The Chordal Fillet Definition dialog box will appear as shown in Figure 5.88.

Next, select the edge and locate the point on the edge. Enter the Chordal length value as 30, 10, and 40mm for points 1, 2, and 3, respectively, in Chordal Fillet Definition dialog box as shown in Figure 5.89.

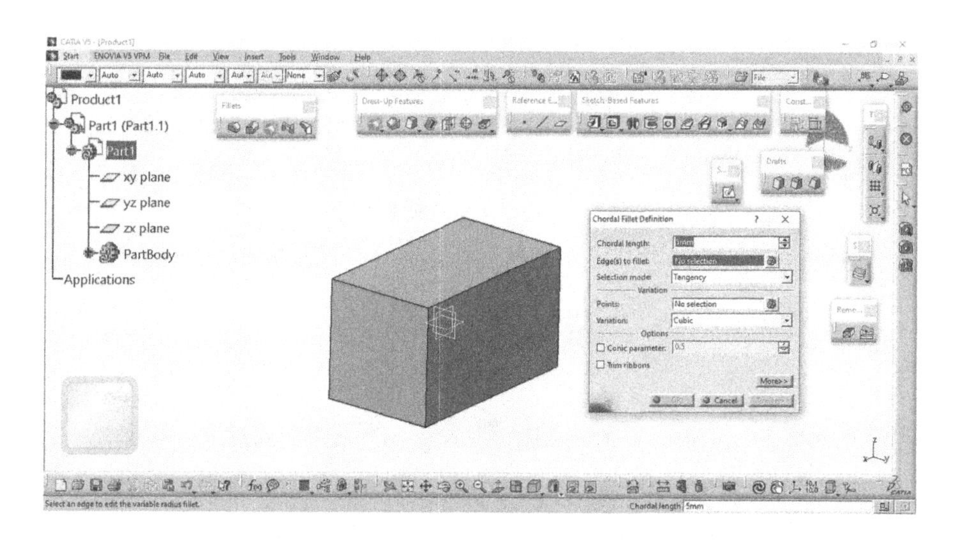

FIGURE 5.88 Selection of Chordal Fillet tool.

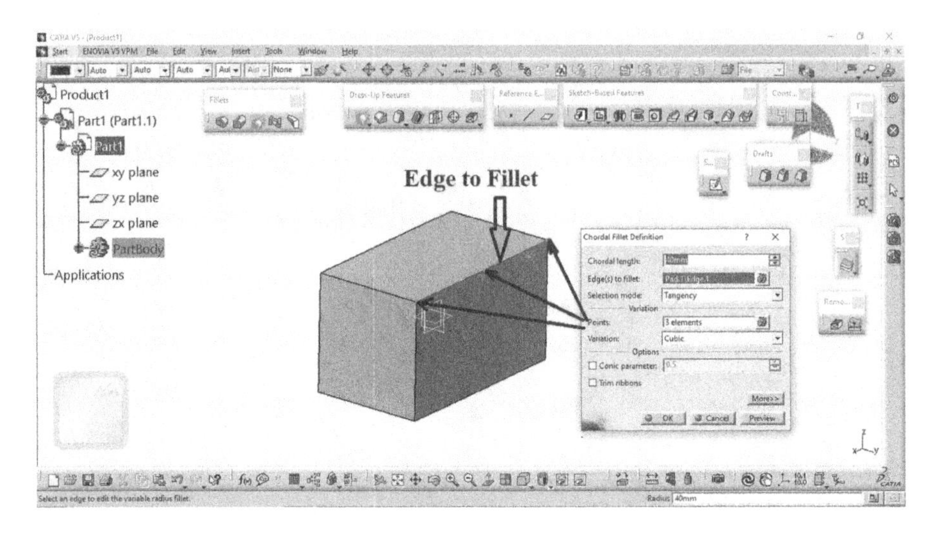

FIGURE 5.89 Selection of Edge and Chordal length value filling in Chordal Fillet Definition dialog box.

After selecting and filling, click the OK button and then you will get the Chordal Fillet, as shown in Figure 5.90.

5.4.1.4 Face-Face Fillet

The Face-Face Fillet tool creates smooth fillet between two edges. Draw the padded object as shown in Figure 5.91.

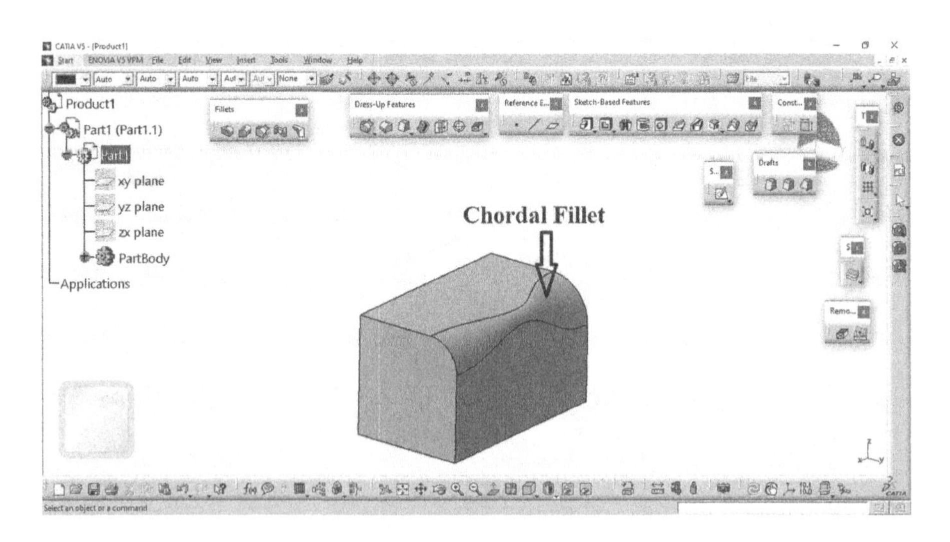

FIGURE 5.90 Solid Edge after the use of Chordal Fillet tool.

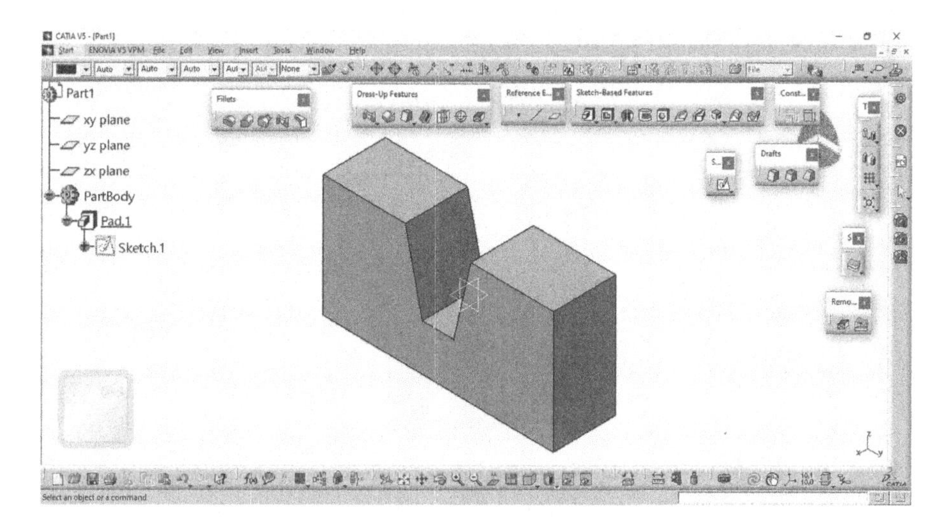

FIGURE 5.91 Padded object.

Next, choose the Face-Face Fillet tool from the Fillets toolbar, as shown in Figure 5.80. The Face-Face Fillet Definition dialog box will be displayed, as shown in Figure 5.92.

Enter the Radius value as 15mm and select 2 elements under Radius and Faces to fillet spanner in Face-Face Fillet Definition dialog box as shown in Figure 5.93.

After selecting and filling, click the OK button and then you will get the Face-Face fillet, as shown in Figure 5.94.

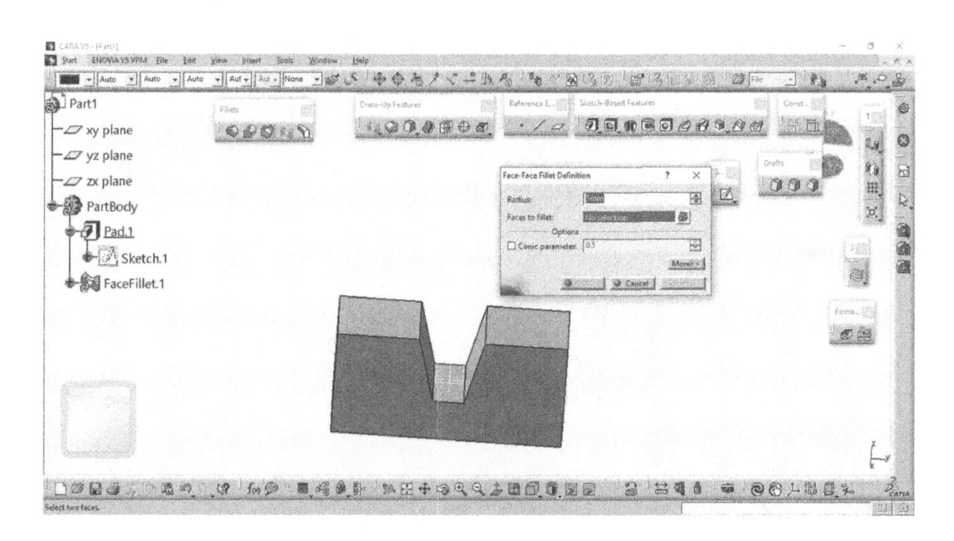

FIGURE 5.92 Selection of Face-Face Fillet tool.

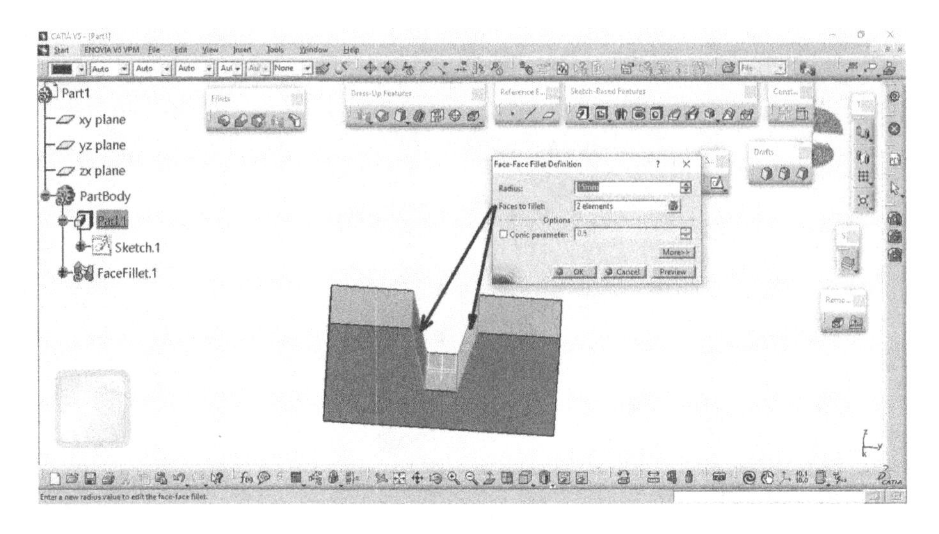

FIGURE 5.93 Selection of Face and Radius value filling in Face-Face Fillet Definition dialog box.

5.4.1.5 Tritangent Fillet

Draw the padded object as shown in Figure 5.95.

Next, choose the Tritangent Fillet tool from the Fillets toolbar as shown in Figure 5.80. The Tritangent Fillet Definition dialog box will appear. Select face to be removed and face to filleted under Face to remove and Faces to fillet spanner in Tritangent Fillet Definition dialog box (Figure 5.96).

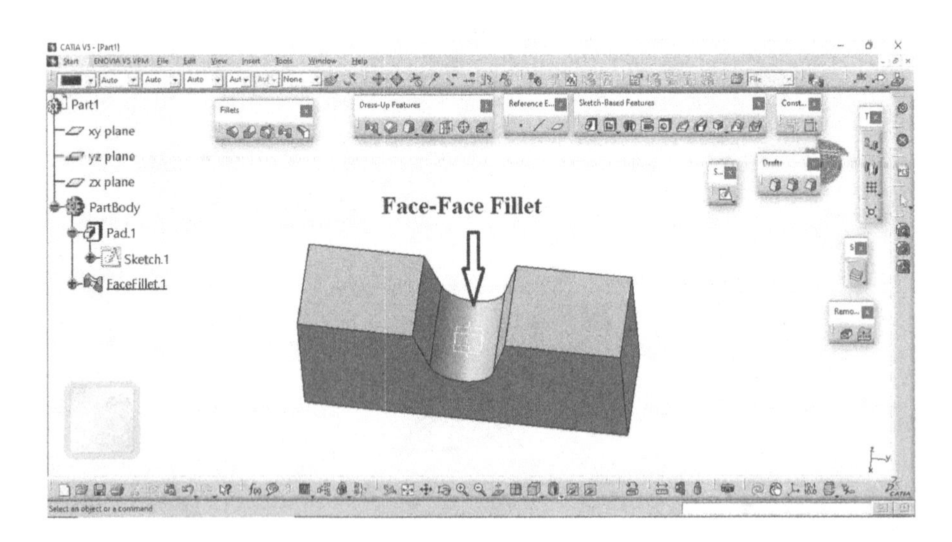

FIGURE 5.94 Solid Edge after the use of Face-Face Fillet tool.

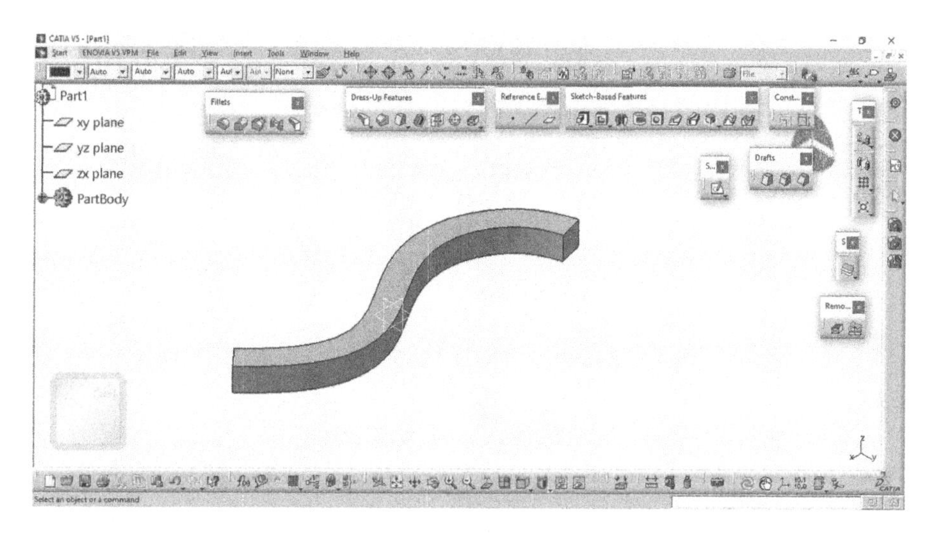

FIGURE 5.95 Padded object.

After selecting faces, click the OK button and then you will get the Tritangent Fillet Definition, as shown in Figure 5.97.

5.4.2 Draft Toolbar

The Draft toolbar creates Face angle. The Draft tool can add or remove material in created solid surface. The material addition or removal totally depends upon Draft Angle and surface of solid. The tools in this toolbar are Draft Angle, Draft Reflect Line and Variable Angle Draft. The Draft toolbar will appear as shown in Figure 5.98.

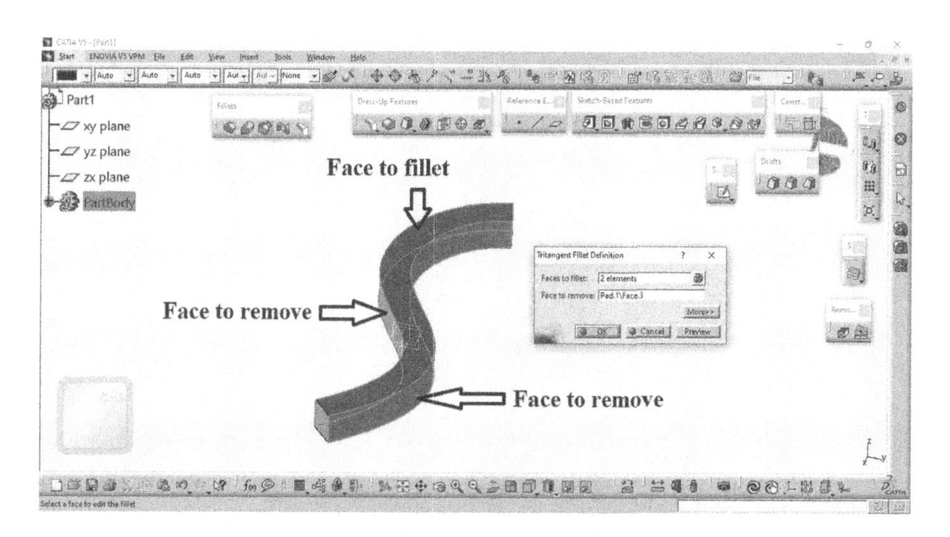

FIGURE 5.96 Selection of faces in Tritangent Fillet Definition dialog box.

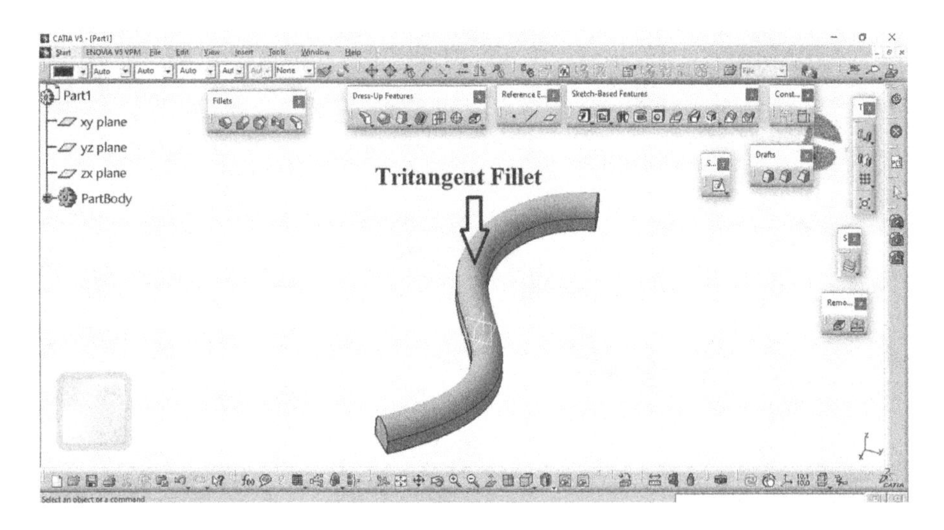

FIGURE 5.97 Solid Edge after the use of Tritangent Fillet Definition tool.

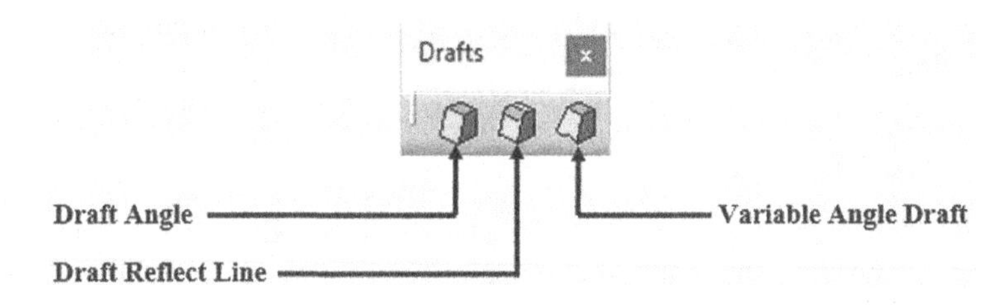

FIGURE 5.98 Draft Toolbar.

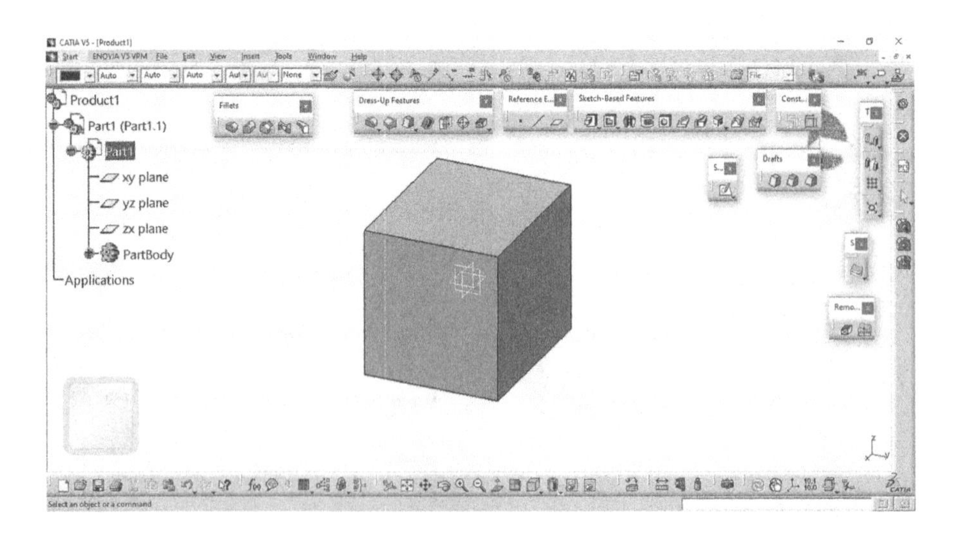

FIGURE 5.99 Padded object.

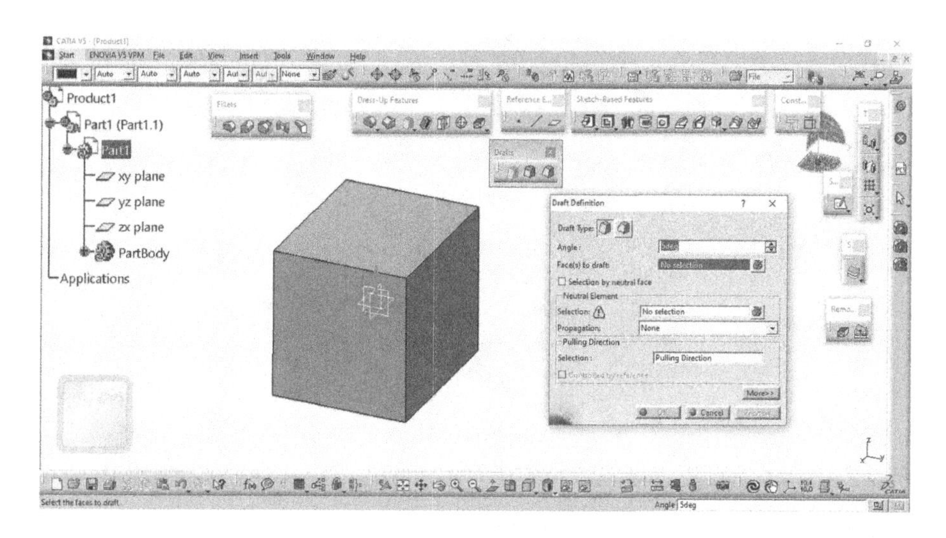

FIGURE 5.100 Selection of Draft tool.

5.4.2.1 Draft Angle

Draw the padded object as shown in Figure 5.99.

Next, choose the Draft Angle tool from the Draft Toolbar (Figure 5.98). The Draft Definition dialog box will appear as shown in Figure 5.100.

Enter the Angle value as 20deg and select two faces under Angle and Face to draft and selection spanner in Draft Definition dialog box (Figure 5.101).

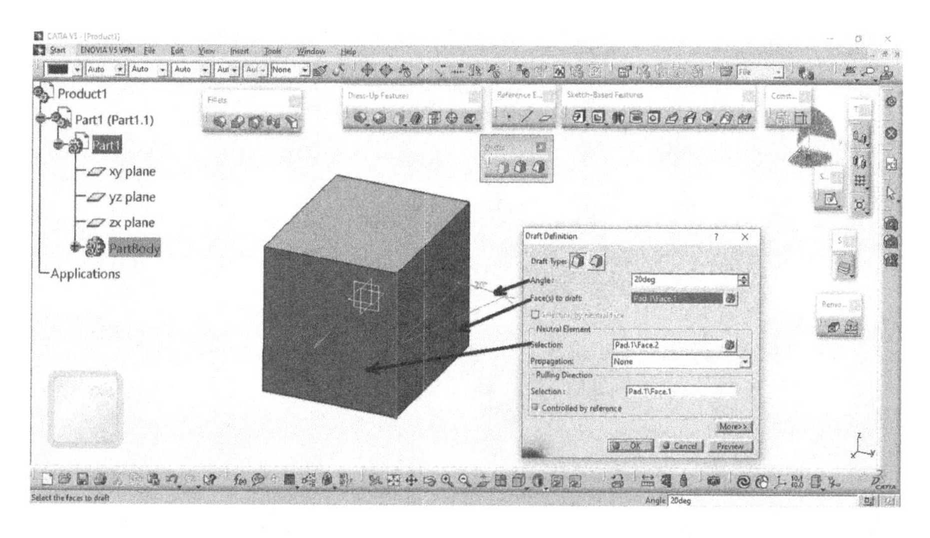

FIGURE 5.101 Selection of faces and angle in Draft Definition dialog box.

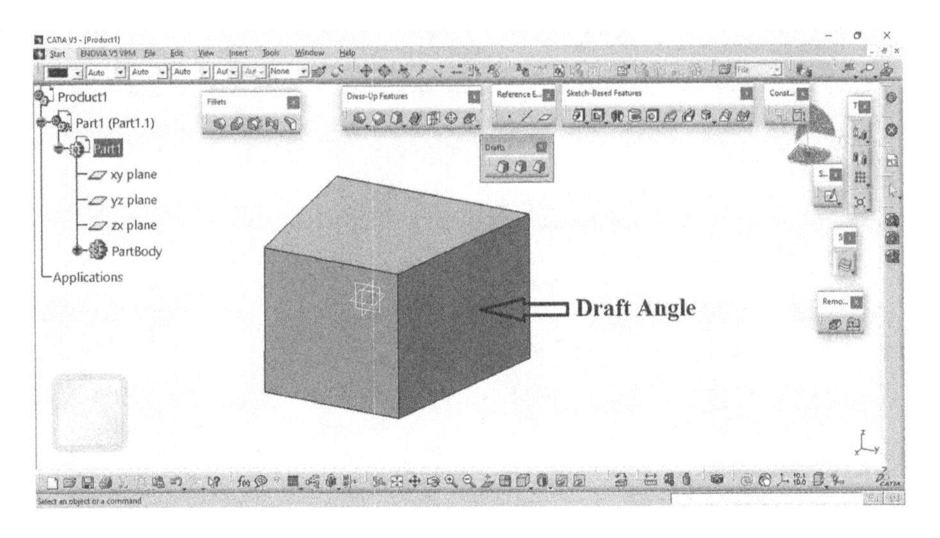

FIGURE 5.102 Solid Edge after the use of Draft Angle tool.

After selecting faces and entering the value of angle, click the OK button and then you will get the Draft Angle, as shown in Figure 5.102.

5.4.2.2 Draft Reflect Line

Draw the padded object as shown in Figure 5.103.

Next, choose the Draft Reflect Line tool from the Draft toolbar as shown in Figure 5.98. The Draft Reflect Line Definition dialog box will appear as shown in Figure 5.104.

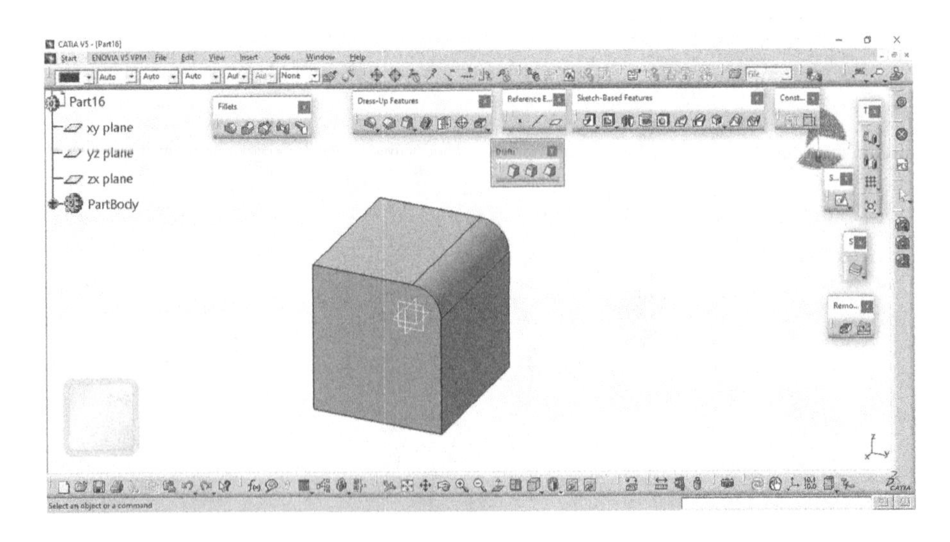

FIGURE 5.103 Padded object.

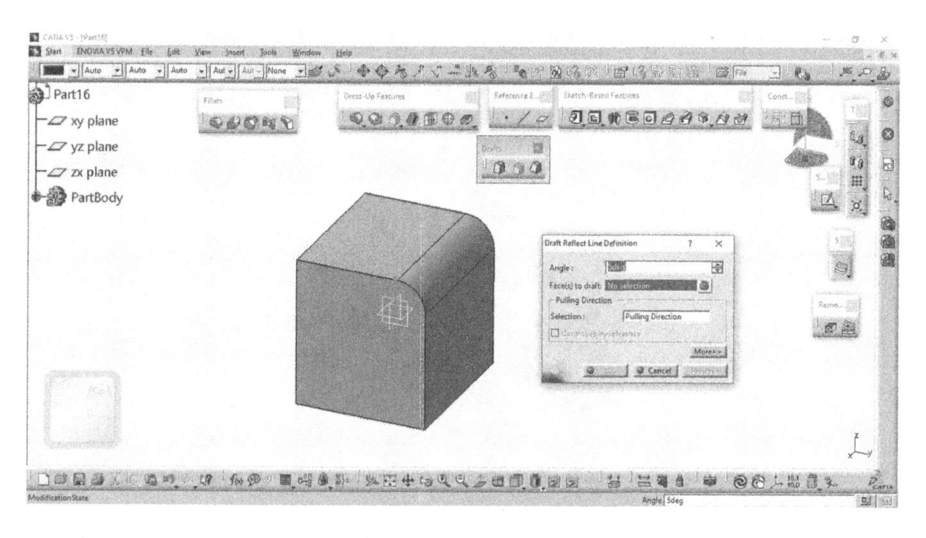

FIGURE 5.104 Selection of Draft Reflect Line Definition tool.

Enter the Angle value as 20deg and select two faces under Angle and Face to draft spanner in Draft Reflect Line Definition dialog box as shown in Figure 5.105.

After selecting faces and entering the value of angle, click the OK button and then you will get the Draft Reflect Line, as shown in Figure 5.106.

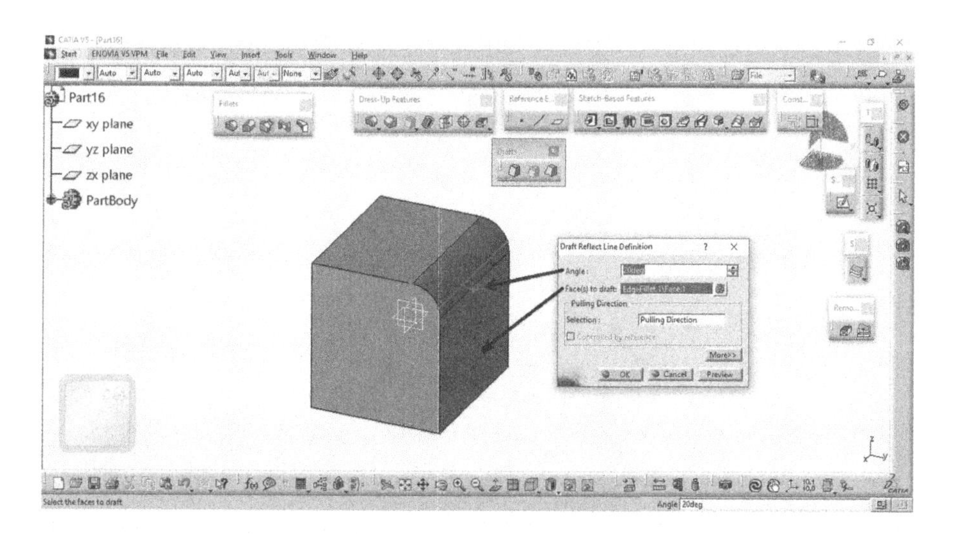

FIGURE 5.105 Selection of faces and angle in Draft Reflect Line Definition dialog box.

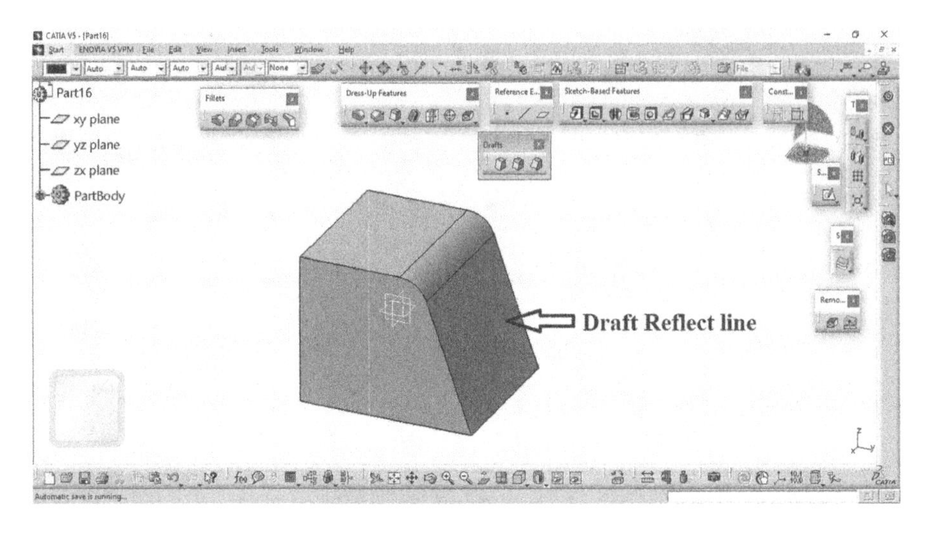

FIGURE 5.106 Solid Edge after the use of Draft Reflect Line tool.

5.4.2.3 Variable Angle Draft

Draw the padded object as shown in Figure 5.107.

Next, choose the Variable Angle Draft tool from the Draft toolbar (Figure 5.98). The Draft Definition dialog box will be displayed. Enter the Angle value as 5deg, 10deg and 30deg in created points and select face under Angle and Face to draft spanner, select other faces under selection spanner in Draft Definition dialog box (Figure 5.108).

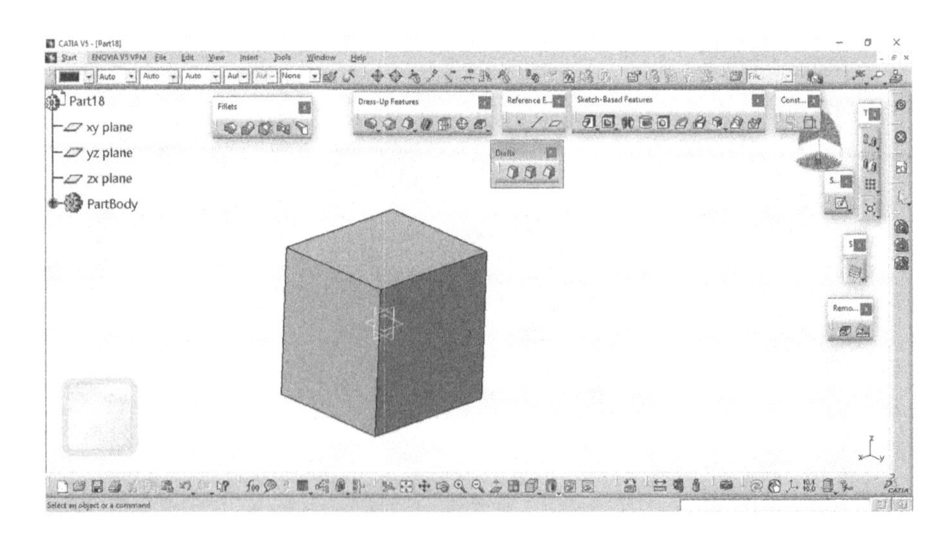

FIGURE 5.107 Padded object.

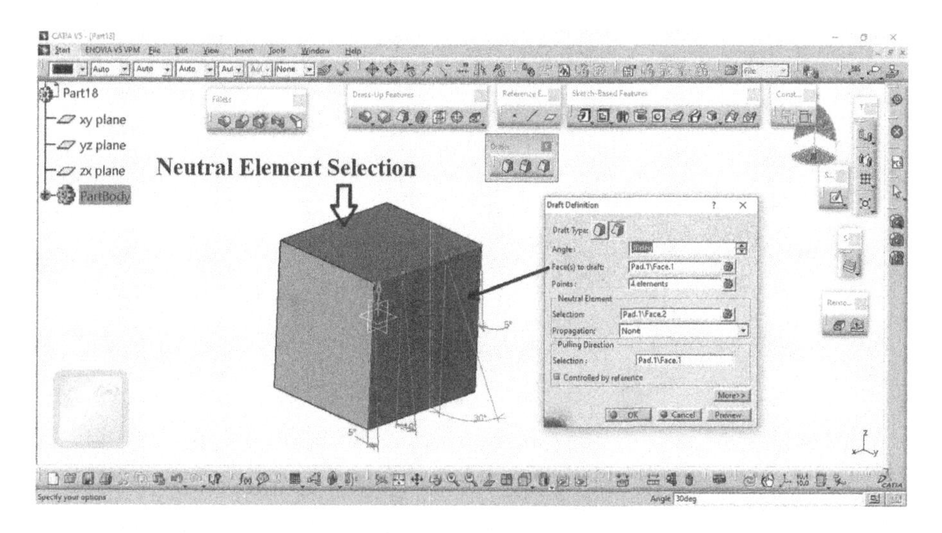

FIGURE 5.108 Selection of face and angle value entry under Draft Definition dialog box.

After selecting faces and entering the value of angle, click on the OK button and then you will get the Variable Angle draft, as shown in Figure 5.109.

5.4.3 Shell

Draw the padded object as shown in Figure 5.110.

Next, choose the Shell tool from the Dress-Up Features toolbar as shown in Figure 5.79. The Shell Definition dialog box will appear as shown in Figure 5.111.

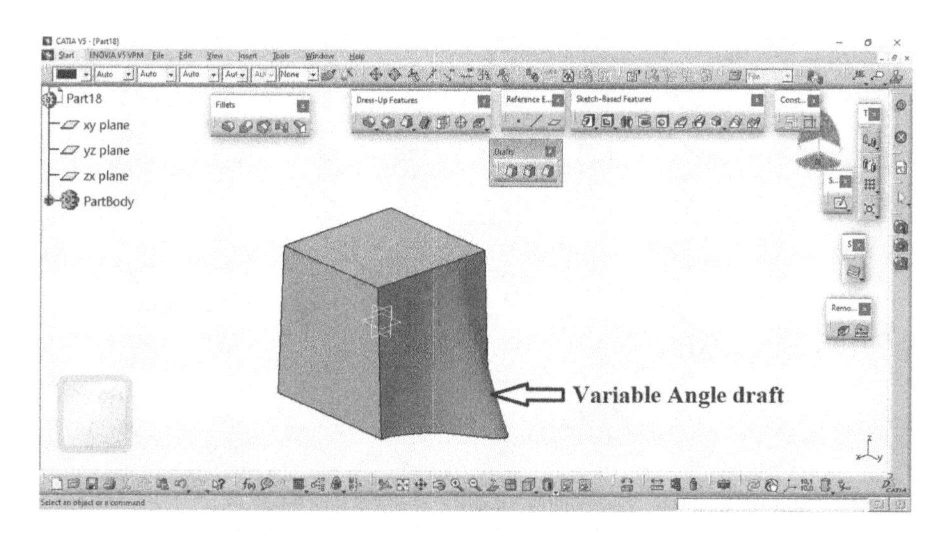

FIGURE 5.109 Solid Edge after the use of Variable Angle Draft tool.

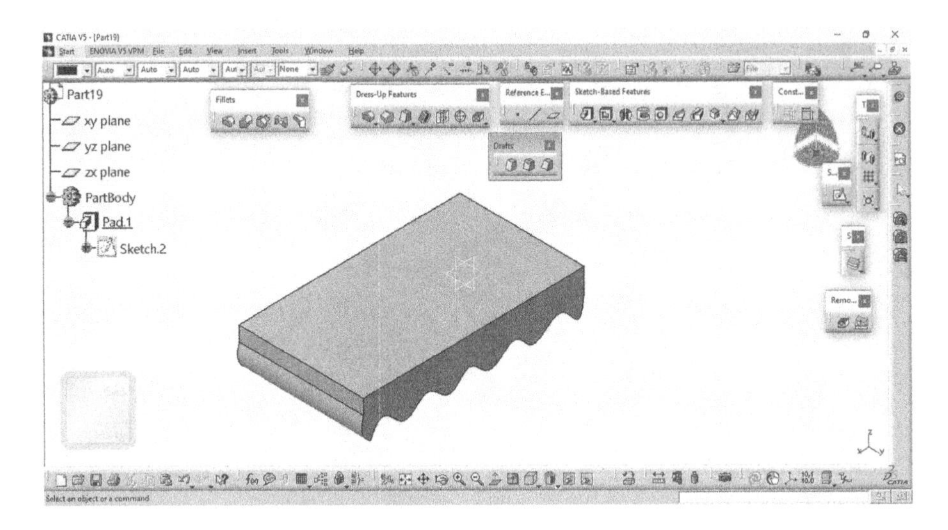

FIGURE 5.110 Padded object.

Enter the inside thickness value as 1mm and select Faces to remove and Other thickness faces in Shell Definition dialog box as shown in Figure 5.112.

After selecting faces and entering the value of thickness, click the OK button and then you will get the shell as shown in Figure 5.113.

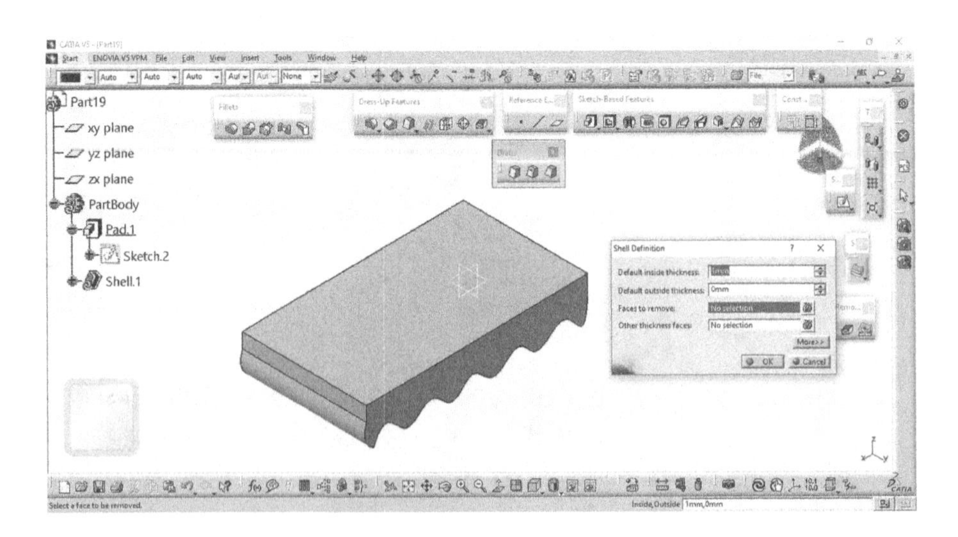

FIGURE 5.111 Selection of Shell Definition tool.

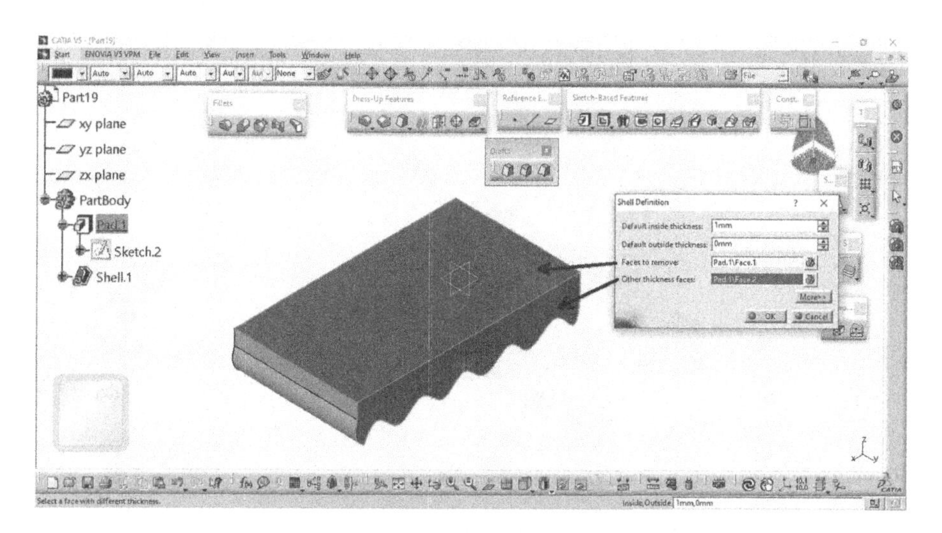

FIGURE 5.112 Selection of faces and thickness in Shell Definition dialog box.

5.4.4 Thickness

The Thickness tool adds thickness to created face solid. Choose the Thickness tool from the Dress-Up Features toolbar, as shown in Figure 5.79. The Thickness Definition dialog box will appear as shown in Figure 5.114.

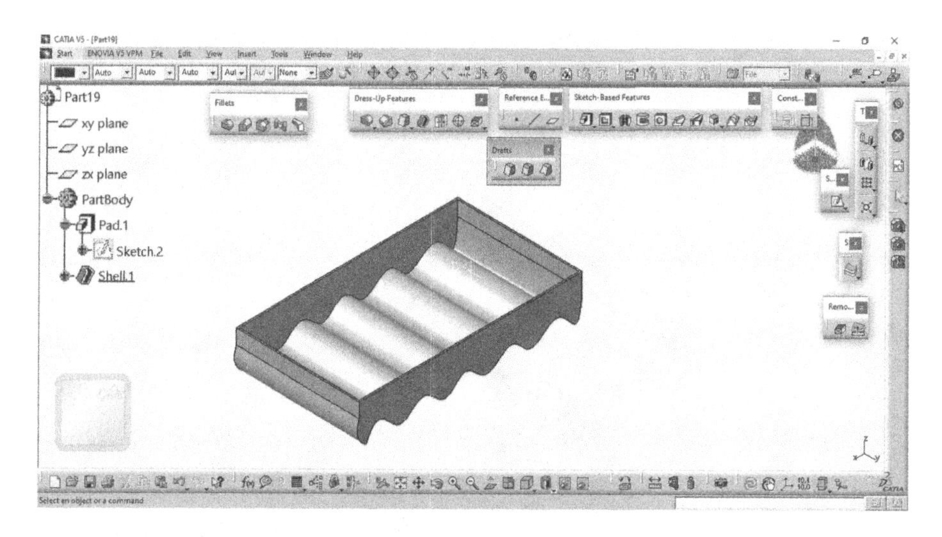

FIGURE 5.113 Solid after the use of Shell tool.

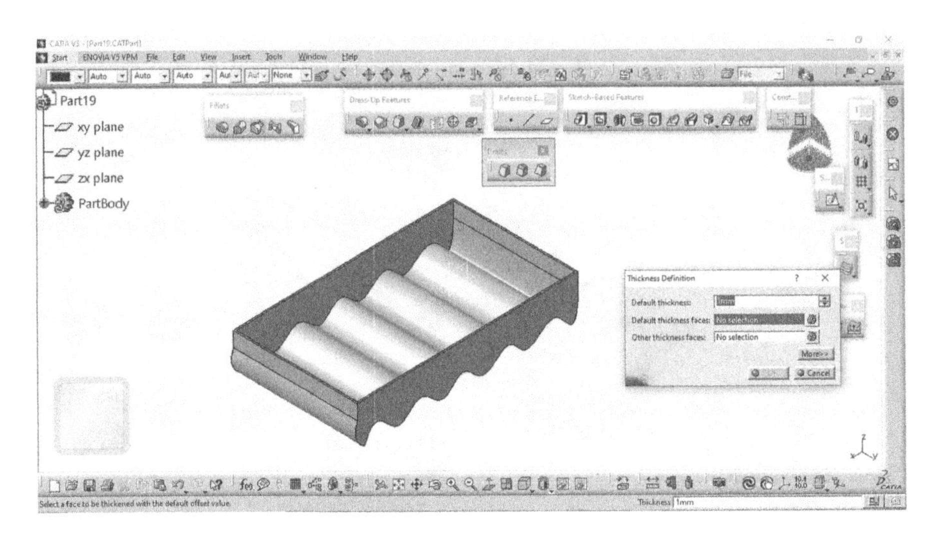

FIGURE 5.114 Selection of Thickness tool.

Enter the thickness value as 10mm and select thickness face to add thickness in Thickness Definition dialog box, as shown in Figure 5.115.

After selecting faces and entering the value of thickness, click the OK button and then you will get the thickness, as shown in Figure 5.116.

5.5 TRANSFORMATION FEATURES TOOLBAR

The Transformation Features toolbar contains commands that transform, move or create a pattern or array of a feature. The tools in this toolbar are Transformations, Mirror, Patterns and Scale as shown in Figure 5.117.

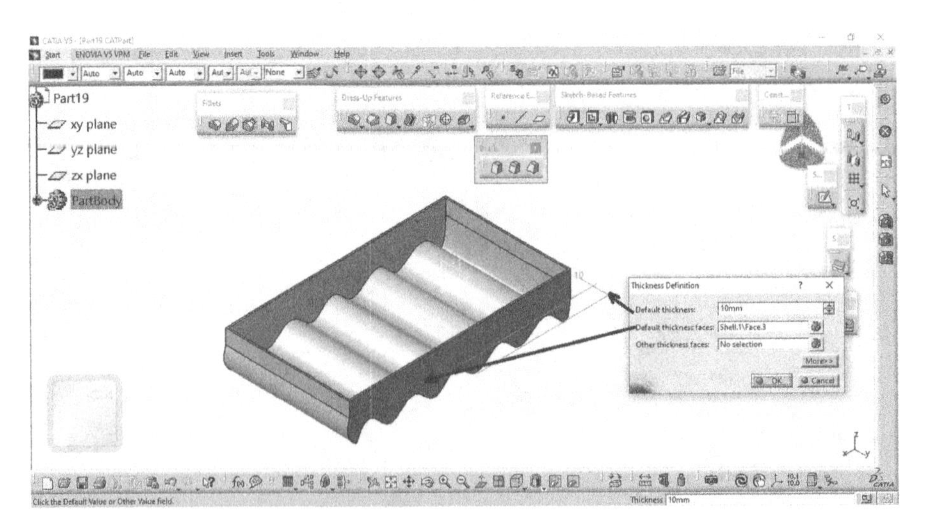

FIGURE 5.115 Selection of faces and thickness value in Thickness Definition dialog box.

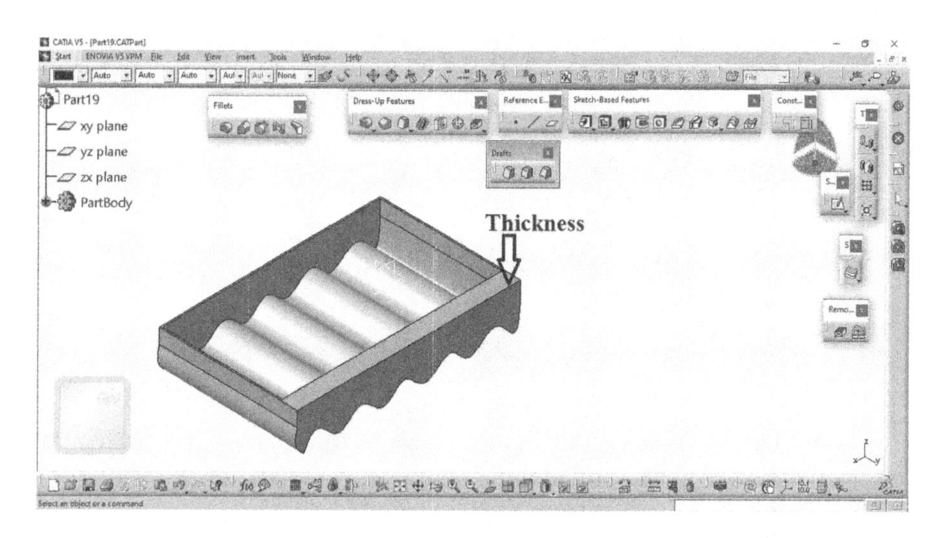

FIGURE 5.116 Solid after the use of Shell tool.

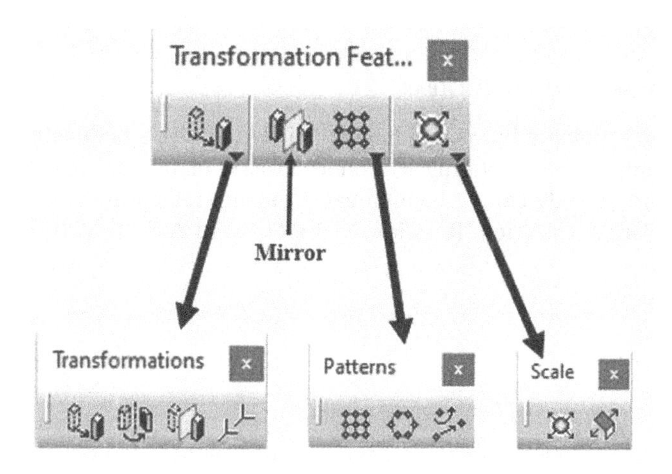

FIGURE 5.117 Transformation Features toolbar.

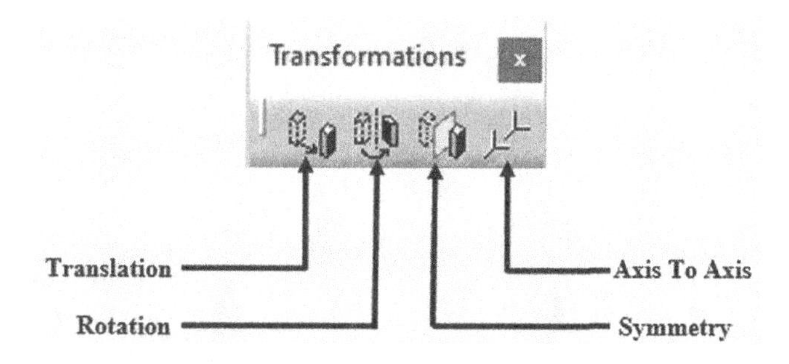

FIGURE 5.118 Transformation toolbar.

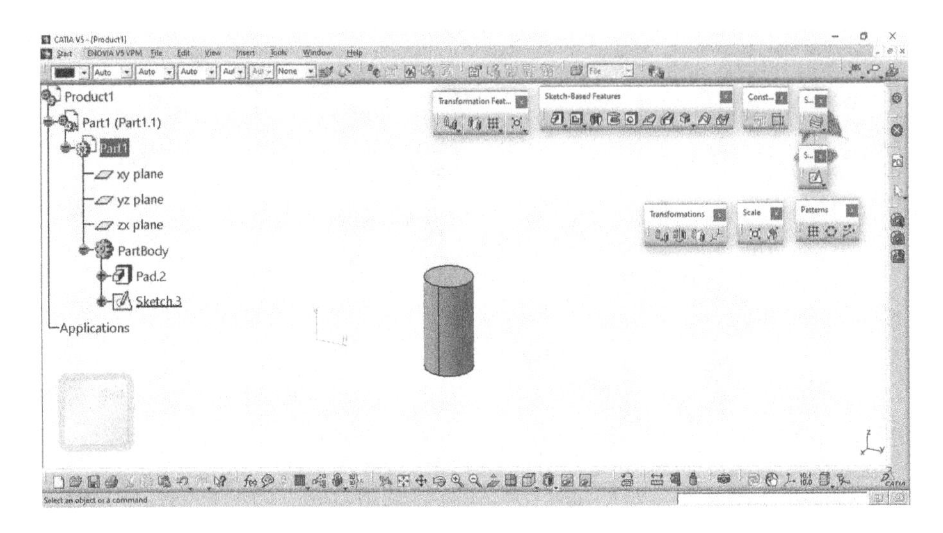

FIGURE 5.119 Padded object.

5.5.1 Transformation Toolbar

The transformation toolbar is tool having ability to move a body either by translating it along an axis, rotating it around an axis or moving it symmetrically about a plane. A body cannot be duplicated during transformation. The tools in this toolbar are Translation, Rotation, Symmetry and Axis To Axis as shown in Figure 5.118.

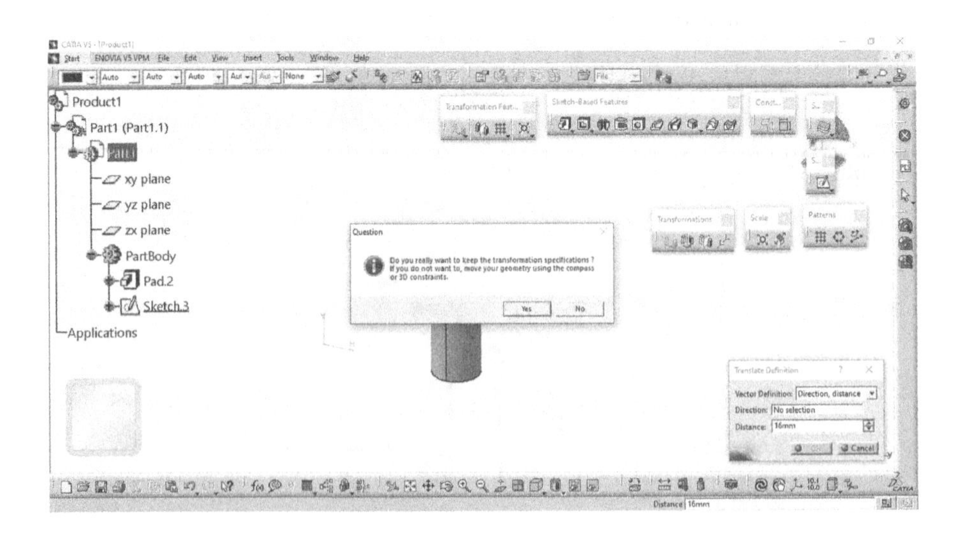

FIGURE 5.120 Selection of Translation tool.

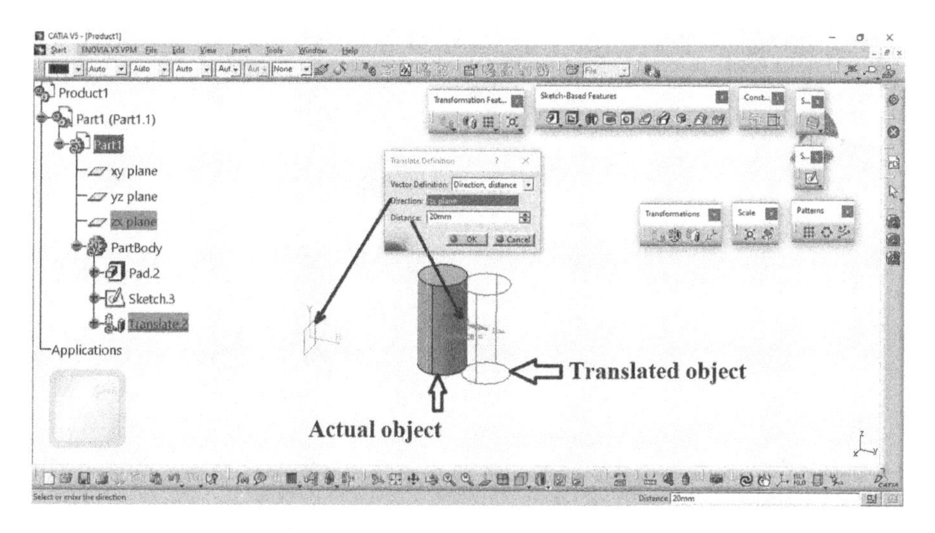

FIGURE 5.121 Selection of plane and distance value in Translate Definition dialog box.

5.5.1.1 Translation

The Translation tool is used for moving the part in selected linear direction. Draw the padded object as shown in Figure 5.119.

Next, choose the Translation tool from the Transformation toolbar, as shown in Figure 5.118. A Translate Definition dialog box and a Question window will appear, *make sure it is okay* to violate the constraints that were imposed in the transformation, click on Yes as shown in Figure 5.120.

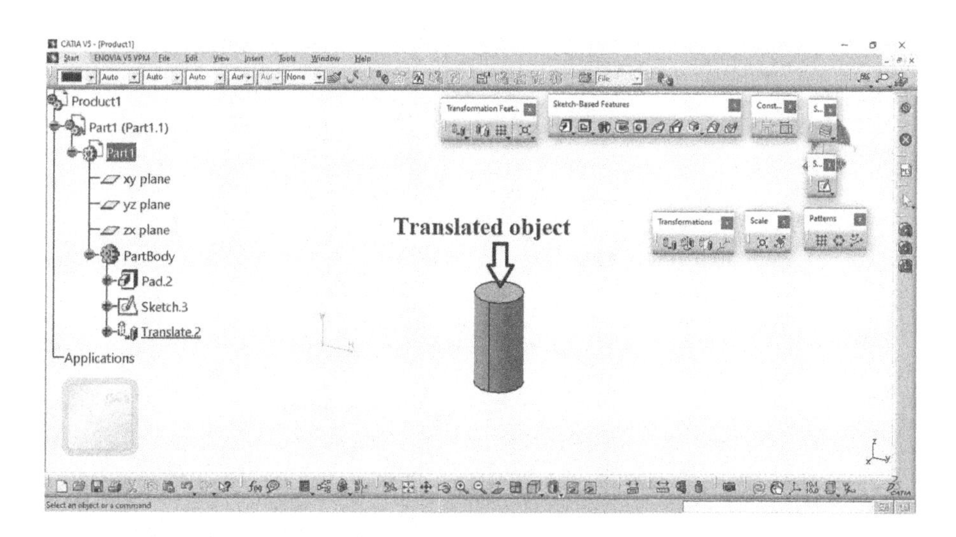

FIGURE 5.122 Solid after the use of Translation tool.

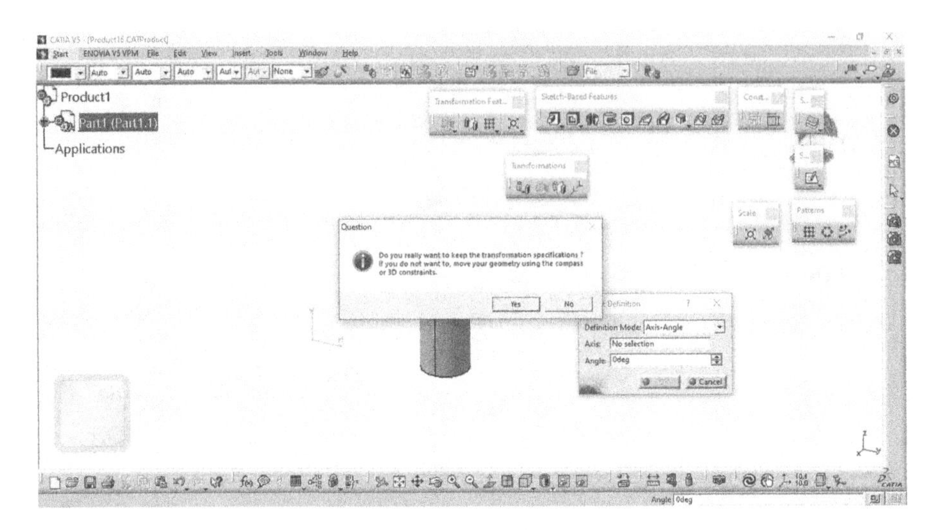

FIGURE 5.123 Selection of Rotation tool.

Enter the Distance value as 20mm and select zx plane under Distance and Direction spanner in Translate Definition dialog box, as shown in Figure 5.121.

After selecting plane and entering the value of distance, click the OK button and then you will get the Translated object (Figure 5.122).

5.5.1.2 Rotation

The Rotation tool is used for rotating a part around a selected line or axis. Draw the padded object (Figure 5.119). Choose the Rotation tool from the Transformation

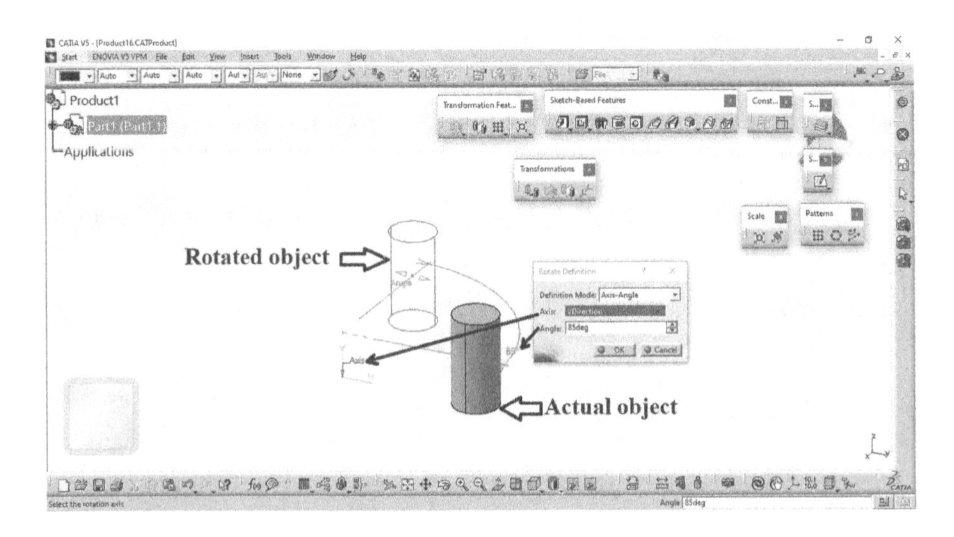

FIGURE 5.124 Selection of Axis and Angle value in Rotate Definition dialog box.

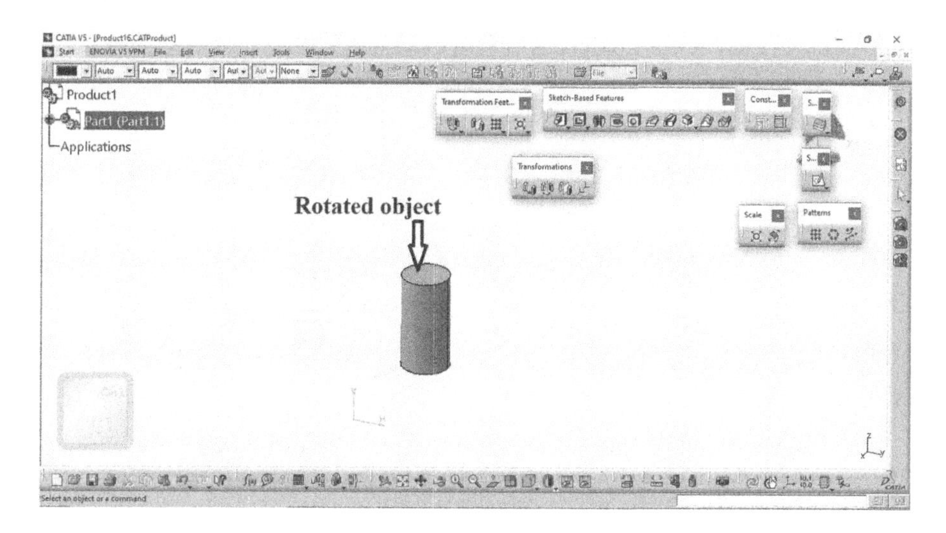

FIGURE 5.125 Solid after the use of Rotation tool.

toolbar as displayed in Figure 5.118. The Rotate Definition dialog box and a Question window will appear, *make sure it is okay* to violate the constraints that were imposed in the transformation, click on Yes as shown in Figure 5.123.

Enter the Angle value as 85deg and select V Direction Axis under Angle and Axis spanner in Rotate Definition dialog box as shown in Figure 5.124.

After selecting axis and entering the value of angle, click on OK button and then you will get the Rotated object as shown in Figure 5.125.

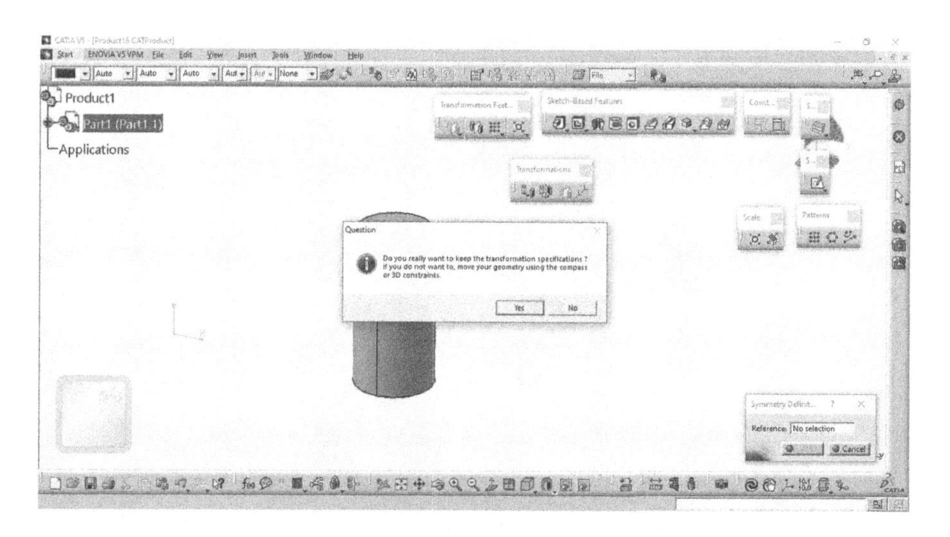

FIGURE 5.126 Selection of Symmetry tool.

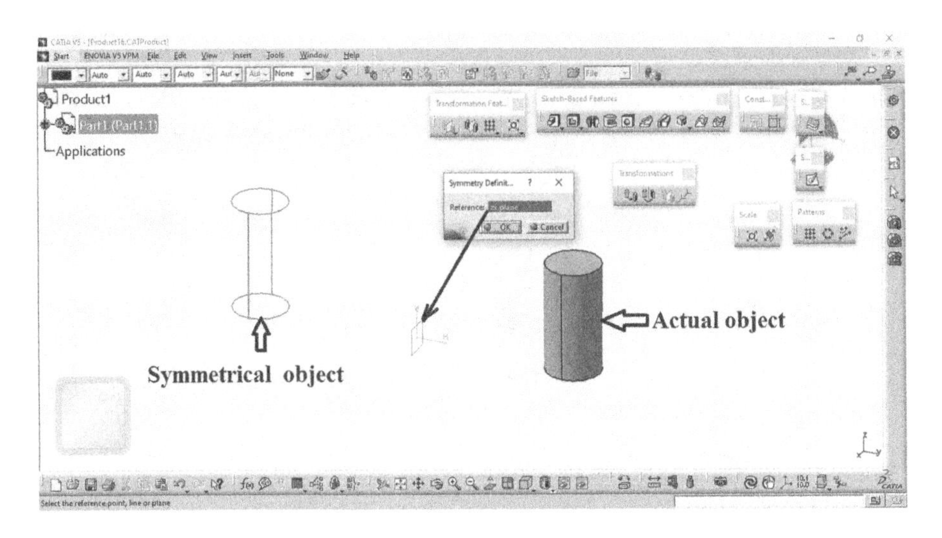

FIGURE 5.127 Selection of plane in Symmetry Definition dialog box.

5.5.1.3 Symmetry

The Symmetry tool is used for replacing the part with its mirror image. Draw the padded object (Figure 5.119). Choose the Symmetry tool from the Transformation toolbar as displayed in Figure 5.118. The Symmetry Definition dialog box and a Question window will appear, *make sure it is okay* to violate the constraints that were imposed in the transformation, click Yes as shown in Figure 5.126.

Select zx plane under Reference spanner in Symmetry Definition dialog box (Figure 5.127).

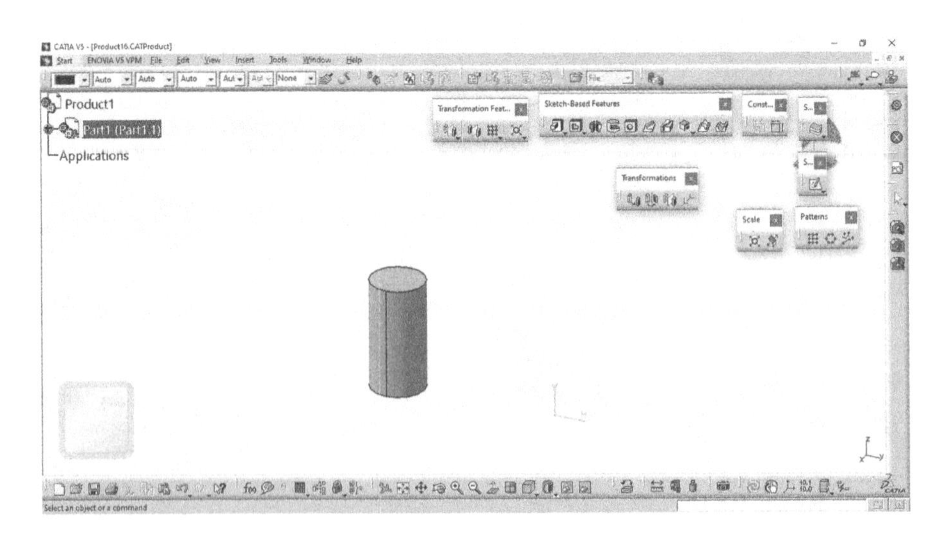

FIGURE 5.128 Solid after the use of Symmetry tool.

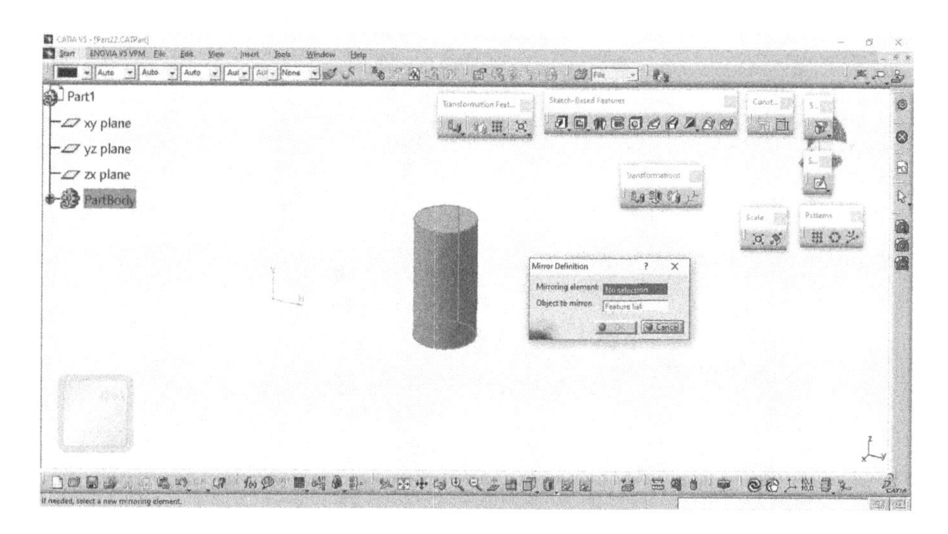

FIGURE 5.129 Selection of Mirror tool.

After selecting plane, click the OK button and then you will get the Symmetry object, as shown in Figure 5.128.

5.5.2 Mirror

The Mirror tool is used to create a duplicate mirror image of a part. Draw the padded object as shown in Figure 5.119. After selecting the object, choose the Mirror tool from the Transformation Features toolbar (Figure 5.117). The Mirror Definition dialog box will appear as shown in Figure 5.129.

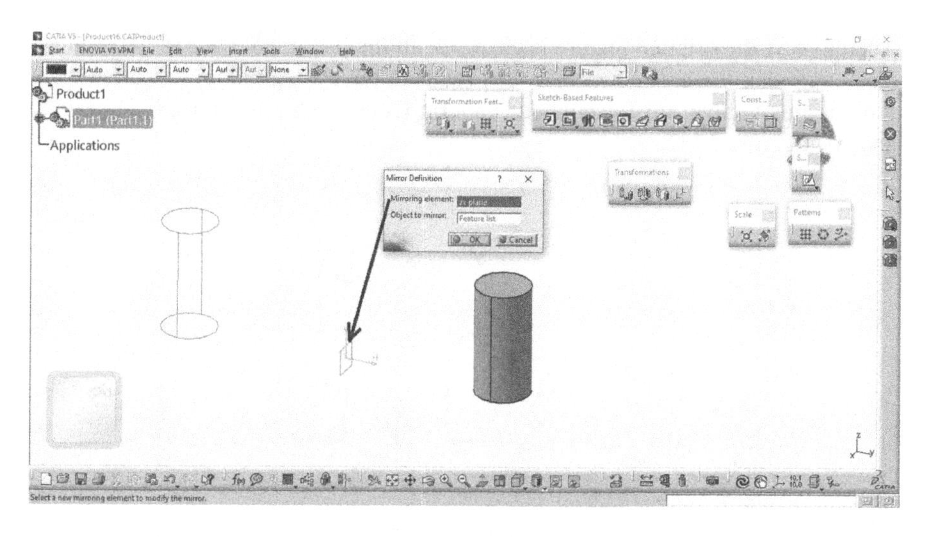

FIGURE 5.130 Selection of plane in Mirror Definition dialog box.

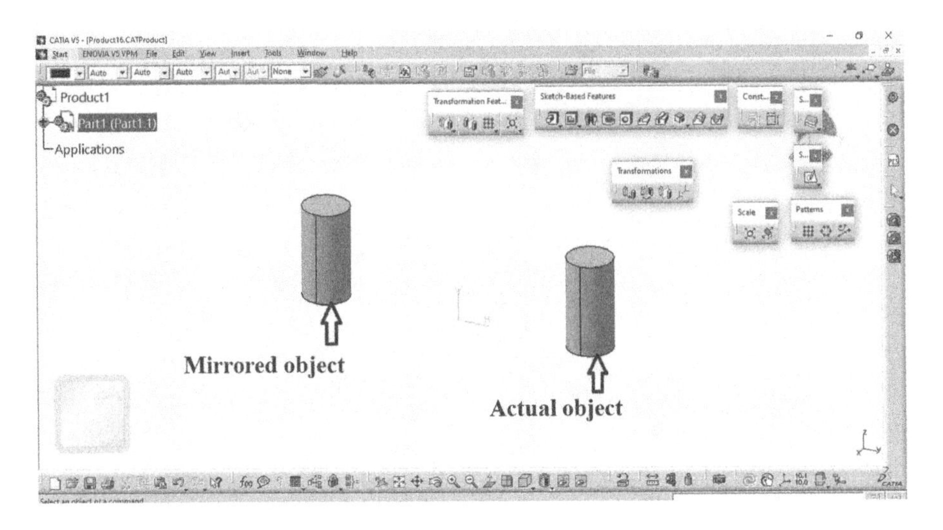

FIGURE 5.131 Solid after the use of Mirror tool.

Select xz plane under Mirroring element spanner in Mirror Definition dialog box as shown in Figure 5.130.

After selecting plane, click the OK button and then you will get the Mirrored object as shown in Figure 5.131.

5.5.3 Patterns Toolbar

The Patterns toolbar allows you to create more than one component of an existing component. You can create rectangular or circular pattern. The positions of the features may also be manually defined. Patterned features are connected or associative. Therefore, if the original feature is changed, all the patterned features will change. You can break the connection between the features by exploding the pattern. The tools in this toolbar are Rectangular Pattern, Circular Pattern and Use Pattern, as shown in Figure 5.132.

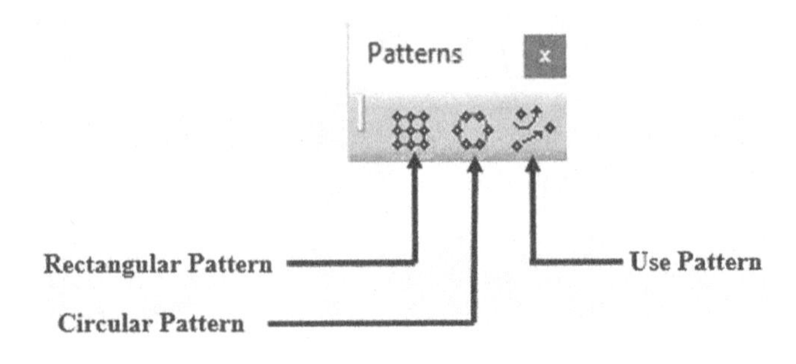

FIGURE 5.132 Patterns toolbar.

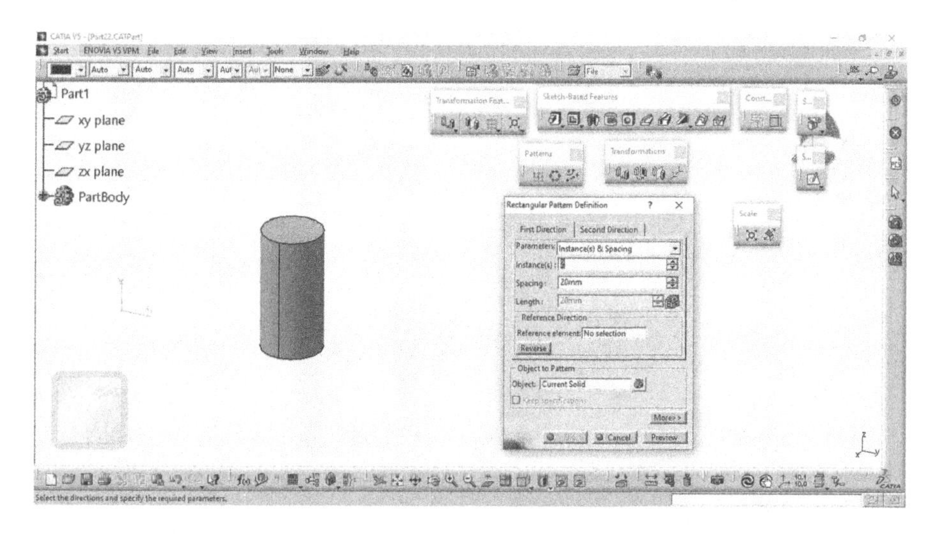

FIGURE 5.133 Selection of Rectangular Pattern tool.

5.5.3.1 Rectangular Pattern

This tool enables you to create rectangular pattern of a feature with a specified number of rows and columns. Draw the one padded object as shown in Figure 5.119. Choose the rectangular pattern tool from the Patterns toolbar as displayed in Figure 5.132. The Rectangular Pattern Definition dialog box window will appear, as shown in Figure 5.133.

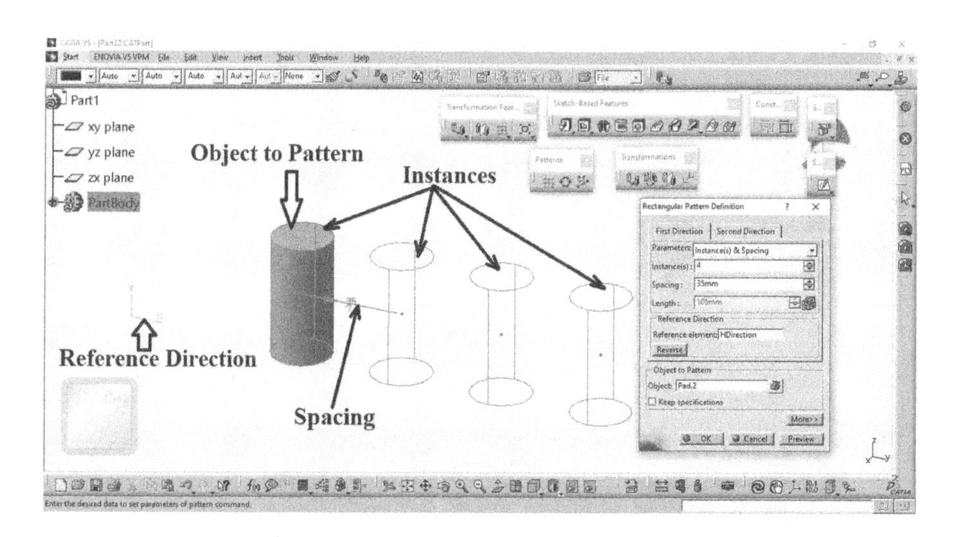

FIGURE 5.134 Selection in Rectangular Pattern dialog box.

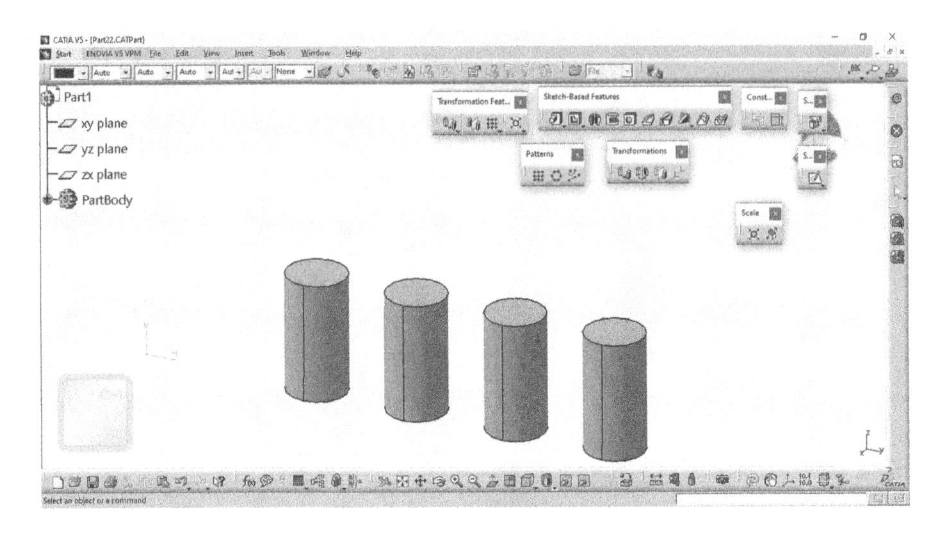

FIGURE 5.135 Solid after the use of Rectangular Pattern tool.

Select padded object under object spanner in Object to pattern area. Select H Direction in Reference element in Reference Direction area. Enter the Instances value as 04 and Spacing distance as 35mm under Instances and Spacing spanner in Rectangular Pattern Definition dialog box as shown in Figure 5.134.

After selecting, click the OK button and then you will get the Patterned object as shown in Figure 5.135.

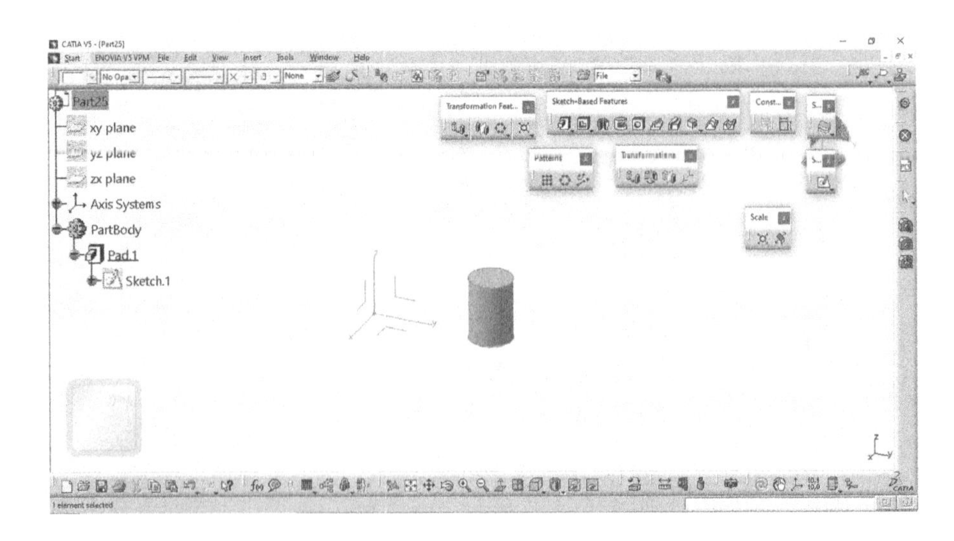

FIGURE 5.136 Padded object.

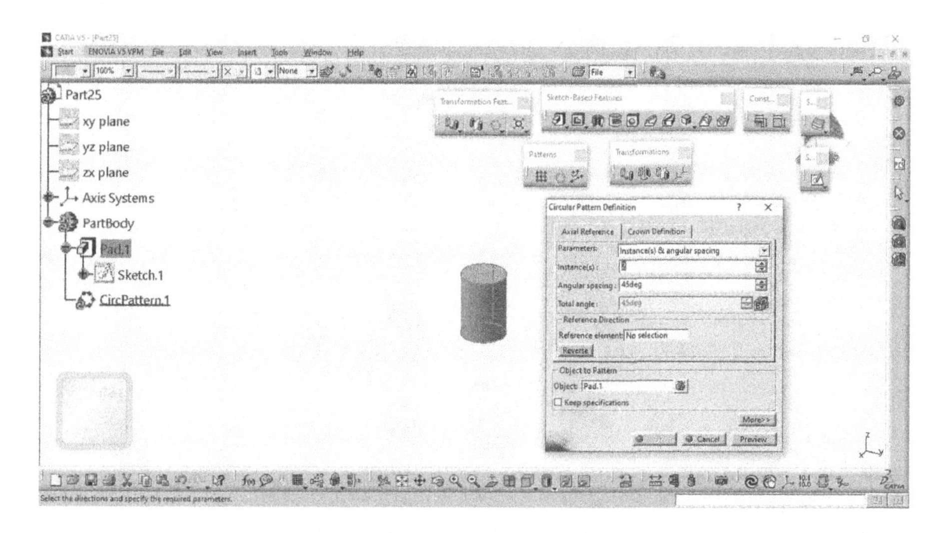

FIGURE 5.137 Selection of Circular Pattern tool.

5.5.3.2 Circular Pattern

This tool enables you to create circular pattern of a feature with a specified number. Draw the one padded object as shown in Figure 5.136.

Next, select the padded object and choose the Circular Pattern tool from the Patterns toolbar, as shown in Figure 5.132. The Circular Pattern Definition dialog box window will appear, as shown in Figure 5.137.

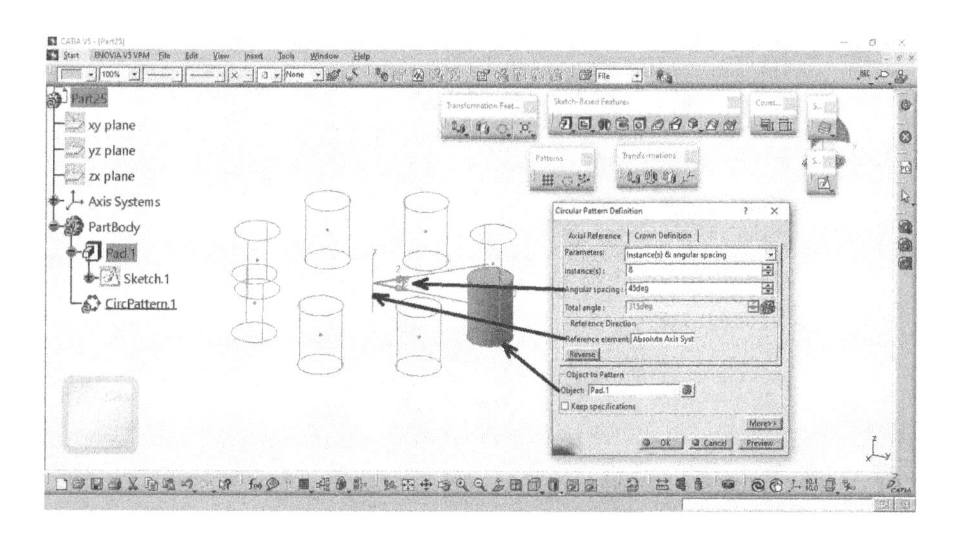

FIGURE 5.138 Selection in Circular Pattern Definition dialog box.

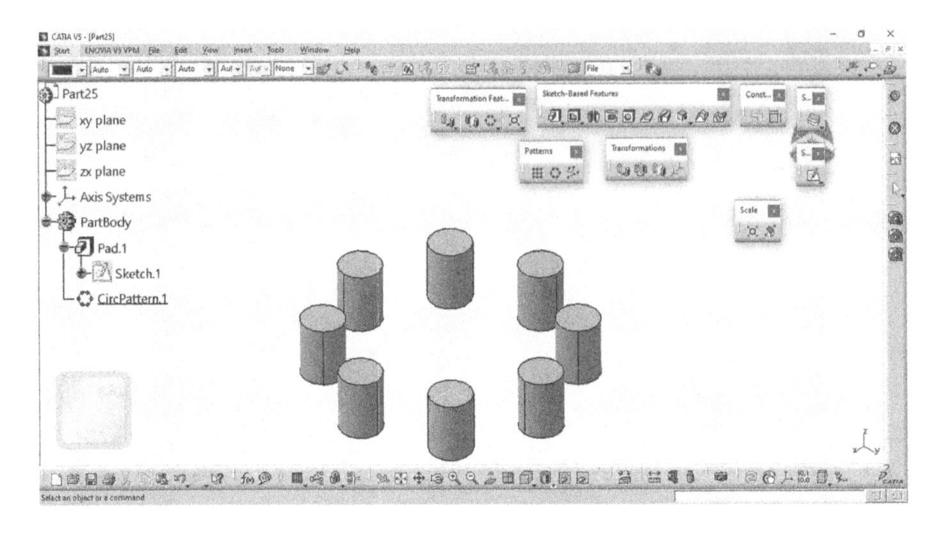

FIGURE 5.139 Solid after the use of Circular Pattern tool.

Select Reference element as Z axis under Reference element spanner in Reference direction area. Enter the Instances value as 08 and Angular spacing as 45deg under Instances and Angular spacing spanner in Circular Pattern Definition dialog box, as shown in Figure 5.138.

After selecting, click the OK button and then you will get the Patterned object, as shown in Figure 5.139.

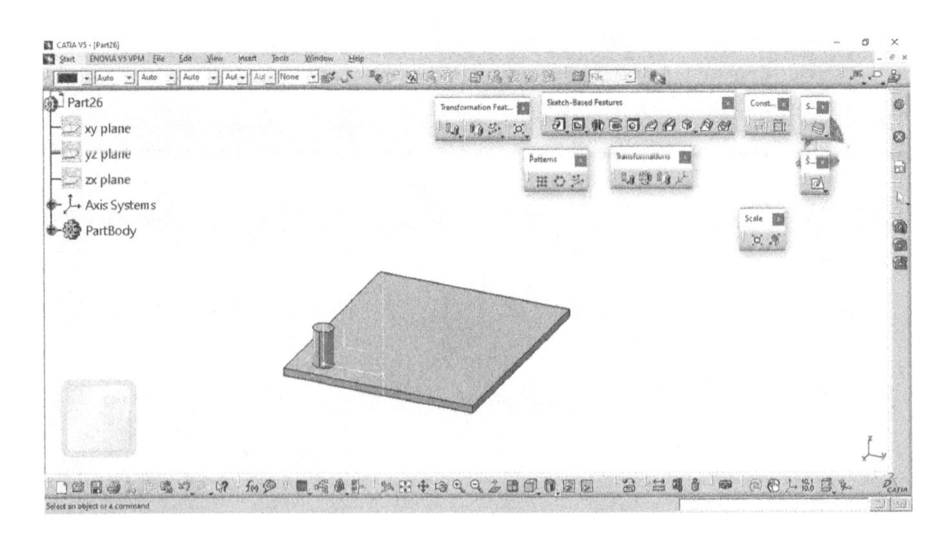

FIGURE 5.140 Padded object.

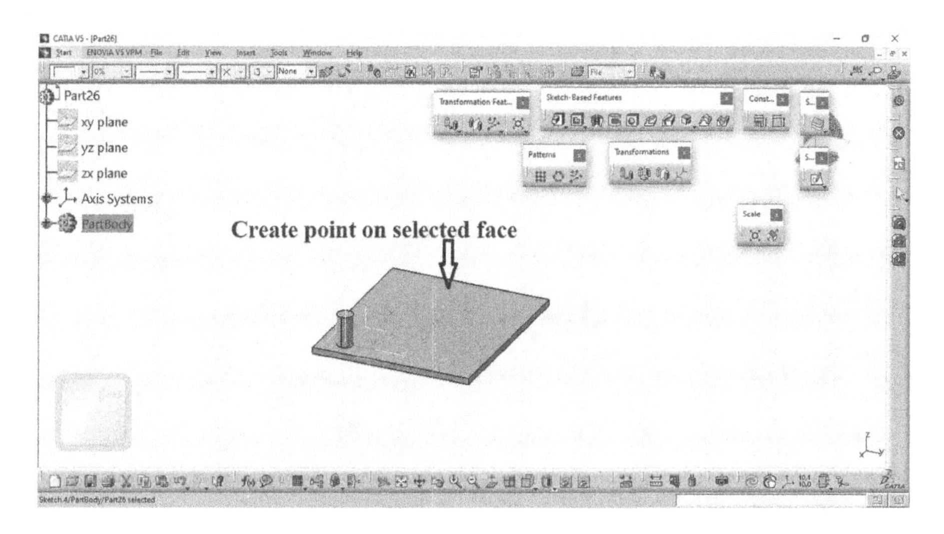

FIGURE 5.141 Point creation for Use Pattern tool.

5.5.3.3 Use Pattern

This tool allows you to create several instances of a feature in predefine locations. The locations are defined by the user in a sketch that usually consists of several points. Draw the one padded object as shown in Figure 5.140.

Create point on selected face as shown in Figure 5.141

Next, select the padded object and choose the Use Pattern tool from the Patterns toolbar, as displayed in Figure 5.132. The User Pattern Definition dialog box window will appear, as shown in Figure 5.142.

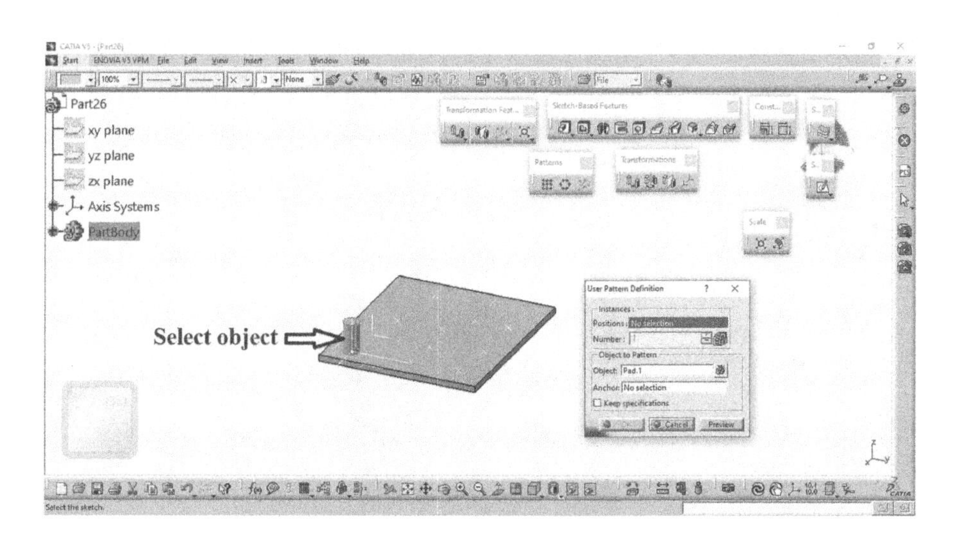

FIGURE 5.142 Selection of object and Use Pattern tool.

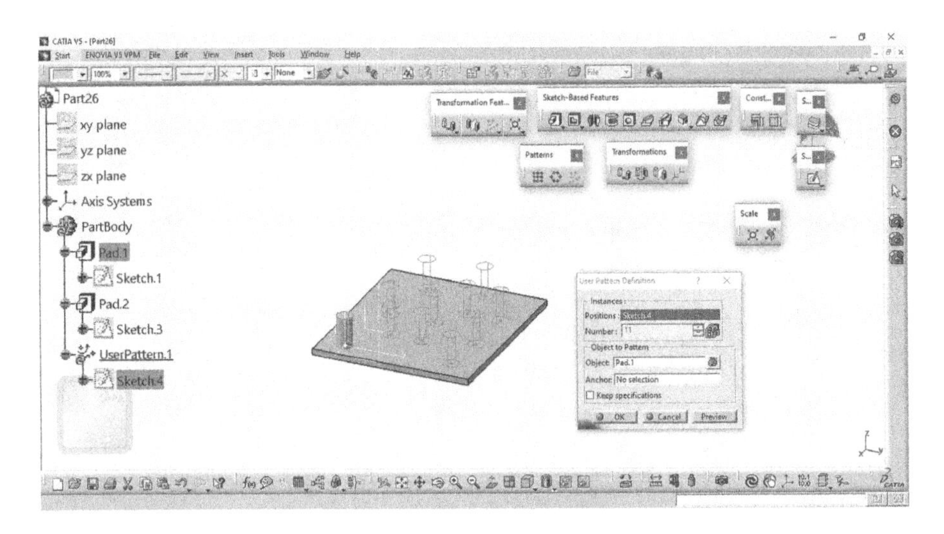

FIGURE 5.143 Selection in Use Patterns Definition dialog box.

Select Point Sketch.4 under Positions spanner in Instances area in Use Patterns Definition dialog box, as shown in Figure 5.143.

After selecting, click the OK button and then you will get the Use Patterned object as shown in Figure 5.144.

5.5.4 Scale Toolbar

The Scale toolbar tools may be used to scale a part. The tools in this toolbar are Transformations, Mirror, Patterns and Scale, as shown in Figure 5.145.

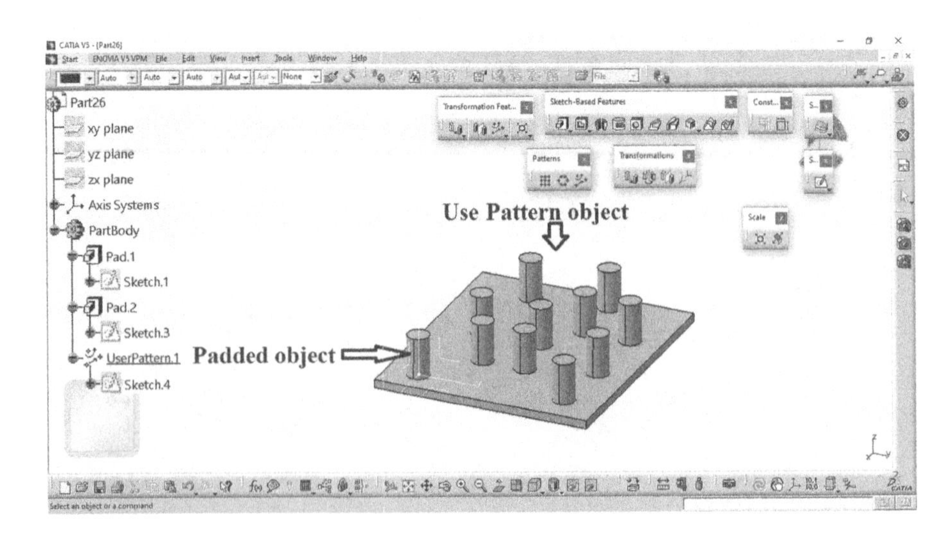

FIGURE 5.144 Solid after the use of Use Pattern tool.

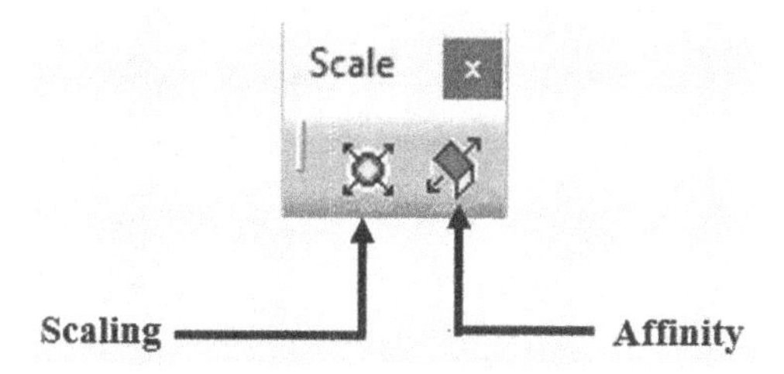

FIGURE 5.145 Transformation Scale toolbar.

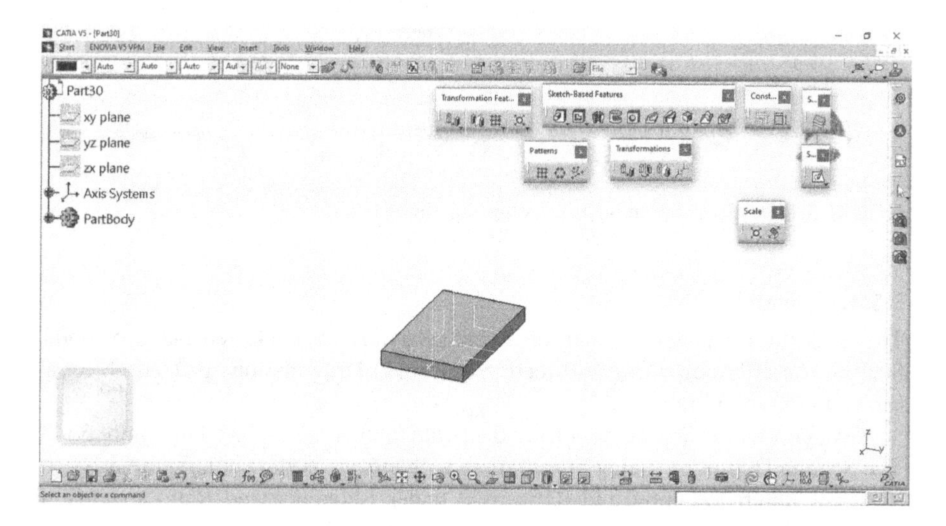

FIGURE 5.146 Padded object.

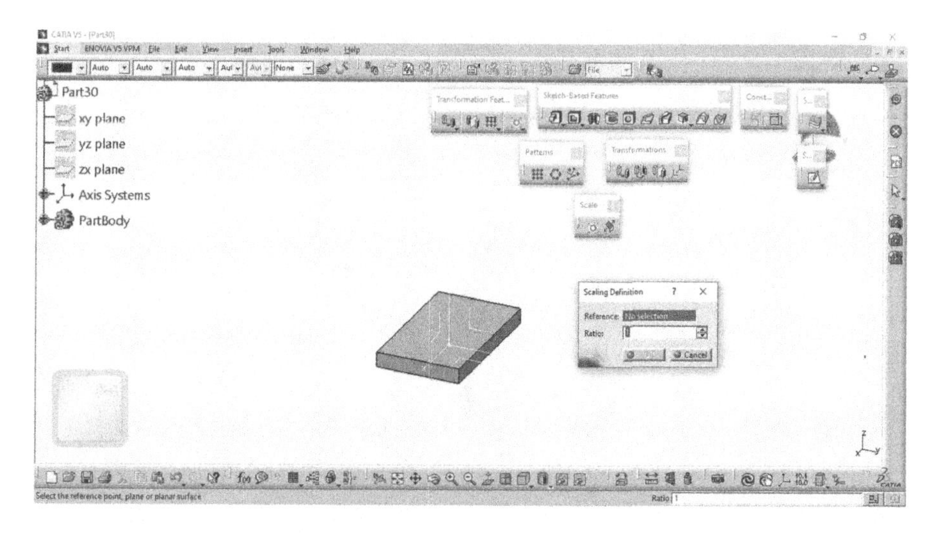

FIGURE 5.147 Selection of Scaling tool.

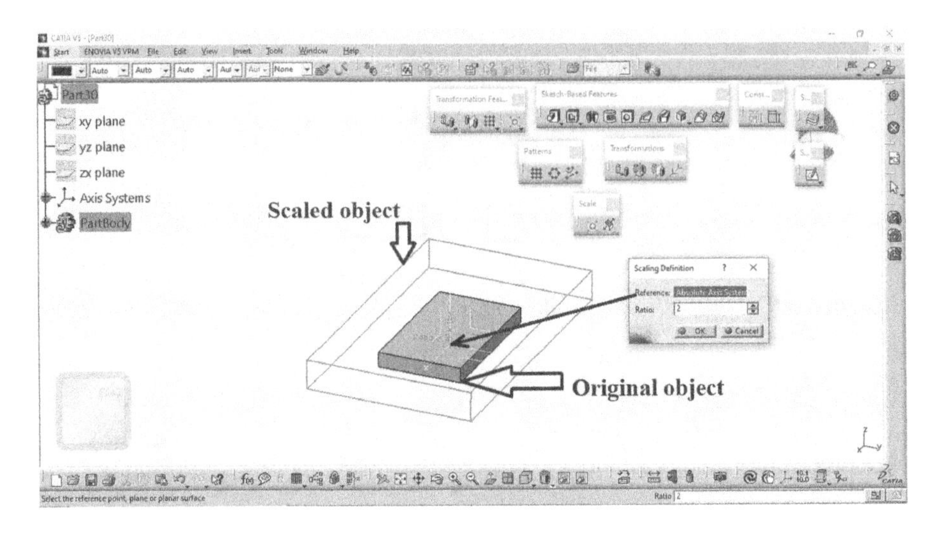

FIGURE 5.148 Selection in Scaling Definition dialog box.

5.5.4.1 Scaling

The Scale tool changes the size of a part relatively to a selected plane or point. Scaling can only occur in one direction at a time. Draw the one padded object as shown in Figure 5.146.

Next, choose the Scaling tool from the Scale toolbar as displayed in Figure 5.145. The Scaling Definition dialog box window will appear, as shown in Figure 5.147.

Select Reference point and Ratio as 2 under Reference and Ratio spanner in Scaling Definition dialog box, as shown in Figure 5.148.

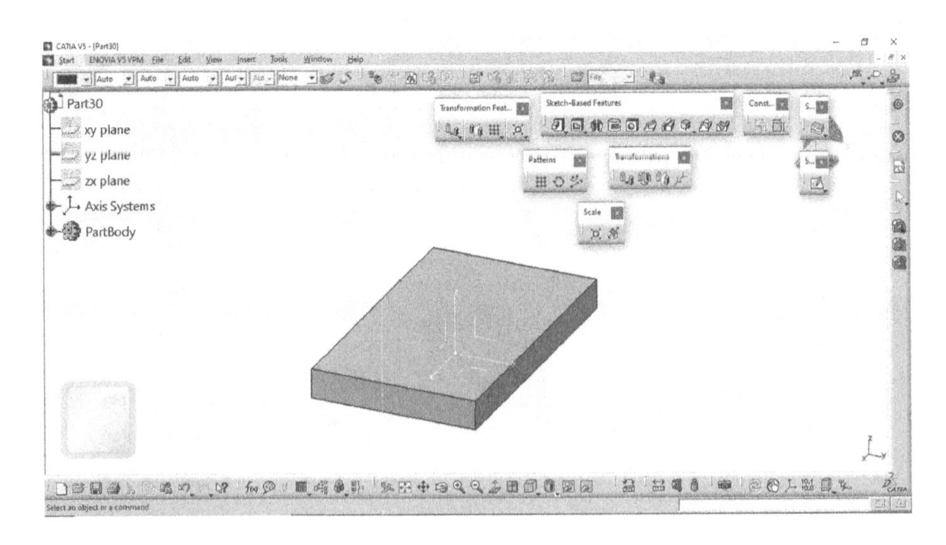

FIGURE 5.149 Solid after the use of Scaling tool.

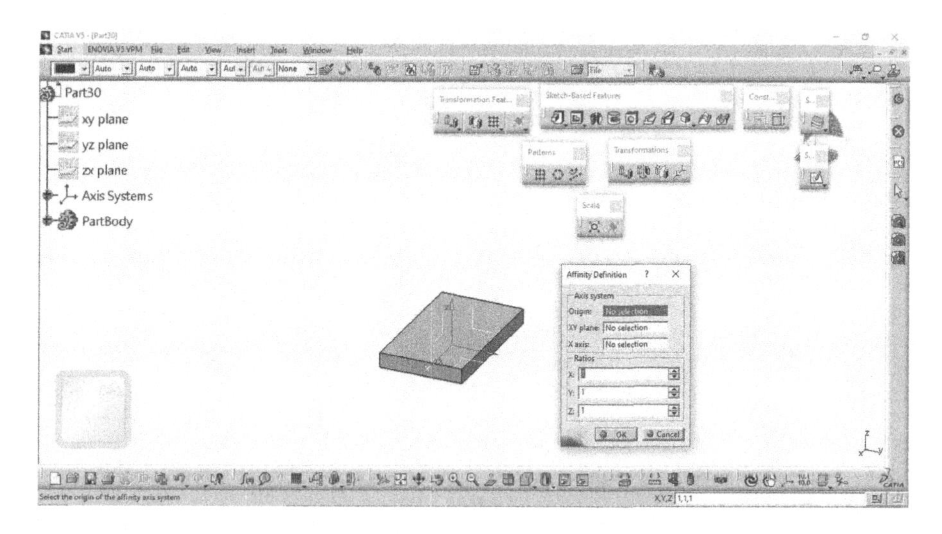

FIGURE 5.150 Selection of Affinity tool.

After selecting, click the OK button and then you will get the scaled object, as shown in Figure 5.149.

5.5.4.2 Affinity

The Affinity tool transforms the absolute size by multiplying a scale ratio. Draw the one padded object as shown in Figure 5.146. Choose the Affinity tool from the Scale toolbar, as displayed in Figure 5.145. The Affinity Definition dialog box window will appear, as shown in Figure 5.150.

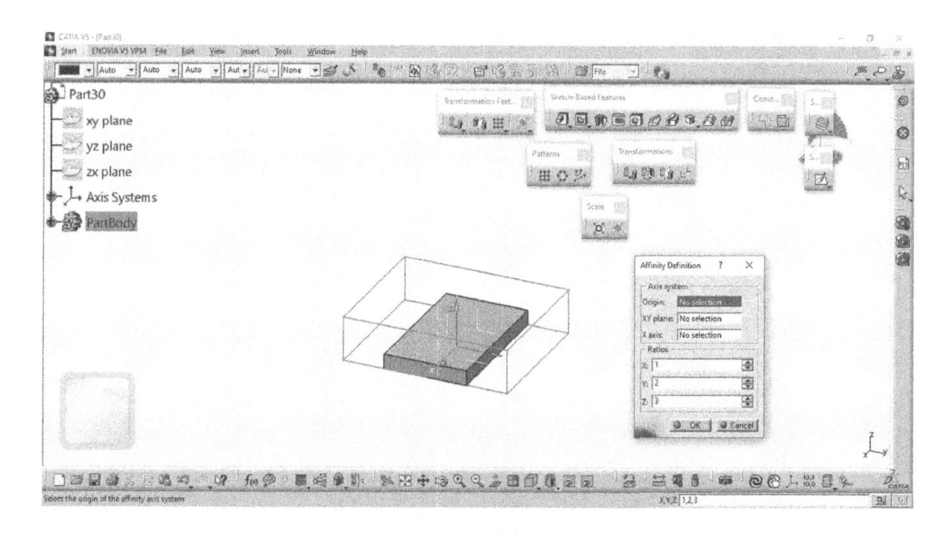

FIGURE 5.151 Selection in Affinity Definition dialog box.

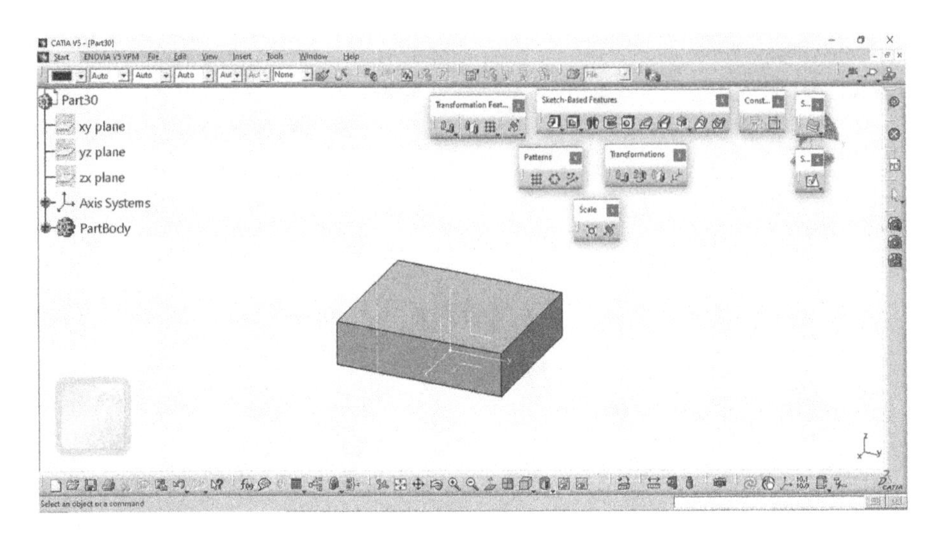

FIGURE 5.152 Solid after the use of Scaling tool.

Enter X as 1, Y as 2 and Z as 3 under Ratios area in Affinity Definition dialog box as shown in Figure 5.151.

After selecting, click the OK button and then you will get the scaled object, as shown in Figure 5.152.

Section IV

Assembly

6 Assembly Design

6.1 INTRODUCTION

This chapter covers assembly of complicated part using different tool in CATIA software. The Assembly Design workbench allows you to design complicated part having more than one component. You can add new or existing parts or subassemblies to the root assembly. Parts can be positioned and constrained within the assembly.

6.2 ACCESSING THE ASSEMBLY DESIGN WORKBENCH

When you start the CATIA software, a new Product file is created and the default name is Product1 and the same comes in the screen as depicted in Figure 6.1.

For accessing the Assembly Design workbench,

Start > Mechanical Design > Assembly Design

from the drop-down menu provided in Start (Figure 6.2).

A new file in Assembly Design workbench comes on the screen similar to as provided in Figure 6.3.

6.3 PRODUCT STRUCTURE TOOLS TOOLBAR

The Product Structure Tools toolbar allows you to insert, replace and reorder components in your product. When you are inserting an existing part or subassembly, their corresponding files are not copied into the assembly. They are just referenced by the assembly. The tools in this toolbar are as shown in Figure 6.4.

FIGURE 6.1 Initial screen after starting CATIA software.

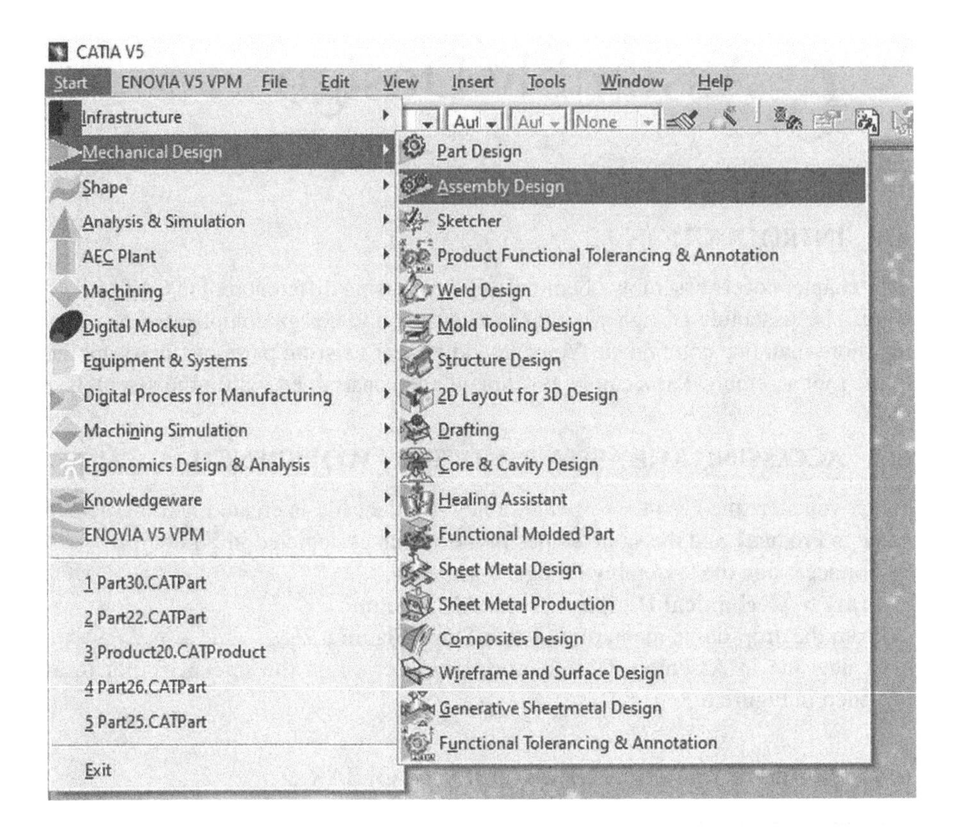

FIGURE 6.2 Screen for access the Assembly Design workbench.

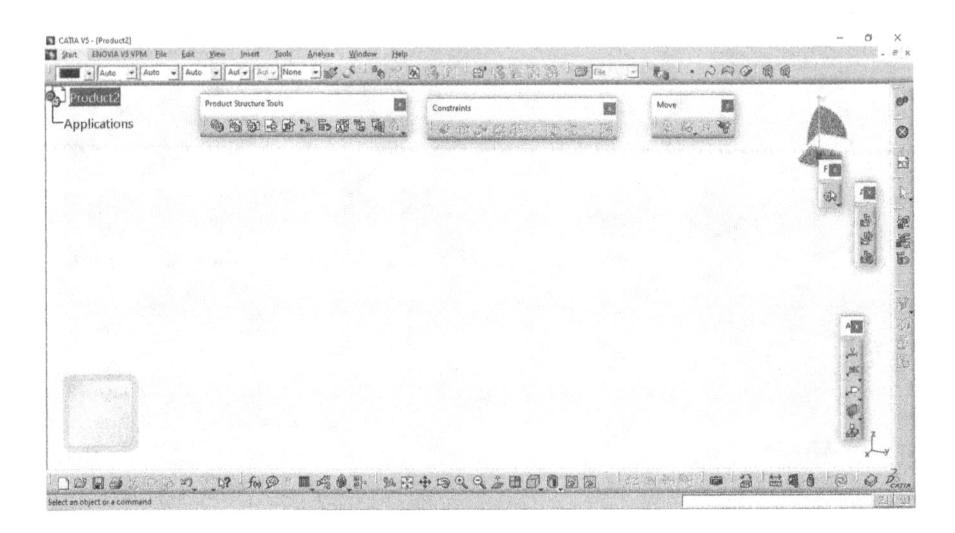

FIGURE 6.3 New Assembly Design dialog box.

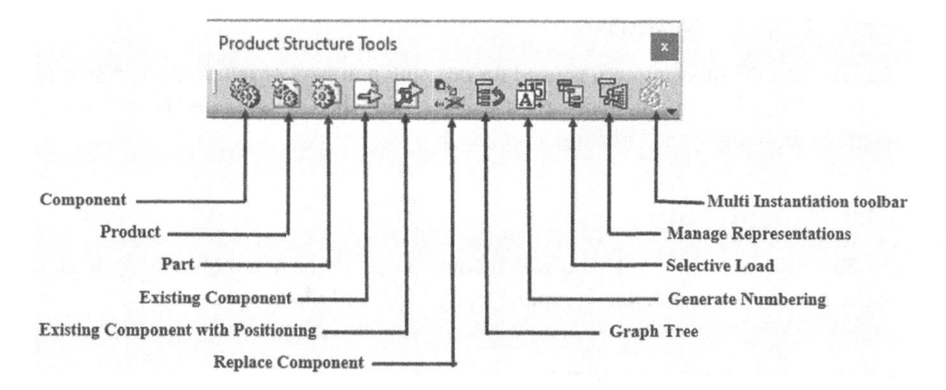

FIGURE 6.4 Product Structure Tools toolbar.

6.3.1 Component

The Component tool is used for inserting a new component that exists only in the root assembly file (CAT Product) and not as its own file.

6.3.2 Product

The Product tool is used for inserting a new product or subassembly (CAT Product) into the root assembly.

6.3.3 Part

The Part tool used is for inserting a new part (CAT Part) that will be designed on the fly while you are designing the assembly.

6.3.4 Existing Component

The Existing Component tool is used for adding an existing part or assembly as a component.

6.3.5 Existing Component With Positioning

The Existing Component with Positioning tool is used for adding an existing part or assembly as a component with a position.

6.3.6 Replace Component

The Replace Component tool is used for replacing a component.

6.3.7 Graph Tree

The Graph Tree tool is used for reordering components in the specification tree.

6.3.8 Generate Numbering

The Generate Numbering tool is used for creating numbers or letters to each part of the assembly. These numbers or letters will be used when creating and blooming the assembly drawing in the Drafting workbench.

6.3.9 Selective Load

The Selective Load tool tells you which components in the assembly will be loaded (opened or active) or not, and components that will be hidden or not.

6.3.10 Manage Representations

The Manage Representations tool is used for managing the representations (name, association etc.) of the components in the assembly.

6.3.11 Multi Instantiation Toolbar

The Multi Instantiation toolbar contains commands that allow you to create instances or duplicates of a component.

6.4 MOVE

Move toolbar (Figure 6.5) is primarily used to place individual components of the complete product in anywhere without any restriction or Constraints. Available tools in this toolbar are as shown in Figure 6.5.

6.4.1 Manipulation

The Manipulation tool is used for moving any part or component randomly with the aid of a mouse. The movement can be in x-axis, y-axis, z-axis or rotation about the axes. When Manipulation tools from Move toolbar are chosen, a Manipulation dialog box appears as provided in Figure 6.6. The dialog box of the same shows all feasible ways of translation and rotation as discussed above. Apart from the above, the users can create their own axes of movement and rotation.

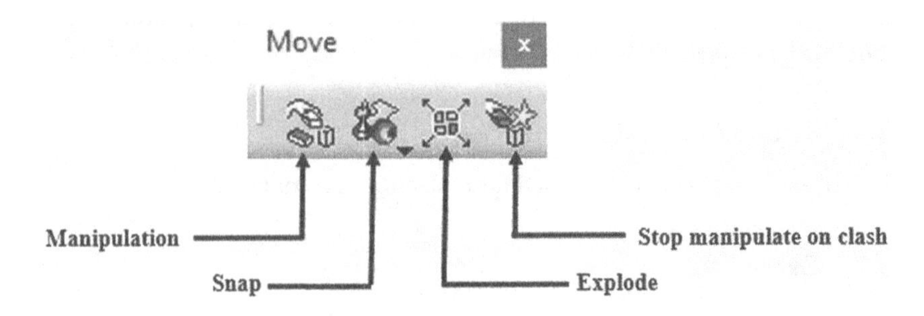

FIGURE 6.5 Move toolbar.

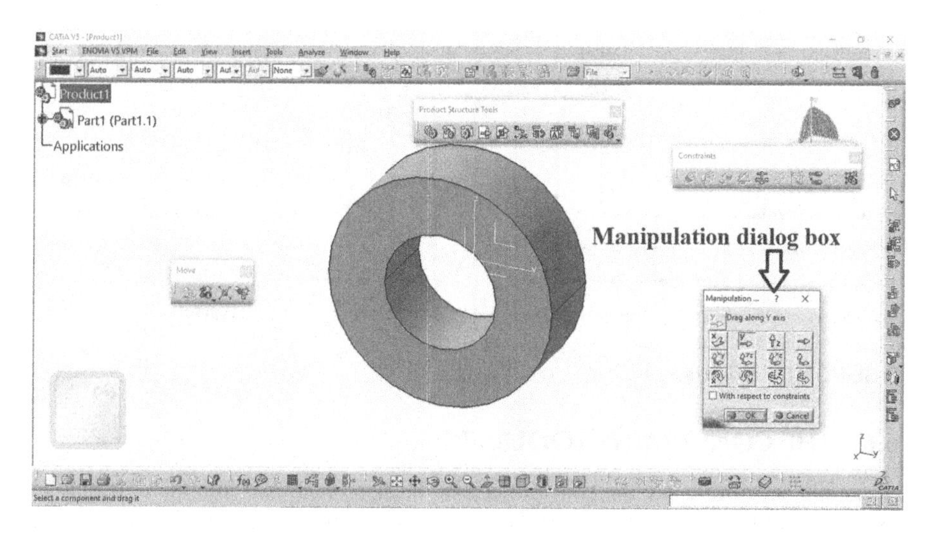

FIGURE 6.6 Manipulation dialog box.

6.4.2 Snap

The Snap tool is used for movement of one part to another by snapping.

6.4.3 Explode

A 3D exploded view can be created using Explode option.

6.4.4 Stop Manipulate on Clash

Whenever there is a clash, Stop Manipulate tool is used to stop the manipulation.

6.5 UPDATE

After moving the parts/product manually, the parts stay at that opposition, but when constraints are applied, the positioning requires an updation. The Update tool updates the product. So with this tool, the parts reset their position from manual status to constrained status as shown in Figure 6.7.

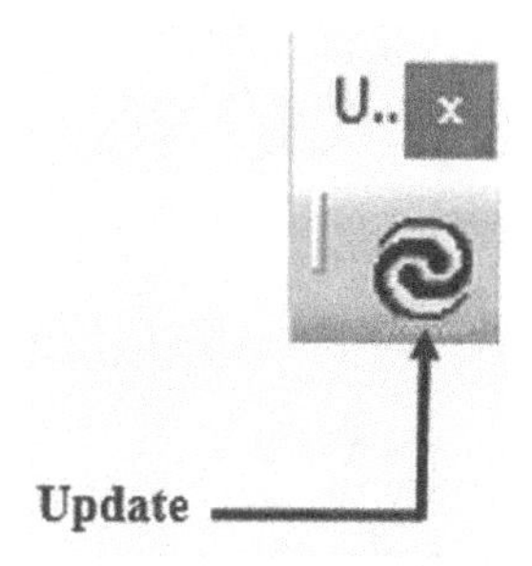

FIGURE 6.7 Update toolbar.

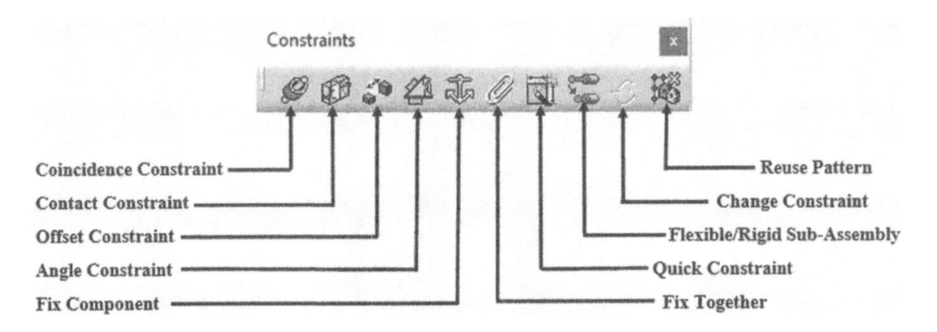

FIGURE 6.8 Constrains toolbar.

6.6 THE CONSTRAINS TOOLBAR

Assembly constraints are the constraints used to place each part into the assembly in its functional position. The general process for constraining the parts in an assembly is, fix one component in space, use the compass to move the other parts into their approximate position, position the parts precisely using the correct constraint(s) and then update the assembly to move the parts into position. If a part(s) has been moved out of position, the update command may be used to move the part(s) back into position. Constraint symbols may be hidden by right clicking on the Constraints inference in the specification tree and selecting Hide/Show. The tools in this toolbar are as shown in Figure 6.8.

6.6.1 Coincidence Constraint

The Coincidence Constraint tool is used to create alignments between axes, planes or points. Draw the padded parts 1 and 2 as shown in Figures 6.9 and 6.10.

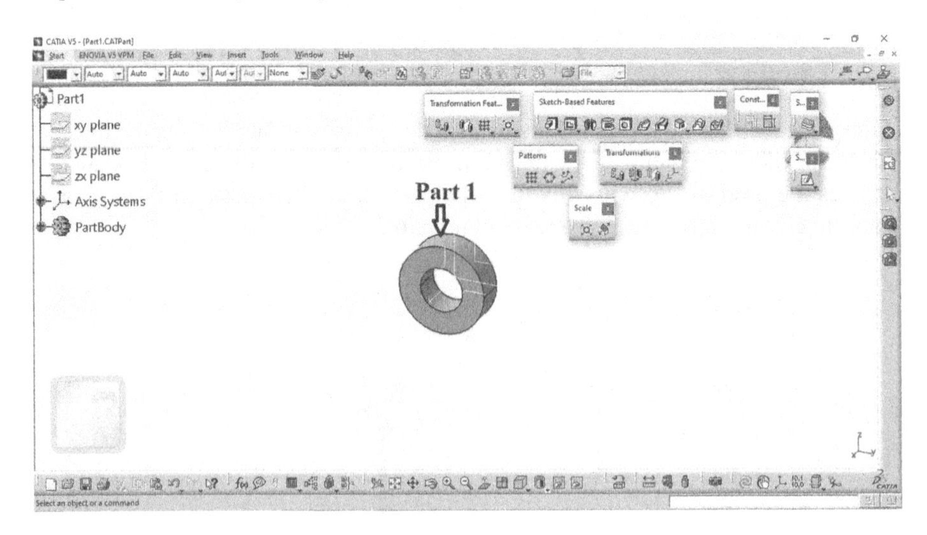

FIGURE 6.9 Padded Part 1.

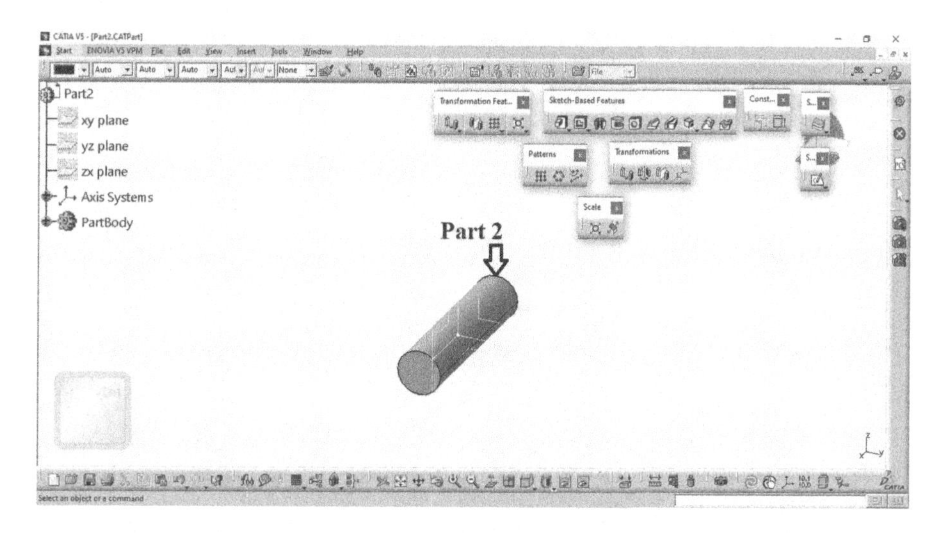

FIGURE 6.10 Padded Part 2.

A new file in the Assembly Design workbench is selected as shown in Figure 6.3. Select Product1 from tree, and select Existing Component tool from Product Structure toolbar as shown in Figure 6.11.

Part 1 is selected from saved folder, and by selecting open button in File Selection dialog box, you will get the Part 1 coming under Assembly Design workbench as shown in Figure 6.12.

Select Product1 from tree, and select Existing Component tool from Product Structure toolbar as shown in Figure 6.11.

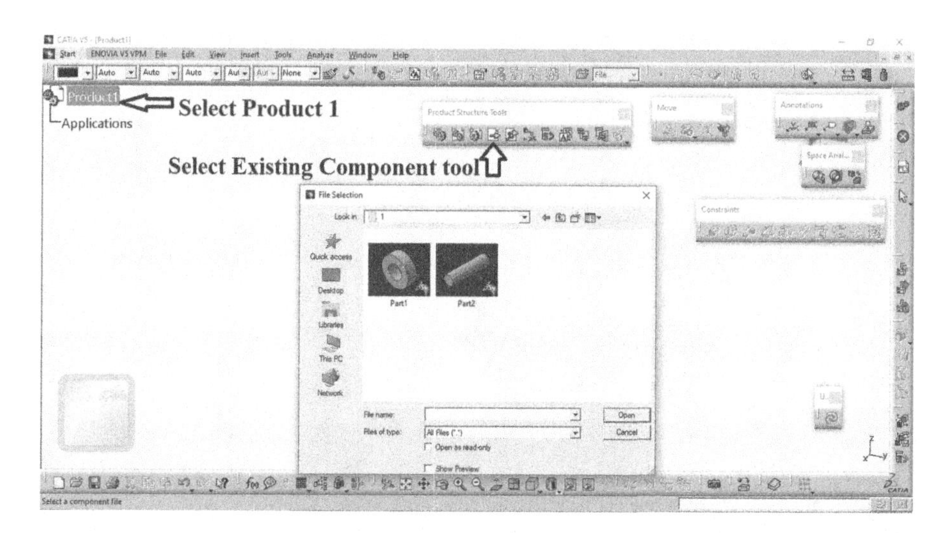

FIGURE 6.11 Selection of Existing Component tool.

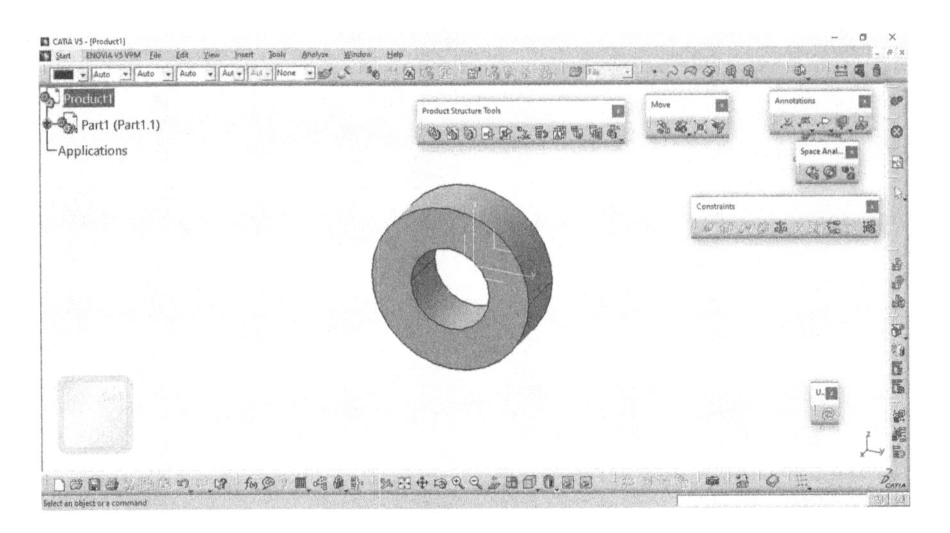

FIGURE 6.12 Part 1 comes under Assembly Design workbench.

Select the Part 2 from saved folder, and by selecting open button in File Selection dialog box, you will get the Part 1 and Part 2 coming under Assembly Design workbench as shown in Figure 6.13.

Select coincidence Constraint tool from Constraint toolbar and after that select axis of parts 1 and 2 as shown in Figure 6.14.

As we have updated the parts with constraints, by clicking on update all tool from Update toolbar, you will get the Assembled Product of parts 1 and 2 using Coincidence Constraint tool as shown in Figure 6.15.

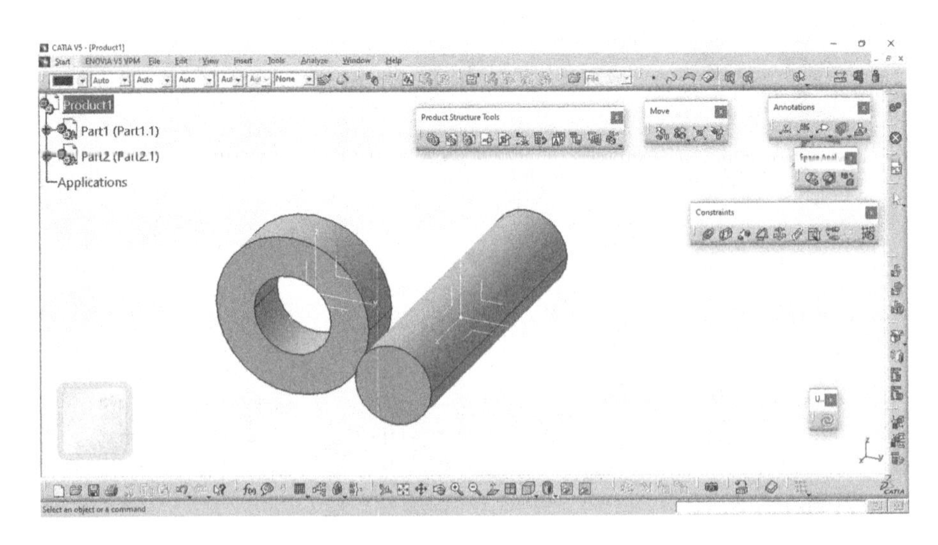

FIGURE 6.13 Part 1 and Part 2 come under Assembly Design workbench.

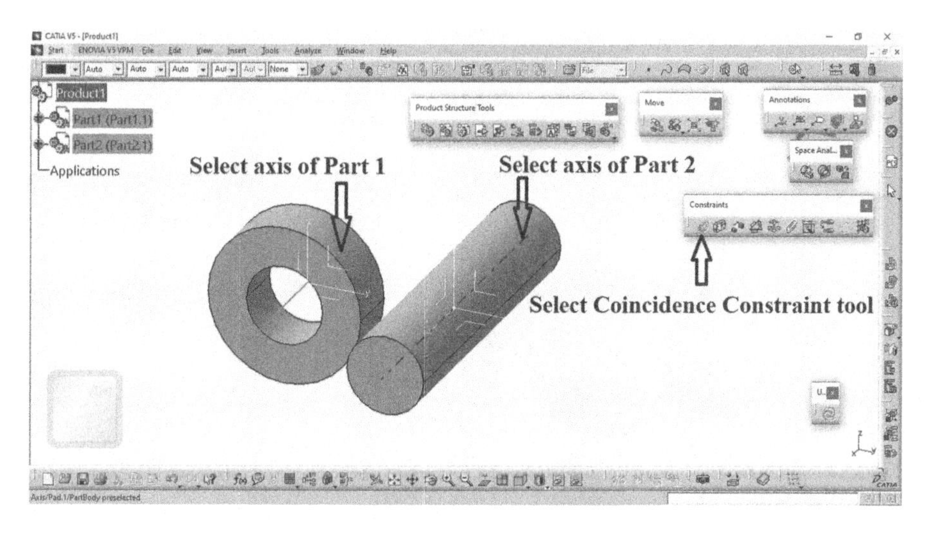

FIGURE 6.14 Selection of Coincidence Constraint tool and axis of parts 1 and 2.

6.6.2 Contact Constraint

The Contact Constraint tool is used to create contact between two planes or faces. Select Contact Constraint tool from constraint toolbar and after that select face of parts 1 and 2 from Assembled Product (Figure 6.15), as shown in Figure 6.16.

Similar to the earlier one, by clicking on update all tool from Update toolbar, you will get the Assembled Product of parts 1 and 2 using Contact Constraint tool as shown in Figure 6.17.

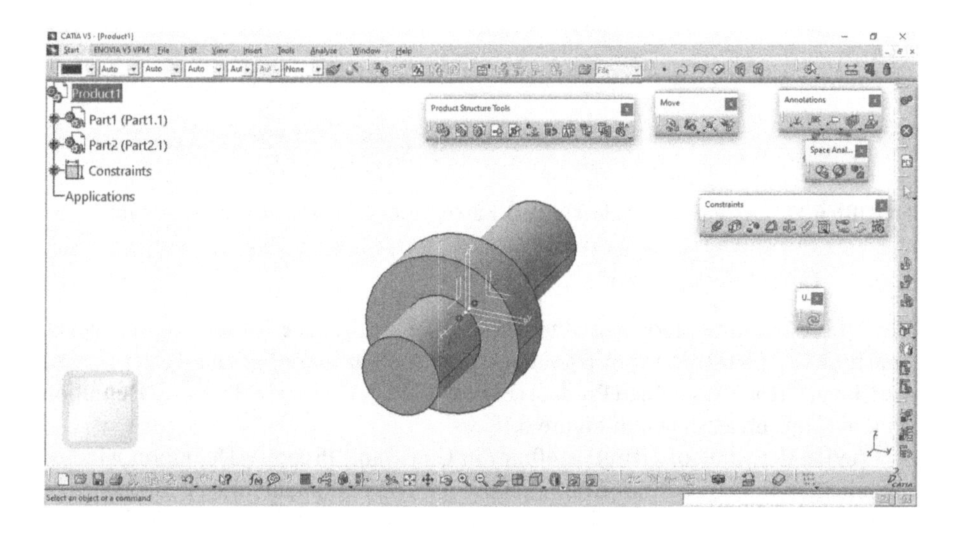

FIGURE 6.15 Assembly Product of parts 1 and 2 after using Coincidence Constraint tool.

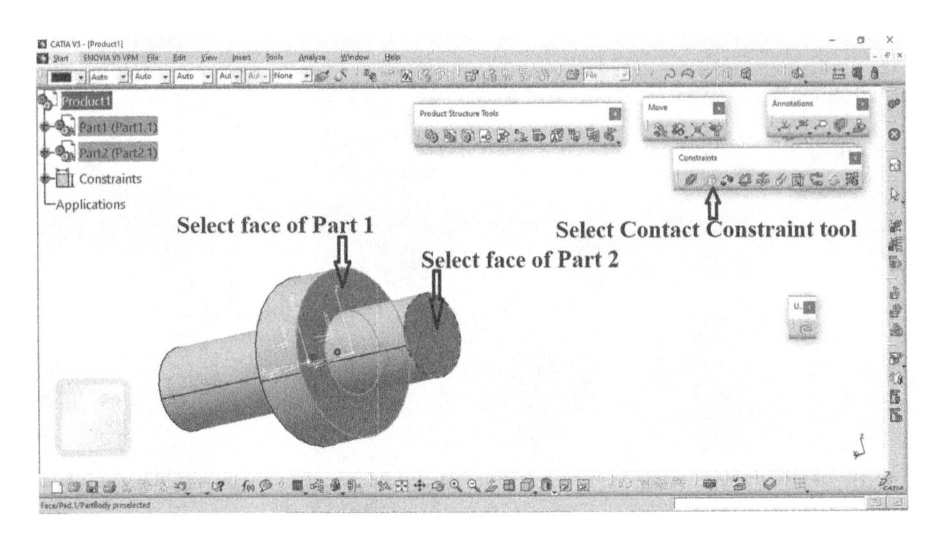

FIGURE 6.16 Selection of Contact Constraint tool and face of parts 1 and 2.

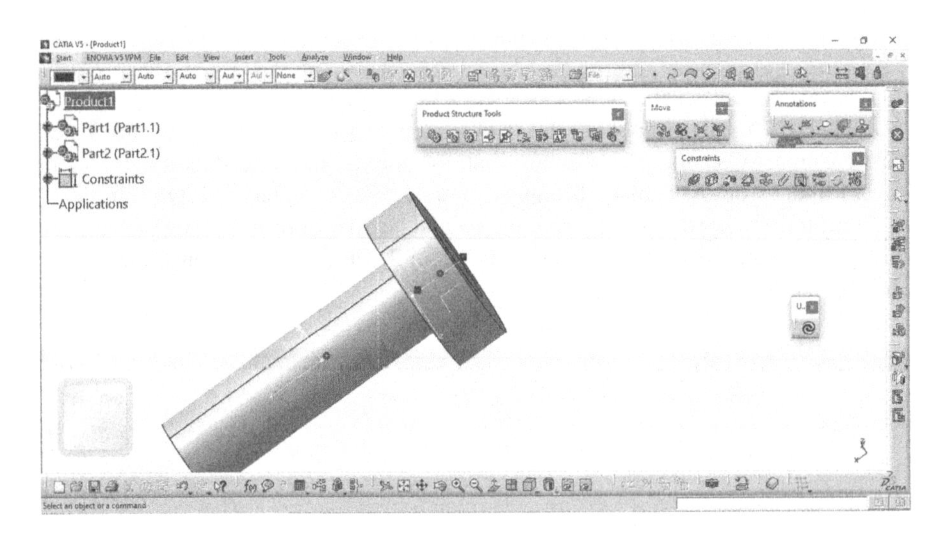

FIGURE 6.17 Assembled Product of Part 1 and 2 after using Contact Constraint tool.

6.6.3 Offset Constraint

The Offset Constraint tool is used to define an offset distance between two elements. Select Offset Constraint tool from constraint toolbar and after that select face of Part 1 and 2 from Assembled Product (Figure 6.15). A Constraint Property Definition window pops up as shown in Figure 6.18.

Provide the value of Offset as 40mm in Constraint Property Definition window and click on the OK button. After clicking the OK button and update all tool from Update toolbar, you will get the Assembled Product of parts 1 and 2 using Offset Constraint tool as shown in Figure 6.19.

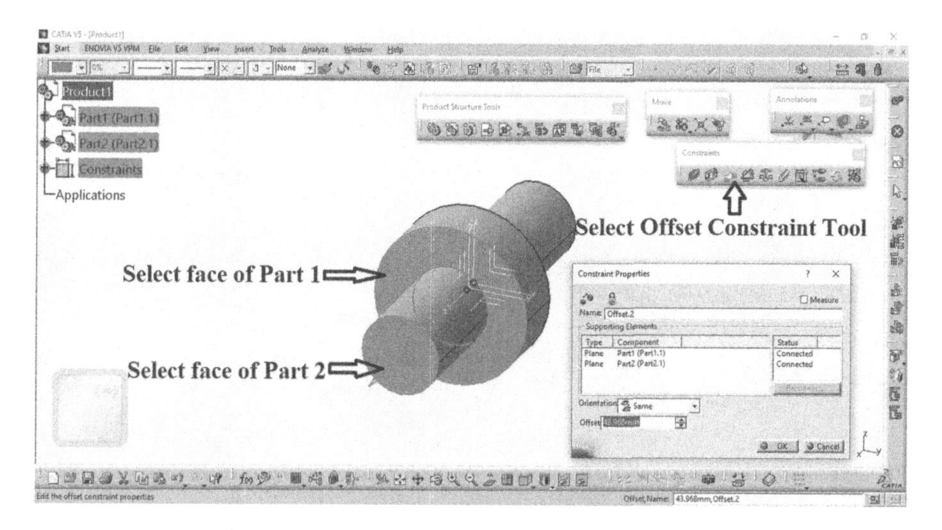

FIGURE 6.18 Selection of Offset Constraint tool and face of parts 1 and 2.

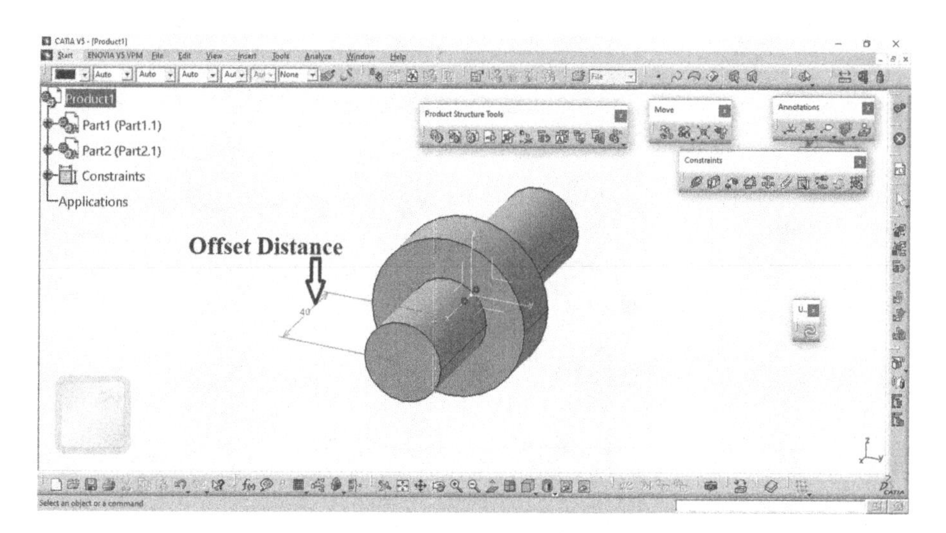

FIGURE 6.19 Assembled Product of Part 1 and 2 after using Offset Constraint tool.

6.6.4 Angle Constraint

The Angle Constraint tool is used to define an angle between two elements. Select Product1 from tree, and select Existing Component tool from Product Structure toolbar. Select the Part from saved folder and select open button in File Selection dialog box, and you will get that it is coming under Assembly Design workbench as shown in Figure 6.20.

Select Angle Constraint tool from Constraint toolbar and after that select face of parts. A Constraint Property Definition window comes as provided in Figure 6.21.

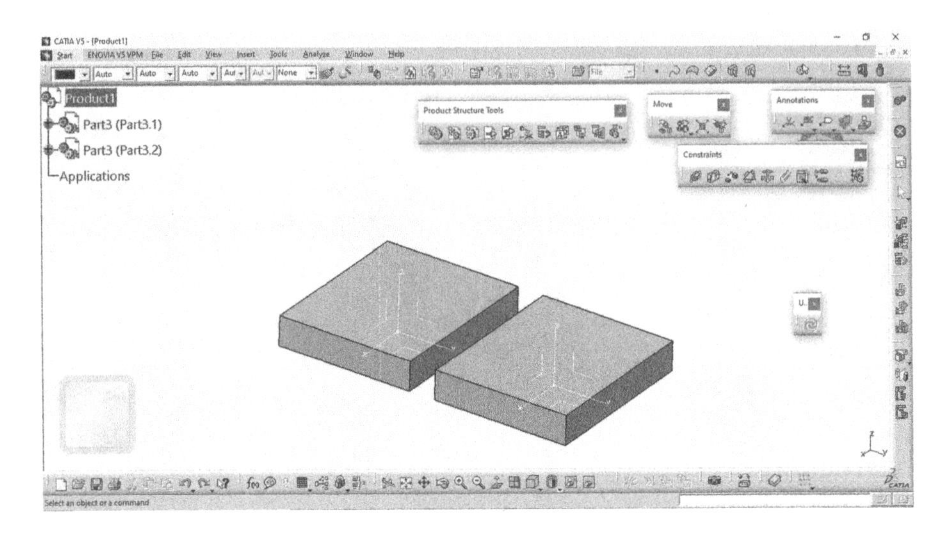

FIGURE 6.20 Parts comes under Assembly Design workbench.

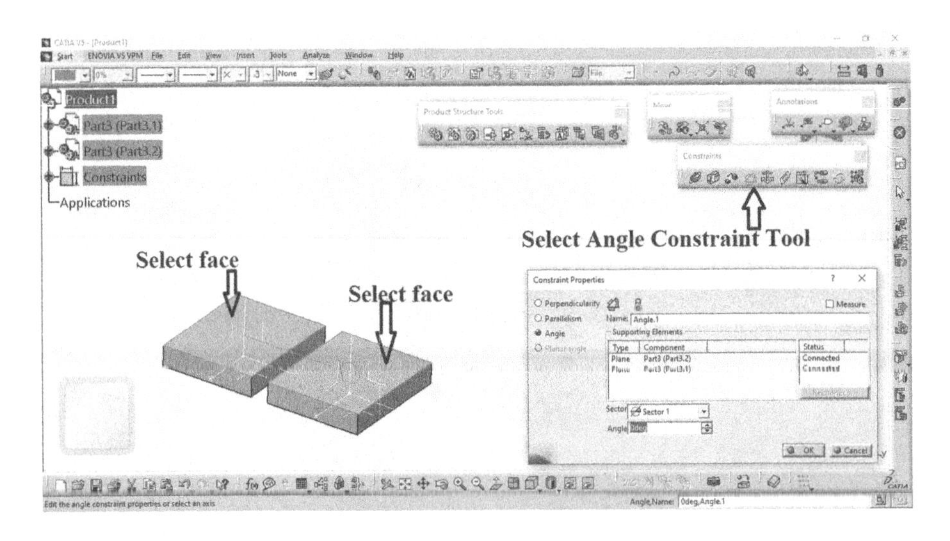

FIGURE 6.21 Selection of Angle Constraint tool and face of parts.

Inter the value of angle as 45deg in Constraint Property Definition window and click OK button. After clicking the OK button and update all tool from Update toolbar, you will get the Assembled Product of Parts using Angle Constraint tool as shown in Figure 6.22.

6.6.5 Fix Component

The Fix Component tool is used to fix a part in space such that it is always located in the same spot. The other parts are constrained relative to the fixed part.

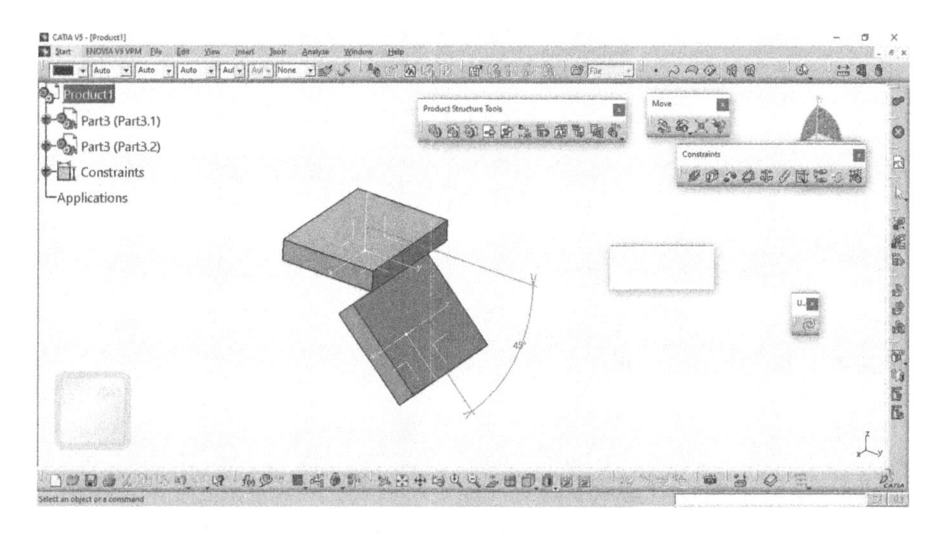

FIGURE 6.22 Assembled Product of Parts after using Angle Constraint tool.

6.6.6 Fix Together

The Fix Together tool is used to fix two parts together.

6.6.7 Quick Constraint

The Quick Constraint tool is used to automatically create what CATIA believes is the correct constraint between two elements.

6.6.8 Flexible/Rigid Sub-Assembly

The Flexible/Rigid Sub-Assembly tool allows you to change a subassembly between flexible and rigid. A flexible subassembly has parts or components that can be moved disregarding the fact that it is not the active component. The positions of the parts can be different than those stored in the reference CAT Product file.

6.6.9 Change Constraint

The Change Constraint tool allows you to change the type of constraint that exists between two elements.

6.6.10 Reuse Pattern

The Reuse Pattern tool allows you to reuse an existing pattern.

6.7 THE MEASURE TOOLBAR

The Measure toolbar contains commands that measure distances, angles, mass, volume, inertia etc. of your assembly or individual geometric element. You can keep the

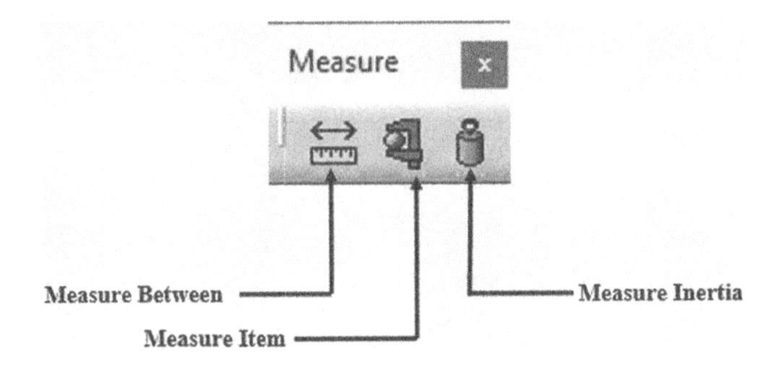

FIGURE 6.23 Measure toolbar.

result of your measurement as an element in your specification tree. The tools in this toolbar are as shown in Figure 6.23.

6.7.1 Measure Between

The Measure Between tool measures between two geometric elements.

6.7.2 Measure Item

The Measure Items tool measures several physical parameters of a demerit such as area and perimeter.

6.7.3 Measure Inertia

The Measure Inertia tool measures physical parameters of a part or assembly. It will calculate the mass, volume, moments of inertia, and center of gravity among other things.

6.8 THE SPACE ANALYSIS TOOLBAR

The tools in this Space Analysis toolbar are as shown in Figure 6.24.

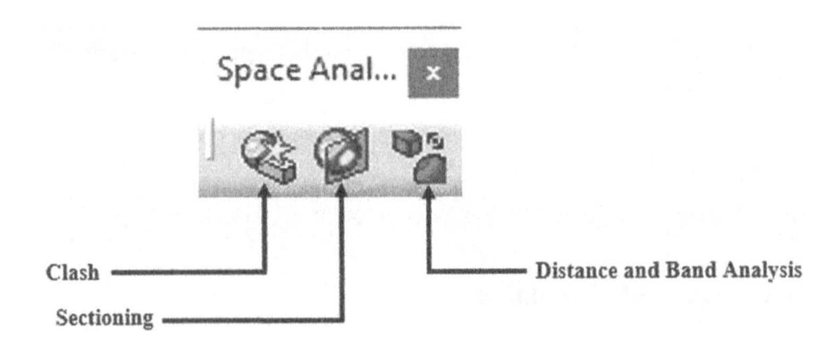

FIGURE 6.24 Space Analysis toolbar.

6.8.1 Clash

The Clash tool analyzes any part interference. If there is an actual material interference between two parts, you will be warned.

6.8.2 Sectioning

The Sectioning tool allows you to see a 2D section and to create a 3D section of your assembly.

6.8.3 Distance and Band Analysis

The Distance and Band Analysis tool analyzes the minimum distance between two sections or between one component and all the other components.

6.9 THE ANNOTATION TOOLBAR

The Annotation toolbar is used for text and symbols that can be seen in the 3D view. The tools in this Annotations toolbar as shown in Figure 6.25.

6.9.1 Weld Feature

The Weld Feature tool allows you to put welding symbols on your part.

6.9.2 Text With Leader

The Text With Leader tool produces a text annotation that points to a selected component or geometric feature.

6.9.3 Flag Note With Leader

The above tool allows you to add hyperlinks to your document that can be used to jump to different locations. For example, you can jump to an Excel spreadsheet or to a HTML page on the Internet.

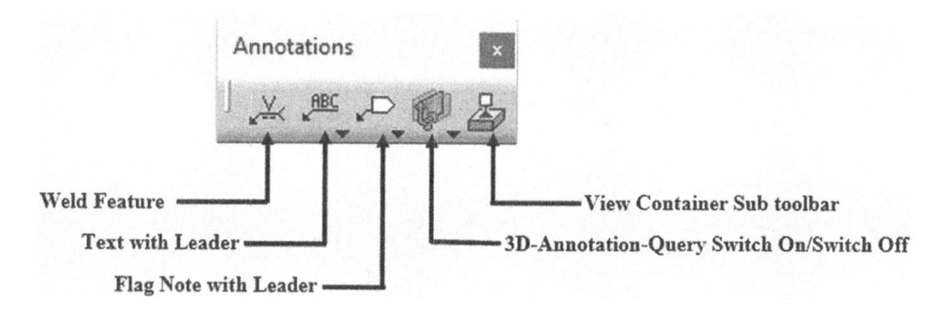

FIGURE 6.25 Annotations toolbar.

6.9.4 View Container Sub Toolbar

The View Container Sub Toolbar tool allows you to define the plane that will contain your annotations.

6.9.5 3D-Annotation-Query Switch On/Switch Off

The above tool enables you to see the relationships between the annotations and the geometries in which they refer to.

Section V

Drafting

7 Drafting

7.1 INTRODUCTION

This chapter covers Drafting of complicated part using different tools in CATIA software. The Drafting workbench allows you to create an orthographic projection or drawing (CAT Drawing) directly from a 3D part (CAT Part) or assembly (CAT Product). A CAT Drawing contains a structure listing like a specification tree. The structure listing shows all the sheets and views contained in the document. CATIA enables you to create generative views that are associative with the 3D part, and to create drawn views that are not associative.

7.2 THE VIEWS TOOLBAR

The tools located in the Views toolbar enable you to create a variety of views and view configurations. The sub-toolbars are located within the Views toolbar as shown in Figure 7.1.

7.2.1 Projections Toolbar

The Projections toolbar contains commands that allow you to create different types of views. The tools in this toolbar appear as shown in Figure 7.2.

7.2.2 Sections Toolbar

This toolbar is used to create a variety of section and cut views. The tools of Sections toolbar appear as shown in Figure 7.3.

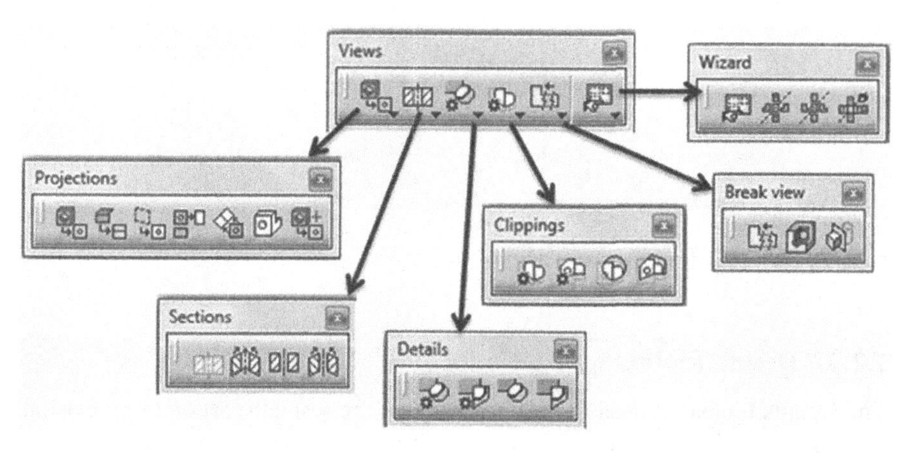

FIGURE 7.1 Views toolbar.

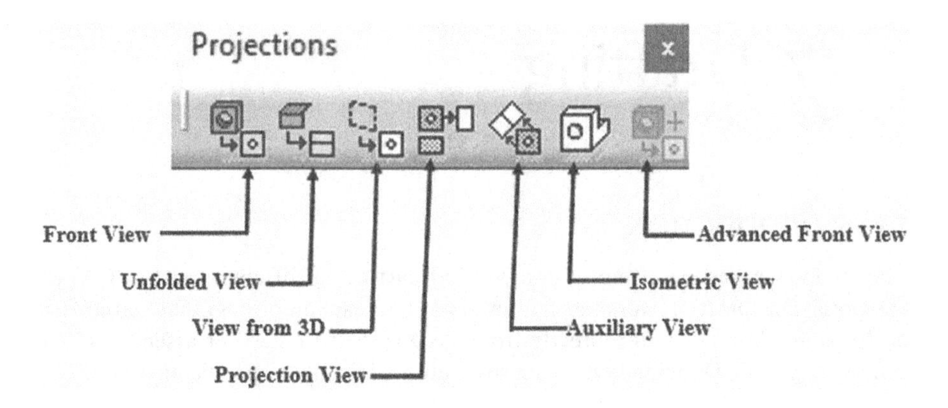

FIGURE 7.2 Projections toolbar.

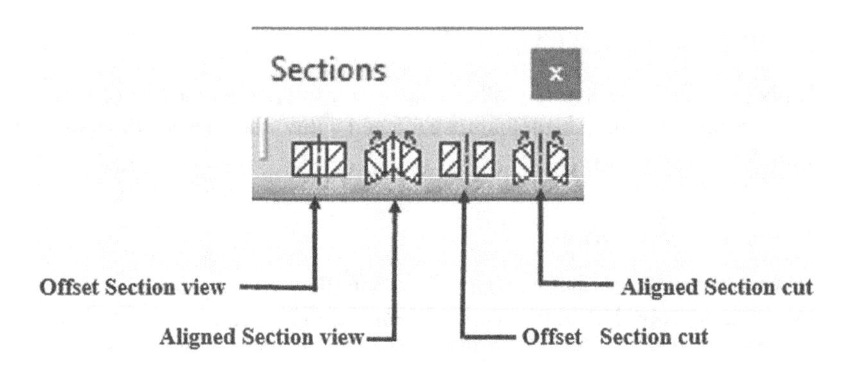

FIGURE 7.3 Sections toolbar.

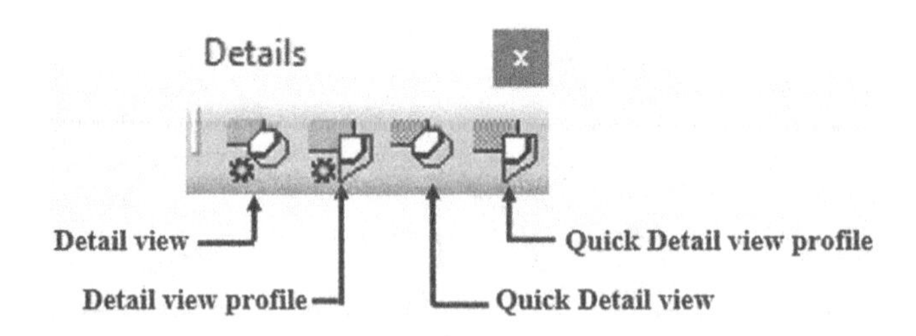

FIGURE 7.4 Details toolbar.

7.2.3 Details Toolbar

The Details toolbar is used to create views that are a small portion of an existing view. The detail view is usually drawn at an increased scale. The tools of Details toolbar appear as shown in Figure 7.4.

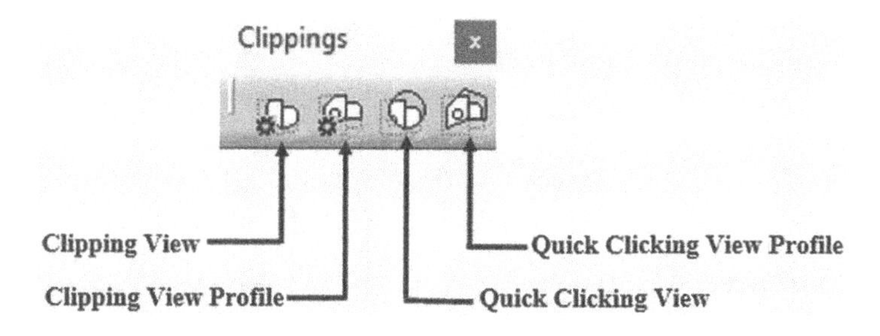

FIGURE 7.5 Clippings toolbar.

7.2.4 Clippings Toolbar

The Clippings toolbar is used to create removed views or removed section views. This is a small area of the part that is shown apart from the original view and is usually shown at an increased scale. The tools of Clippings toolbar appear as shown in Figure 7.5.

7.2.5 Break View Toolbar

The Break View toolbar is used to break the part in a specified location. This is usually done to save drawing space. The missing section of the part is usually uninteresting and not worth showing. There is also a command that enables you to create a broken out section. The tools of Break View toolbar appear as shown in Figure 7.6.

7.2.6 Wizard Toolbar

This toolbar is used to select from several predefined view configurations or to define your own custom configuration. The tools of Wizard toolbar appear as shown in Figure 7.7.

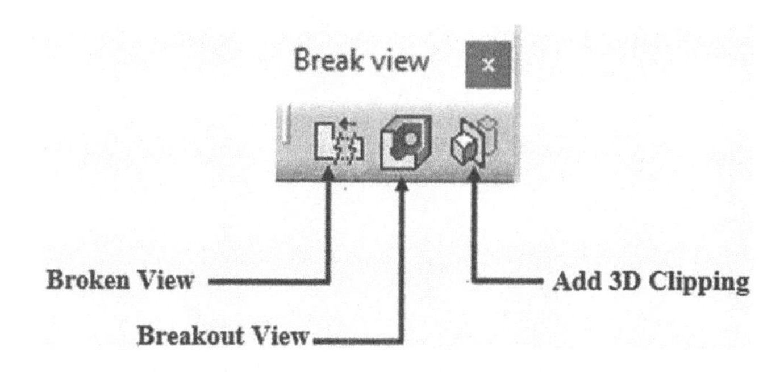

FIGURE 7.6 Break View toolbar.

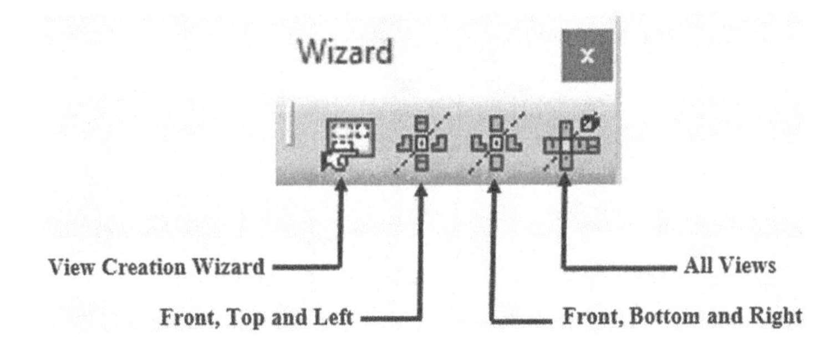

FIGURE 7.7 Wizard toolbar.

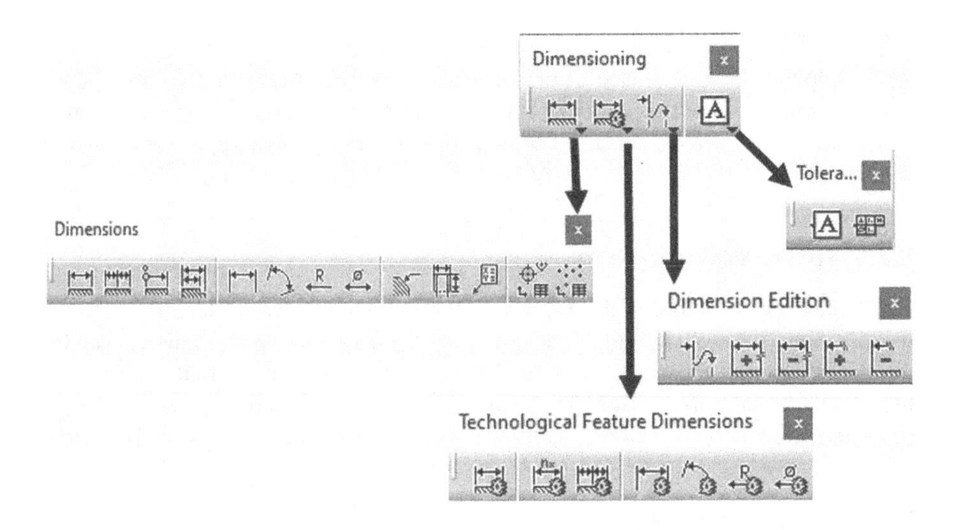

FIGURE 7.8 Dimensioning toolbar.

7.3 THE DIMENSIONING TOOLBAR

The Dimensioning toolbar is used to create dimensions and tolerances manually. The tools and tool bar of Dimensioning toolbar appear as shown in Figure 7.8.

Section VI

Case Study

8 Case Study – 1
Modeling of Oldham's Coupling

8.1 ASSEMBLY

8.1.1 AIM

To create Assembly of Oldham's Coupling by given components 1, 2 and 3 as shown in Figures 8.1, 8.2 and 8.3.

Draw the components 1, 2 and 3 as per dimension shown in Figures 8.1, 8.2 and 8.3 using Sketcher and Part Design workbench as explained in previous chapter. Create components 1, 2 and 3 as shown in Figures 8.4, 8.5 and 8.6.

Select a new file in the Assembly Design workbench as displayed on the screen, as shown in Figure 8.7. Select Product 1 from tree, and select Existing component tool from Product Structure Tool bar. Select the Component 1 from saved folder and

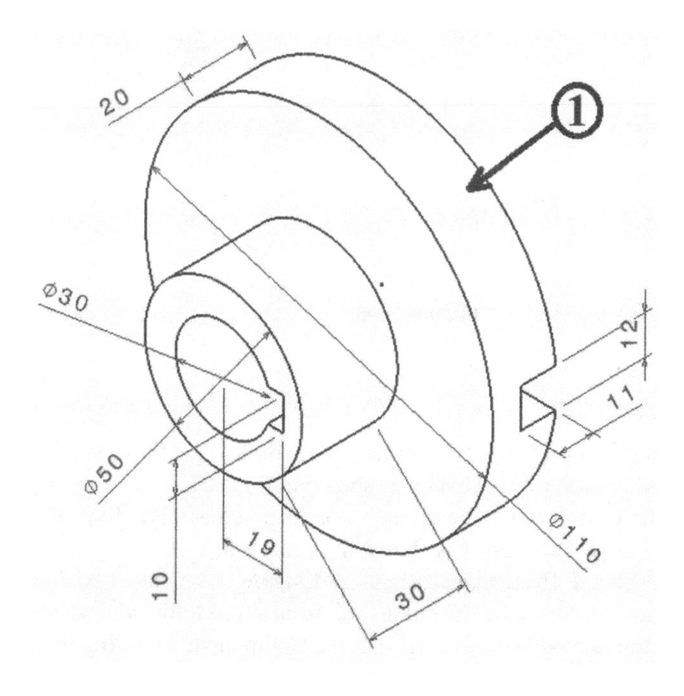

FIGURE 8.1 Component 1 with dimensions.

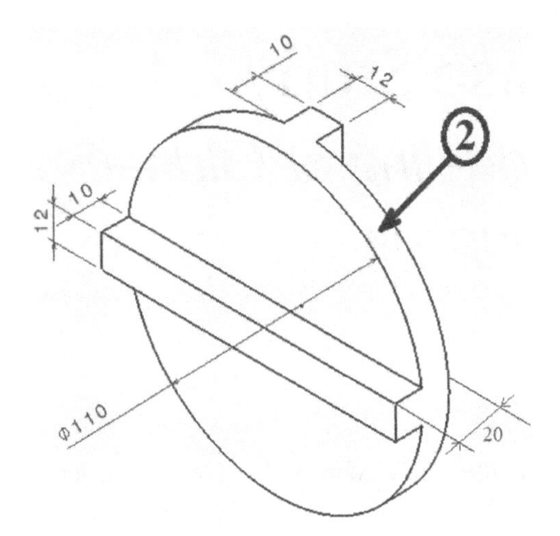

FIGURE 8.2 Component 2 with dimensions.

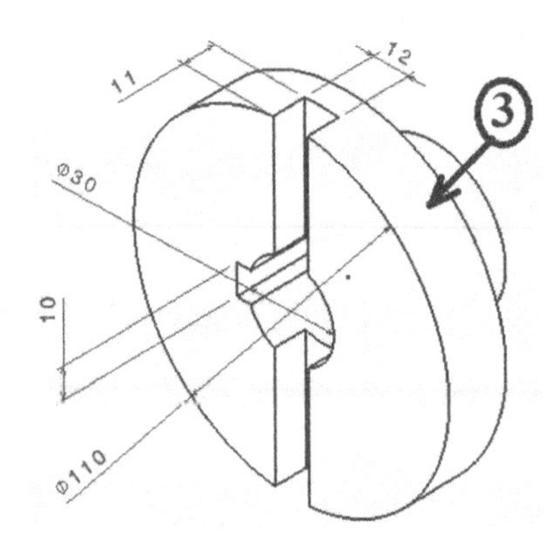

FIGURE 8.3 Component 3 with dimensions.

select open button in File selection dialog box; you will get the Component 1 coming under Assembly Design workbench as shown in Figure 8.7.

Select Fix Component Constraint Tool from constraint Tool Bar and select Product 1. Fixed Component 1 is shown in Figure 8.8.

Select Product 1 from tree, and select Existing component tool from Product Structure Tool bar. Select the Component 2 from saved folder and select open button in File selection dialog box; you will get the Component 2 coming under Assembly Design workbench as shown in Figure 8.9.

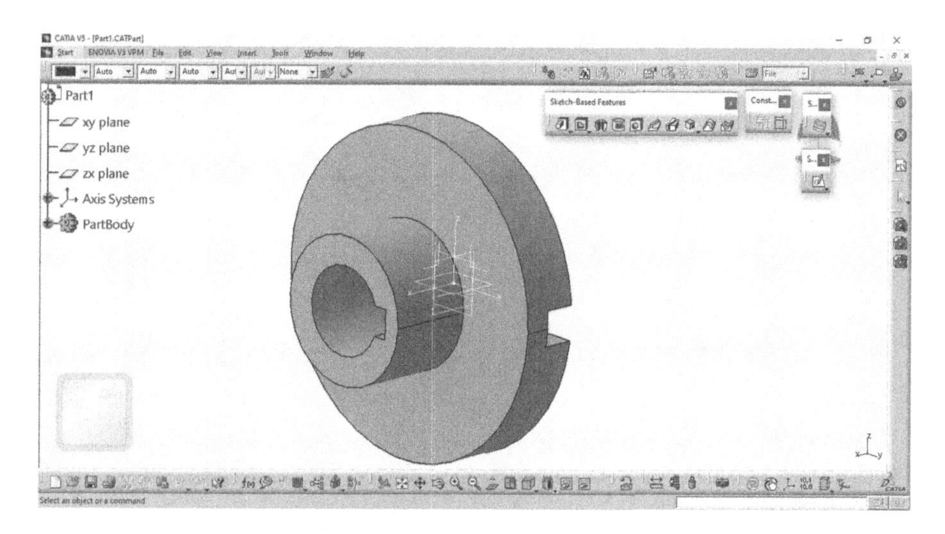

FIGURE 8.4 Created Component 1.

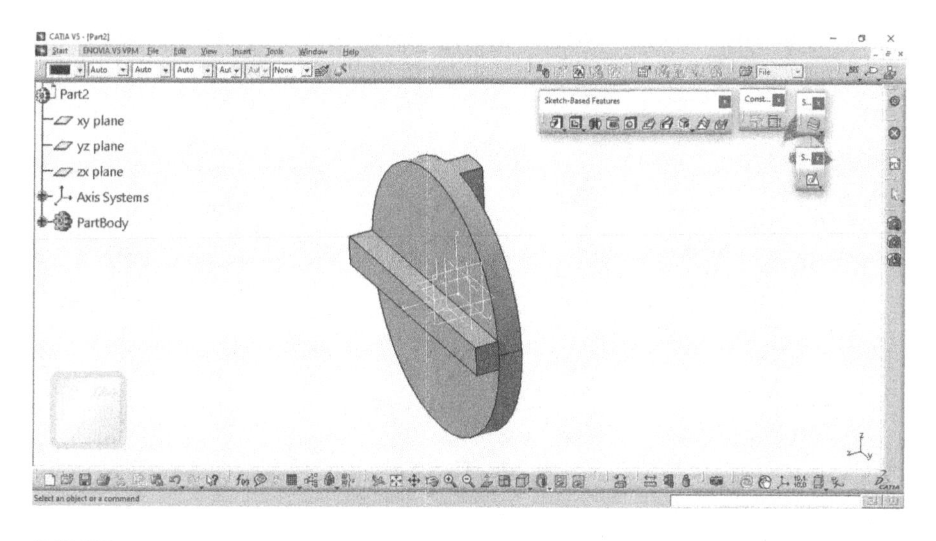

FIGURE 8.5 Created Component 2.

By using coincidence and Contact Constraint Tool from the constraint Tool Bar assemble components 1 and 2. Assembly of components 1 and 2 as shown in Figure 8.10.

Select Product 1 from tree, and select Existing component tool from Product Structure Tool bar. Select the Component 3 from saved folder and select open button

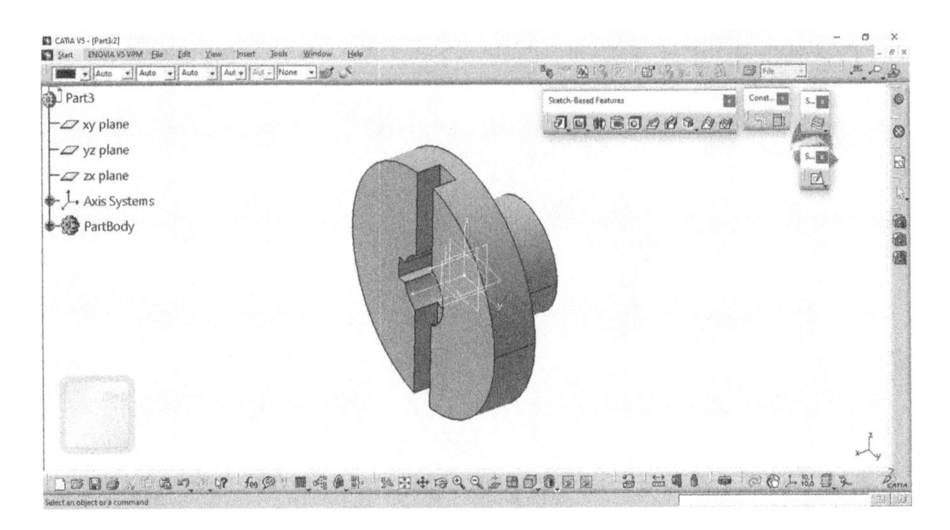

FIGURE 8.6 Created Component 3.

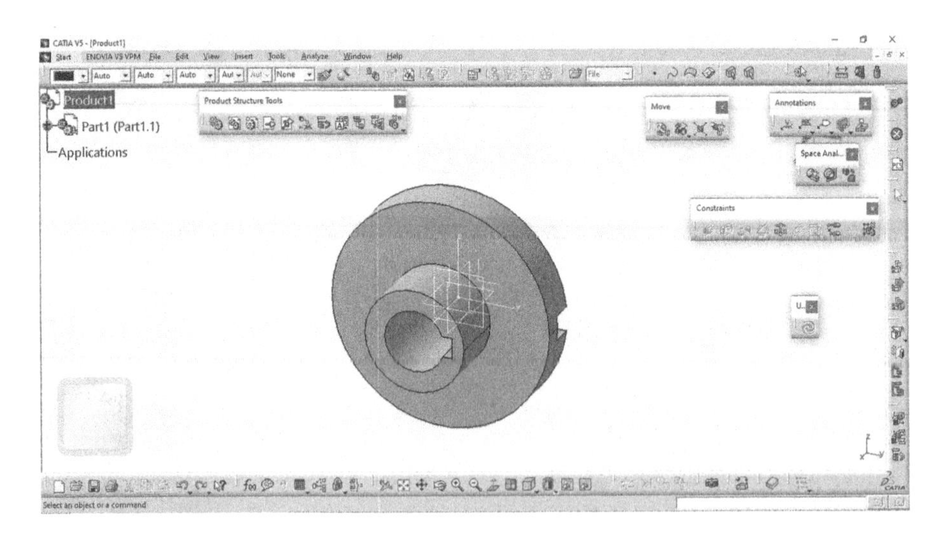

FIGURE 8.7 Component 1 comes under Assembly Design workbench.

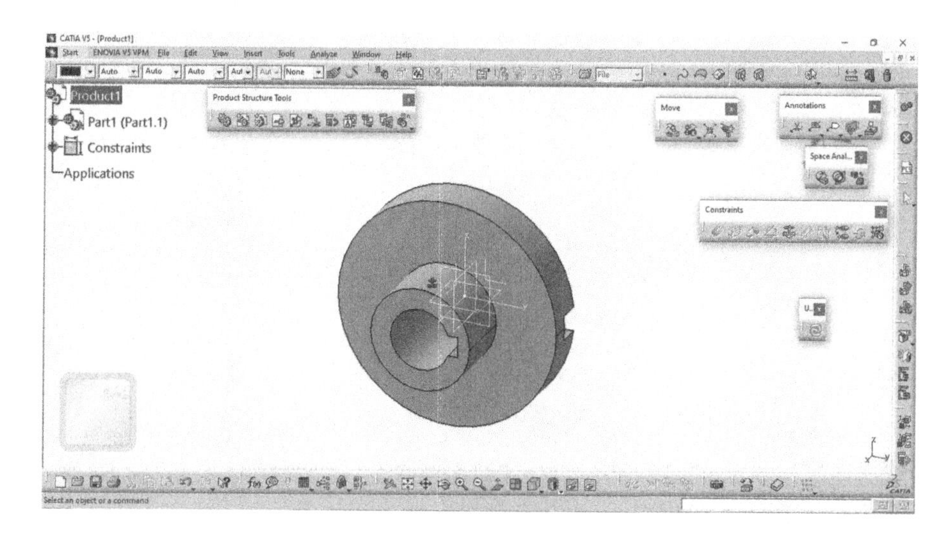

FIGURE 8.8 Component 1 fixed using Fix constraint tool.

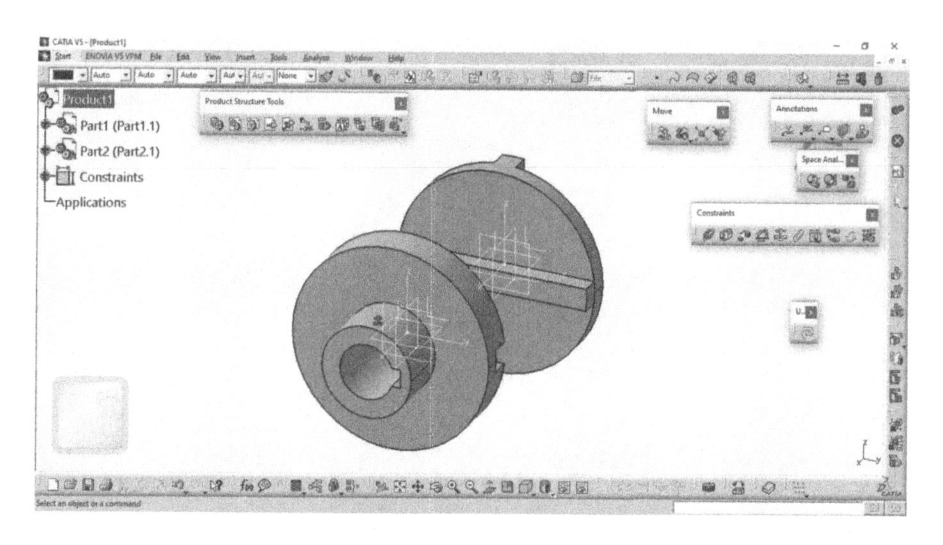

FIGURE 8.9 Components 1 and 2 comes under Assembly Design workbench.

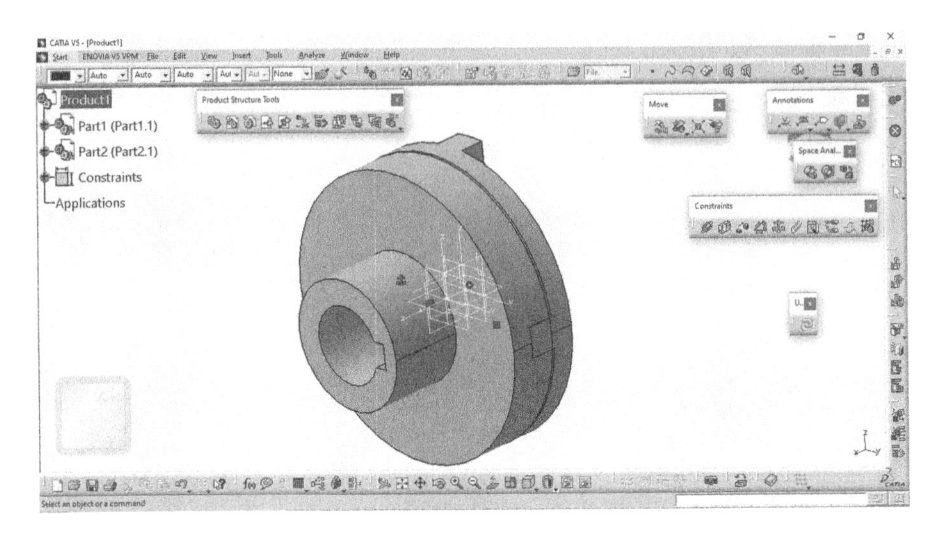

FIGURE 8.10 Assembly of components 1 and 2 using coincidence and Contact constraint tool.

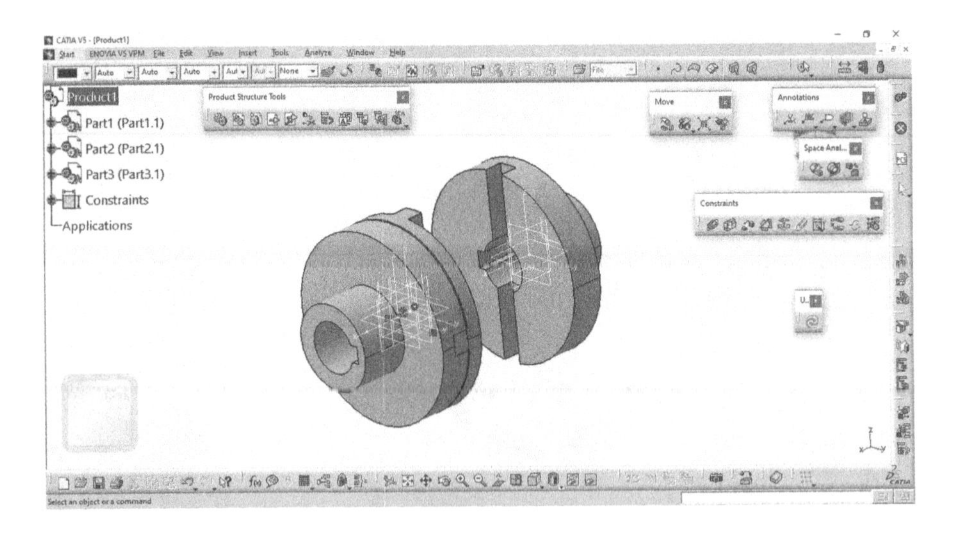

FIGURE 8.11 Components 1, 2 and 3 comes under Assembly Design workbench.

in File selection dialog box; you will get the Component 3 coming under Assembly Design workbench as shown in Figure 8.11.

By using coincidence and Contact Constraint Tool from constraint Tool Bar, assemble Components 1, 2 and 3. Assembly of Oldham's Coupling by using Components 1, 2 and 3 as shown in Figure 8.12.

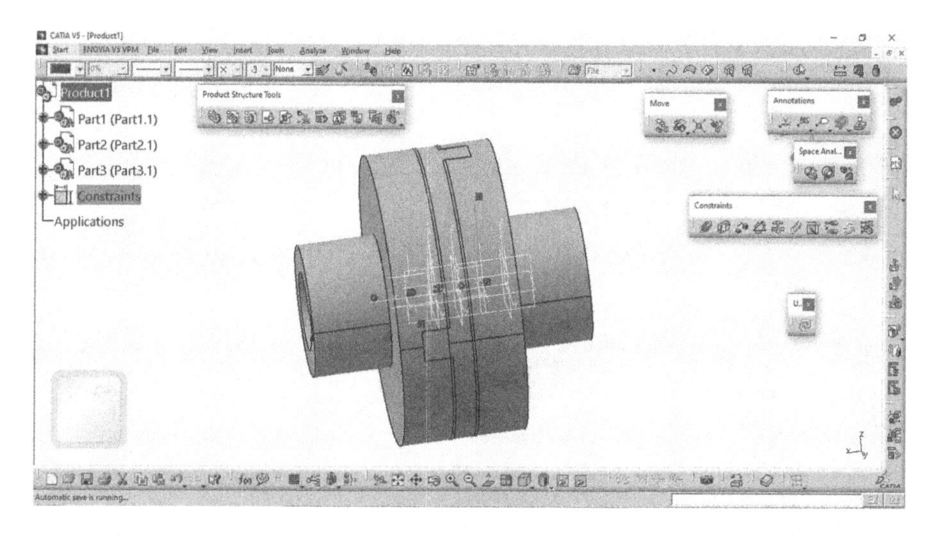

FIGURE 8.12 Assembly of Oldham's coupling.

8.2 DRAFTING

8.2.1 Aim

To create front view, top view, side view and isometric view of Assembled Oldham's Coupling by using drafting.

Go to File, open Assembly Design and select the saved Assembly of Oldham's Coupling, as shown in Figure 8.13.

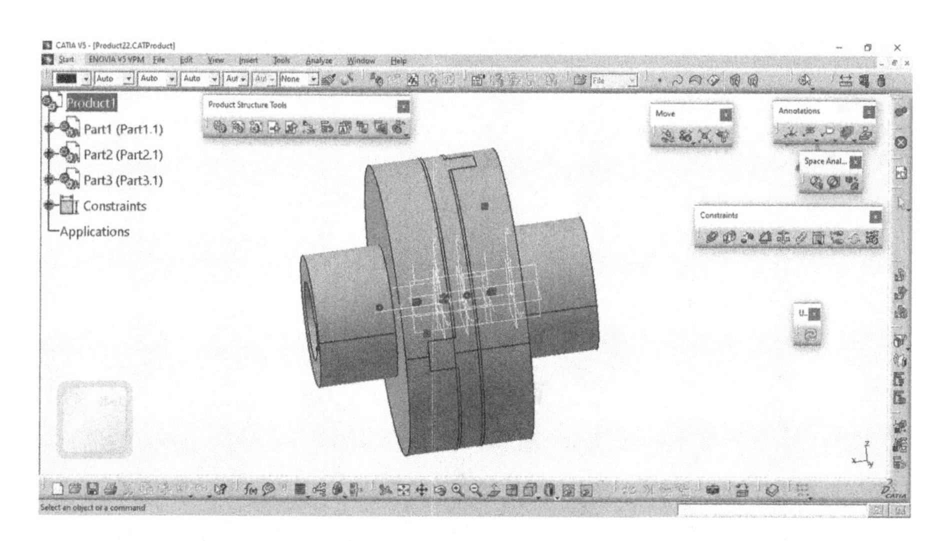

FIGURE 8.13 Selection of Assembled Oldham's coupling.

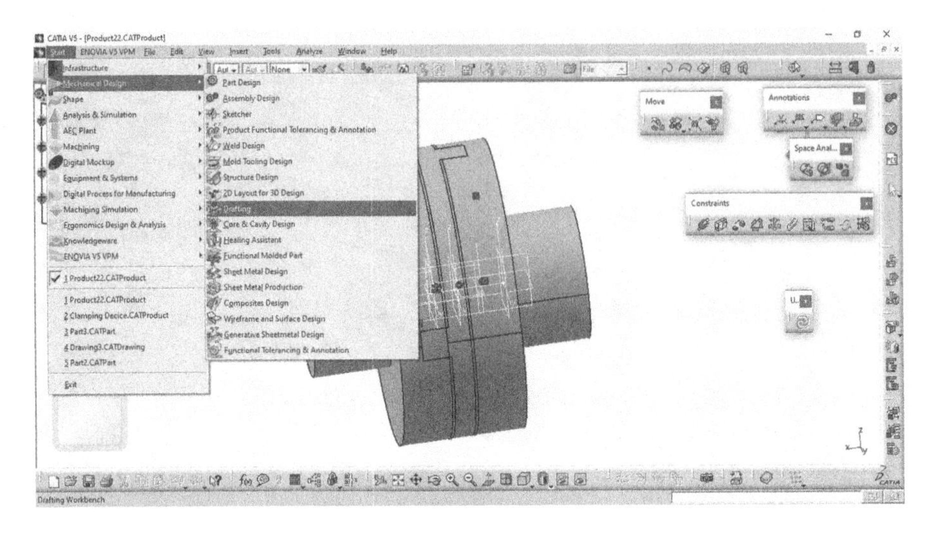

FIGURE 8.14 Selection of drafting.

After selecting file, go to *Start* > *Mechanical Design* > *Drafting* as shown in Figure 8.14.

A New drawing creation dialog box will appear as shown in Figure 8.15.

Select the second view in New drawing creation dialog box and click OK. The front view, top view, side view and isometric view of assembled Oldham's Coupling drafted Drawing will be displayed as shown in Figure 8.16.

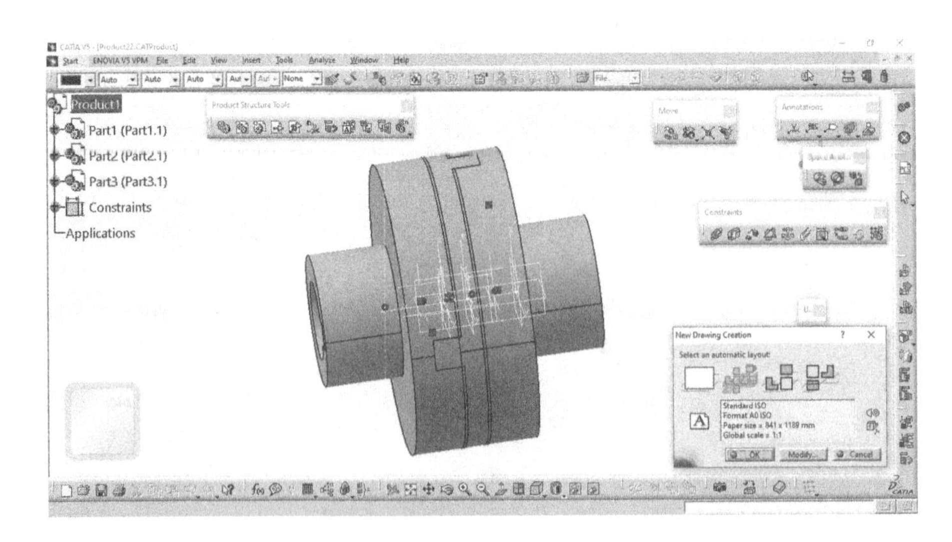

FIGURE 8.15 Selection of New drawing creation dialog box.

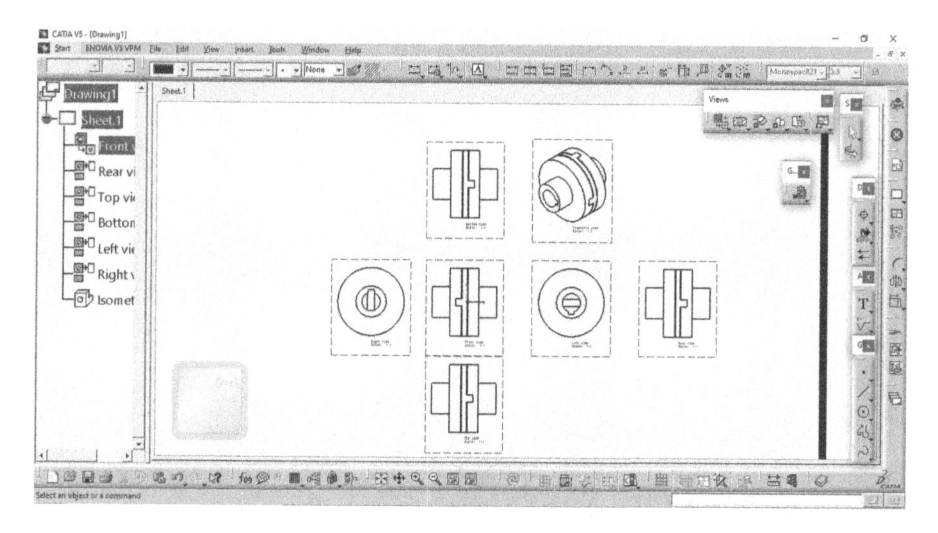

FIGURE 8.16 Drafted drawing of Assembled Oldham's coupling.

Now select the Generate Dimension option to generate the dimensions in Figure 8.16. Now Figure 8.16 will have all the dimensions as shown in Figure 8.17.

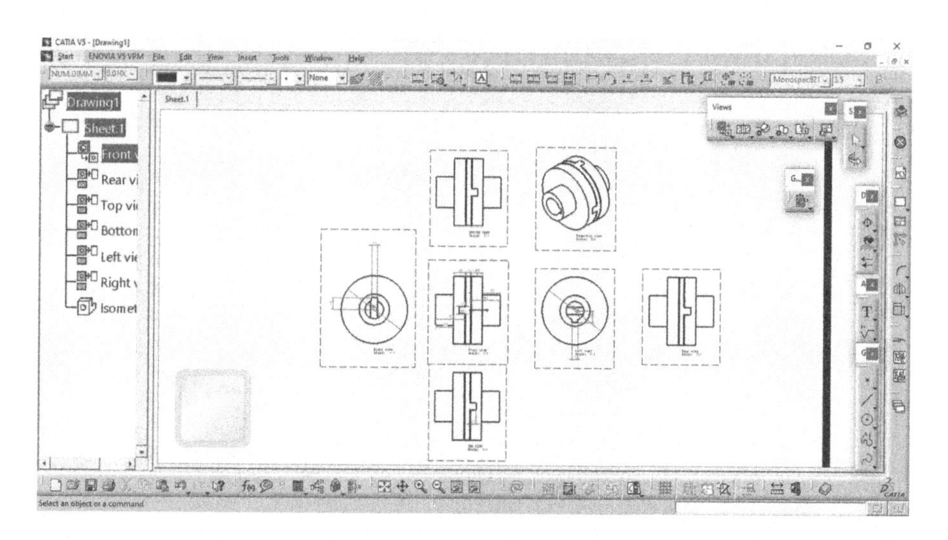

FIGURE 8.17 Dimension generation of Assembled Oldham's coupling.

9 Case Study – 2
Modeling of Clamping Device

9.1 ASSEMBLY

To create Assembly of Clamping Device by given components 1, 2, 3 and 4 as shown in Figures 9.1, 9.2, 9.3 and 9.4.

Draw the components 1, 2, 3 and 4 as per dimension shown in Figures 9.1, 9.2, 9.3 and 9.4 using Sketcher and Part Design workbench as explained in the previous chapter. Create components 1, 2, 3 and 4 as shown in Figures 9.5, 9.6, 9.7 and 9.8.

Select a new file in the Assembly Design workbench as displayed on the screen, as shown in Figure 9.9. Select Product 1 from tree, and select Existing component tool from Product Structure Tool bar. Select the Component 1 from saved folder and select open button in File selection dialog box; you will get the Component 1 coming under Assembly Design workbench as shown in Figure 9.9.

Select Fix Component Constraint Tool from constraint Tool Bar and select Product 1. Fixed Component 1 as shown in Figure 9.10.

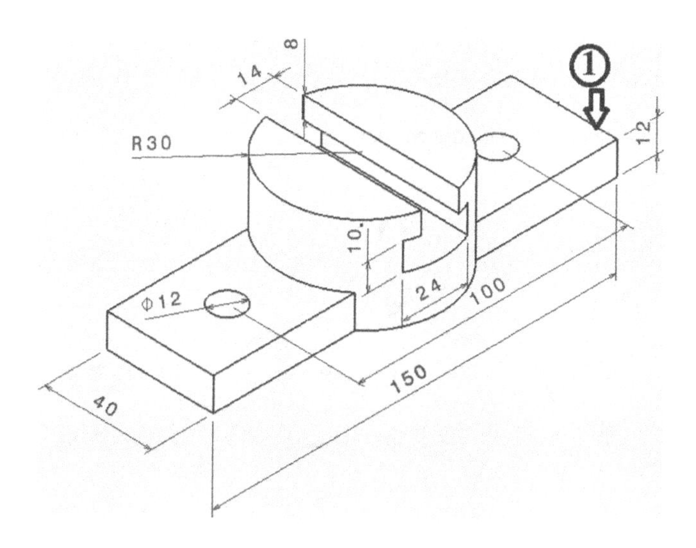

FIGURE 9.1 Component 1 with dimensions.

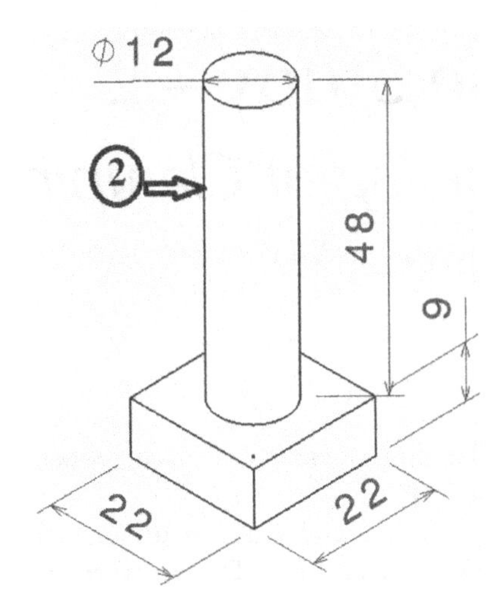

FIGURE 9.2 Component 2 with dimensions.

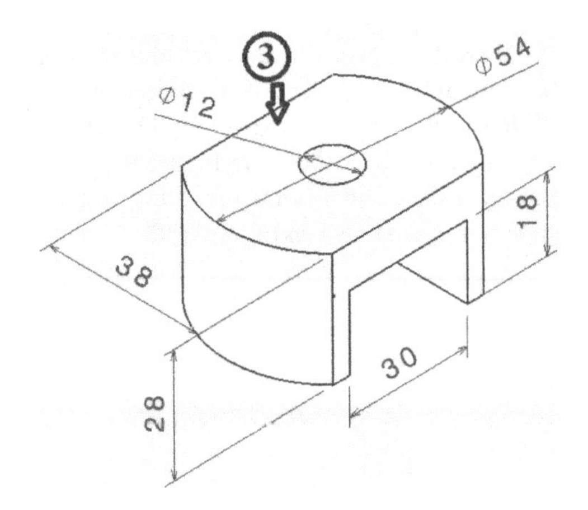

FIGURE 9.3 Component 3 with dimensions.

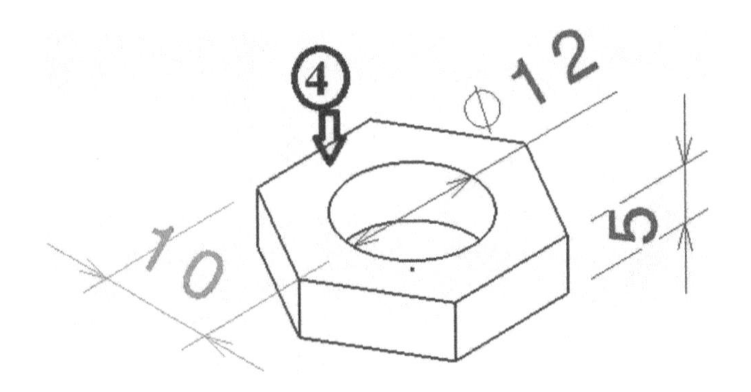

FIGURE 9.4 Component 4 with dimensions.

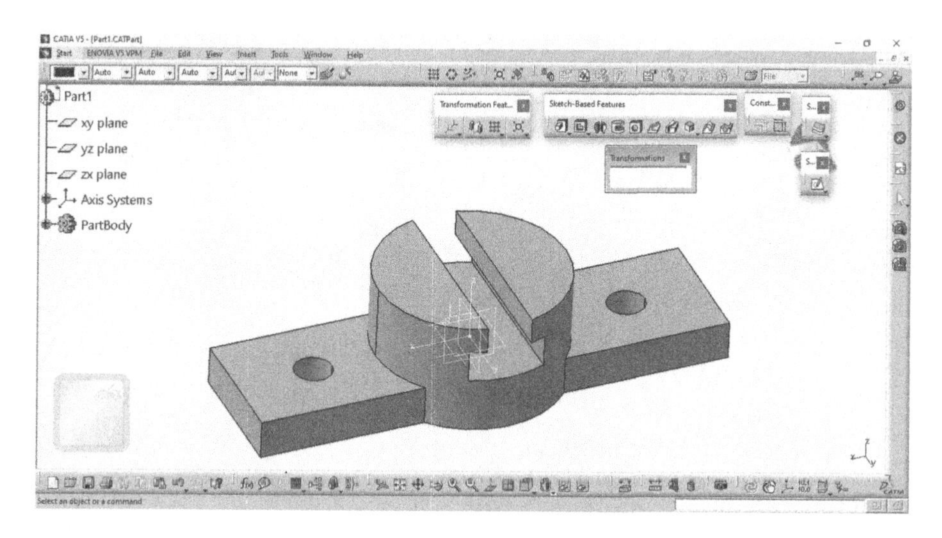

FIGURE 9.5 Created Component 1.

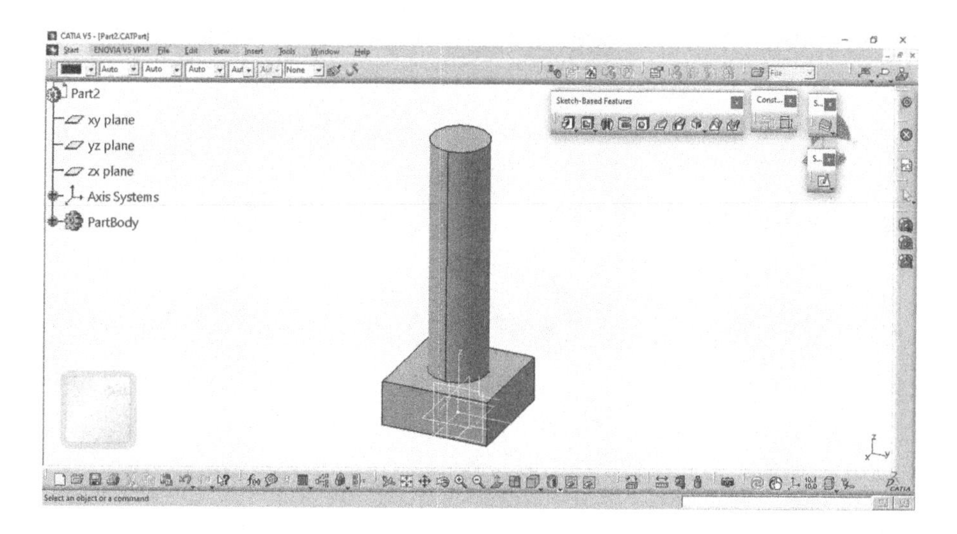

FIGURE 9.6 Created Component 2.

Select Product 1 from tree, and select Existing component tool from Product Structure Tool bar. Select the Component 2 from saved folder and select open button in File selection dialog box; you will get the Component 2 coming under Assembly Design workbench as shown in Figure 9.11.

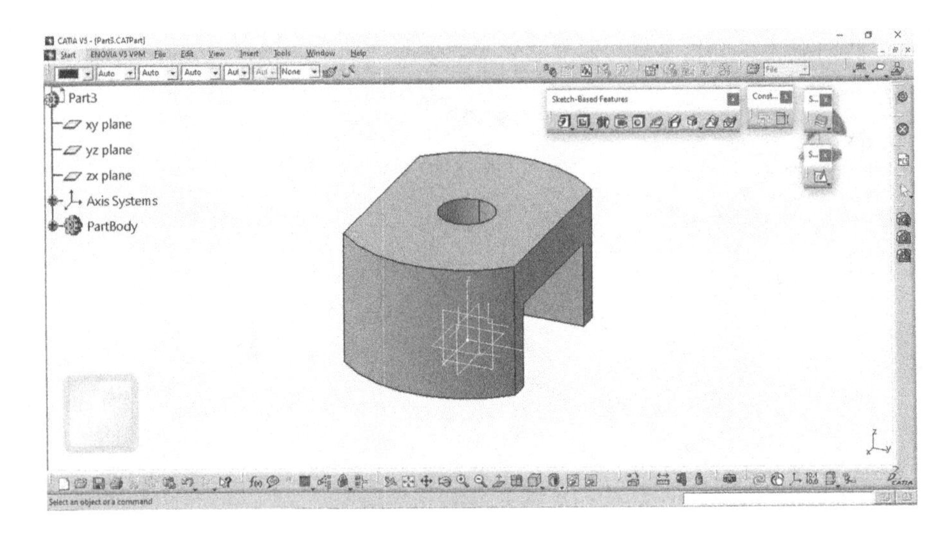

FIGURE 9.7 Created Component 3.

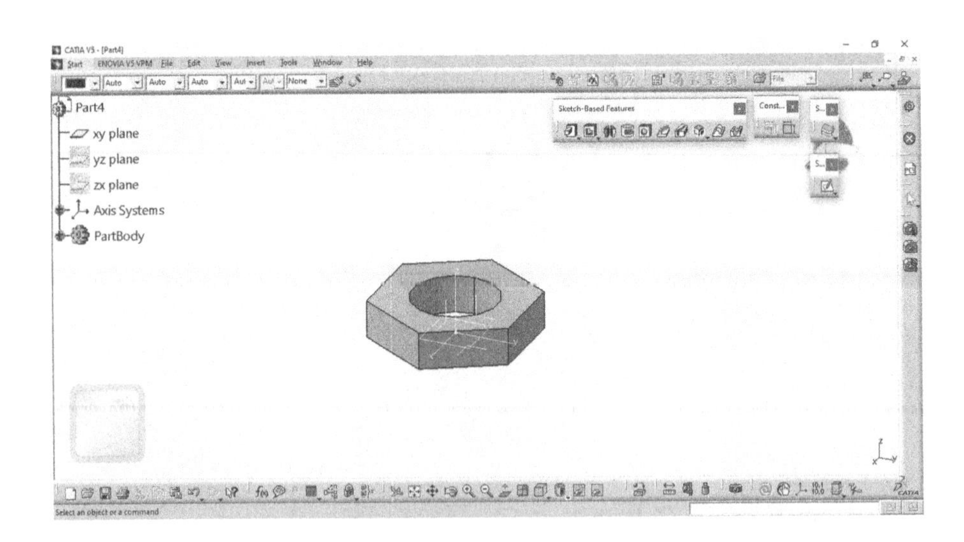

FIGURE 9.8 Created Component 4.

By using coincidence and Contact Constraint Tool from the constraint Tool Bar assemble components 1 and 2. Assembly of components 1 and 2 as shown in Figure 9.12.

Select Product 1 from tree, and select Existing component tool from Product Structure Tool bar. Select the Component 3 from saved folder and select open button

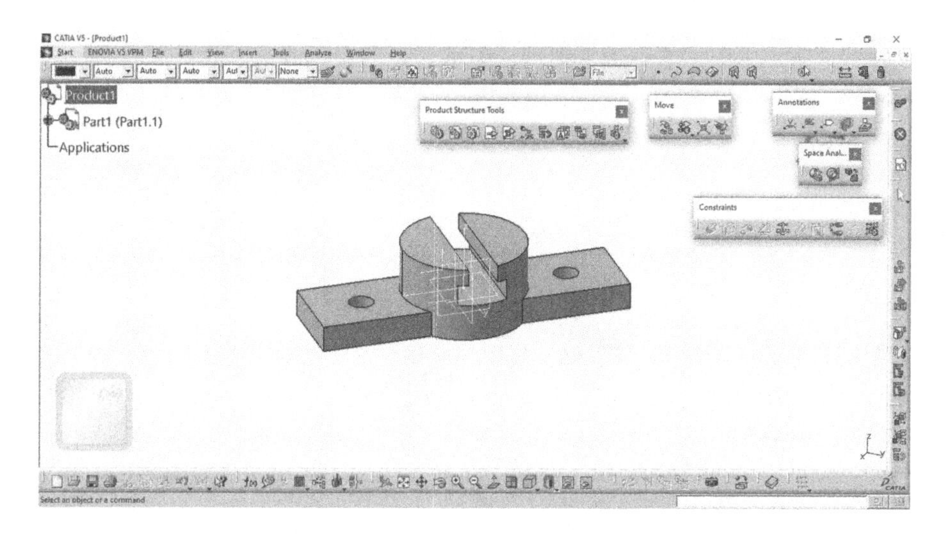

FIGURE 9.9 Component 1 comes under Assembly Design workbench.

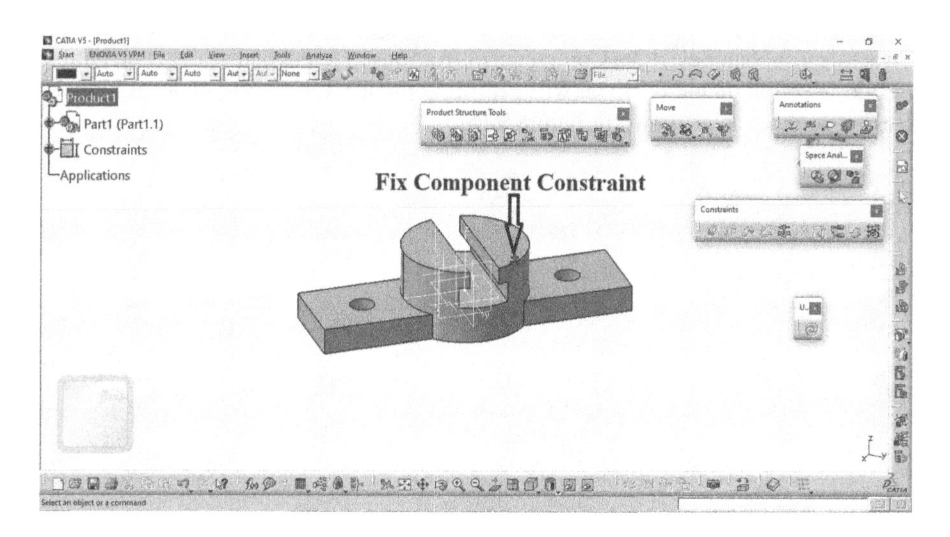

FIGURE 9.10 Component 1 fixed using Fix constraint tool.

in File selection dialog box; you will get the Component 3 coming under Assembly Design workbench as shown in Figure 9.13.

By using coincidence and Contact Constraint Tool from constraint Tool Bar assemble components 1, 2 and 3. Assembly of Component 3 as shown in Figure 9.14.

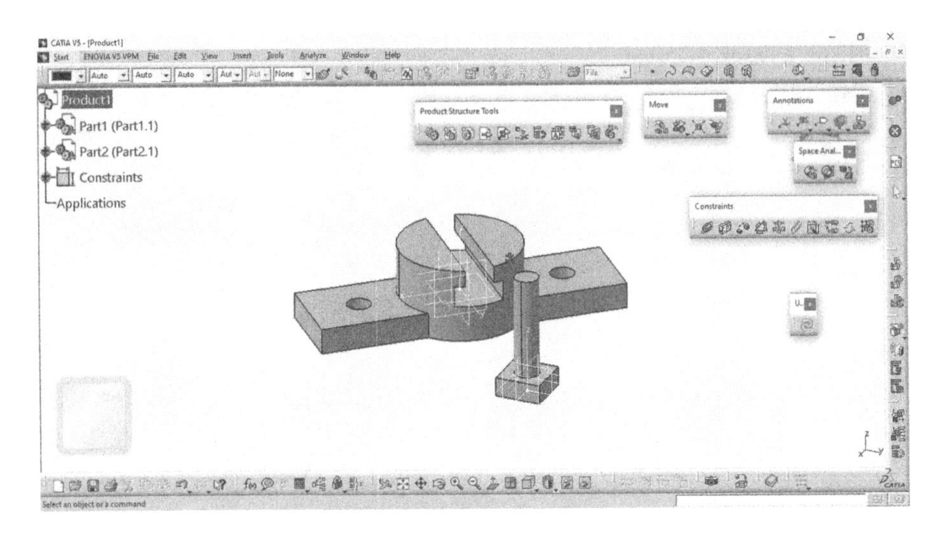

FIGURE 9.11 Components 1 and 2 comes under Assembly Design workbench.

Select Product 1 from tree, and select Existing component tool from Product Structure Tool bar. Select the Component 4 from saved folder and select open button in File selection dialog box; you will get the Component 4 coming under Assembly Design workbench as shown in Figure 9.15.

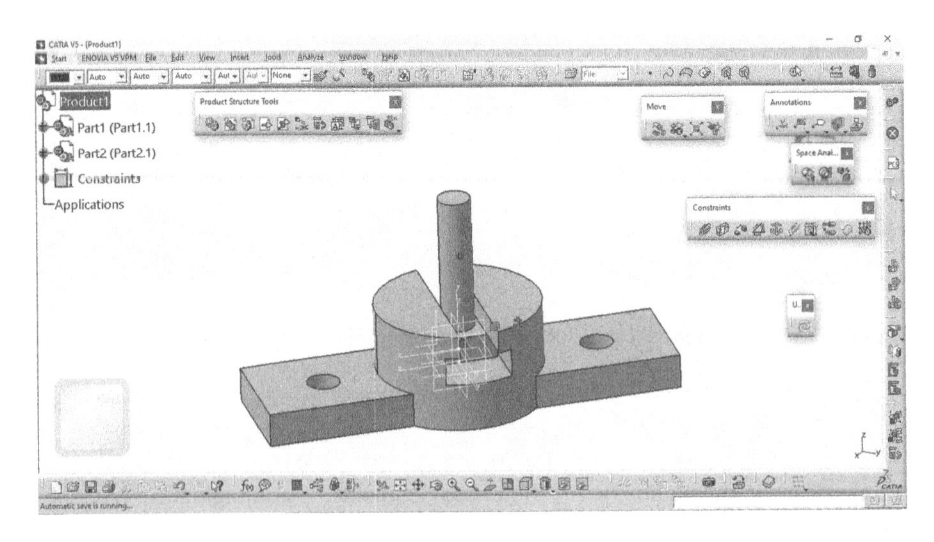

FIGURE 9.12 Assembly of components 1 and 2 using coincidence and Contact constraint tool.

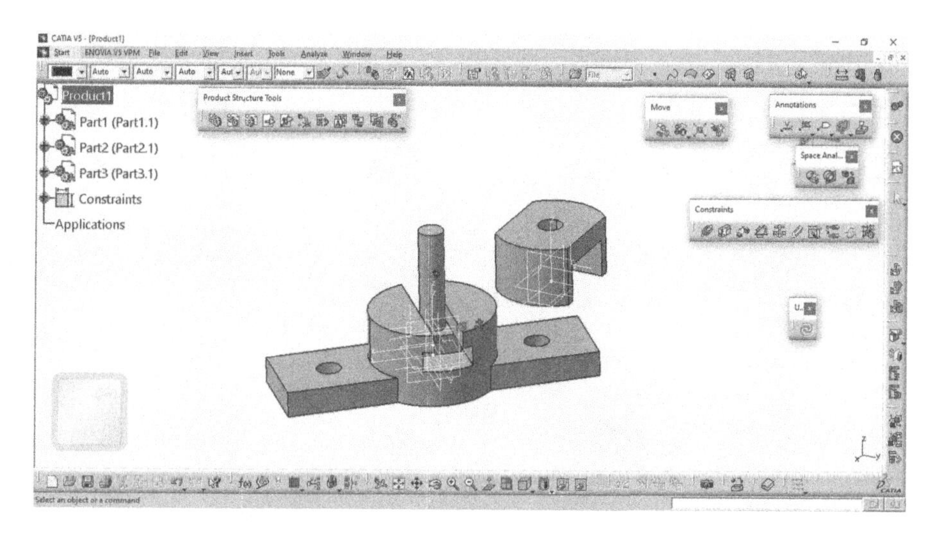

FIGURE 9.13 Components 1, 2 and 3 come under Assembly Design workbench.

By using coincidence and Contact Constraint Tool from the constraint Tool Bar assemble components 1, 2, 3 and 4. Assembly of Clamping Device by using components 1, 2, 3 and 4 as shown in Figure 9.16.

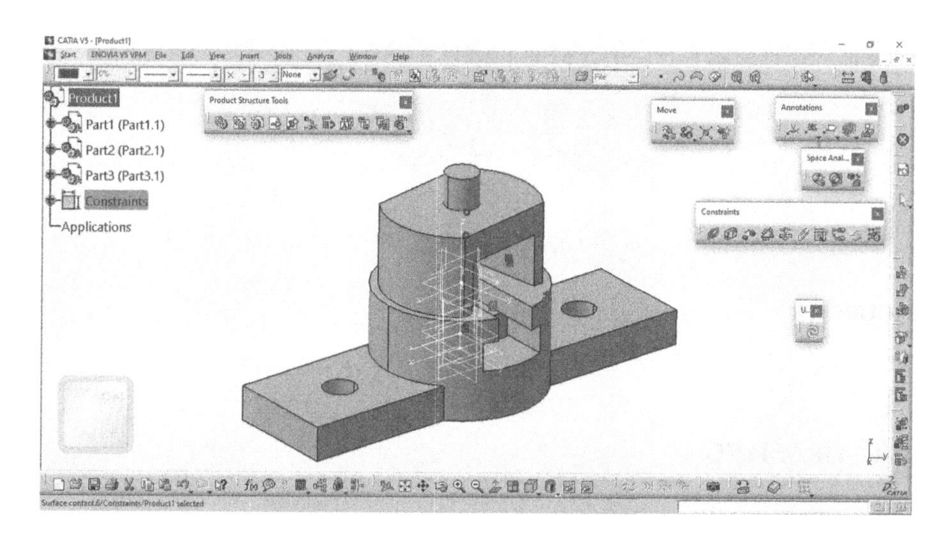

FIGURE 9.14 Assembly of components 1, 2 and 3 using coincidence and Contact constraint tool.

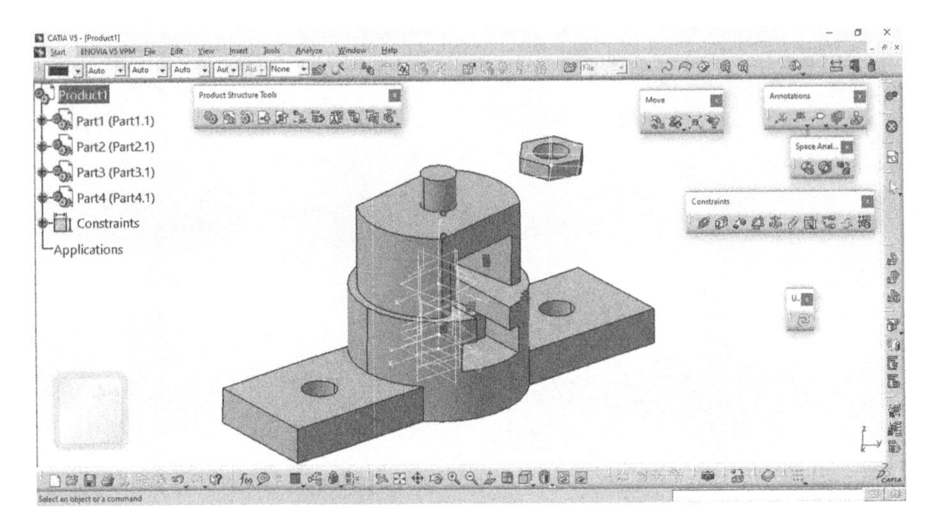

FIGURE 9.15 Components 1, 2, 3 and 4 come under Assembly Design workbench.

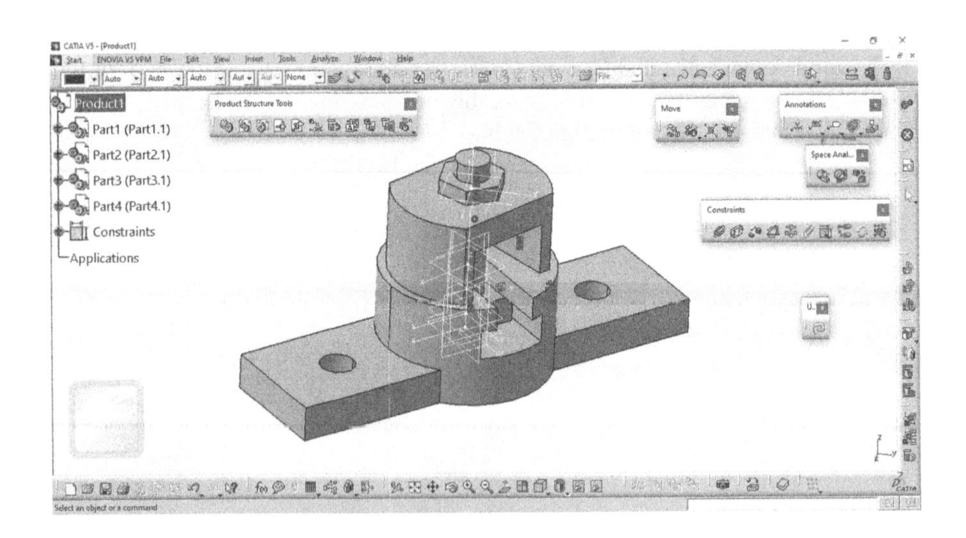

FIGURE 9.16 Assembly of clamping device.

9.2 DRAFTING

9.2.1 Aim

To create front view, top view, side view and isometric view of Assembled Clamping Device by using drafting.

Go to File, open Assembly Design and select the saved Assembly of Clamping Device, as shown in Figure 9.17.

After selecting file, go to *Start > Mechanical Design > Drafting* as shown in Figure 9.18.

A New drawing creation dialog box will appear as depicted in Figure 9.19.

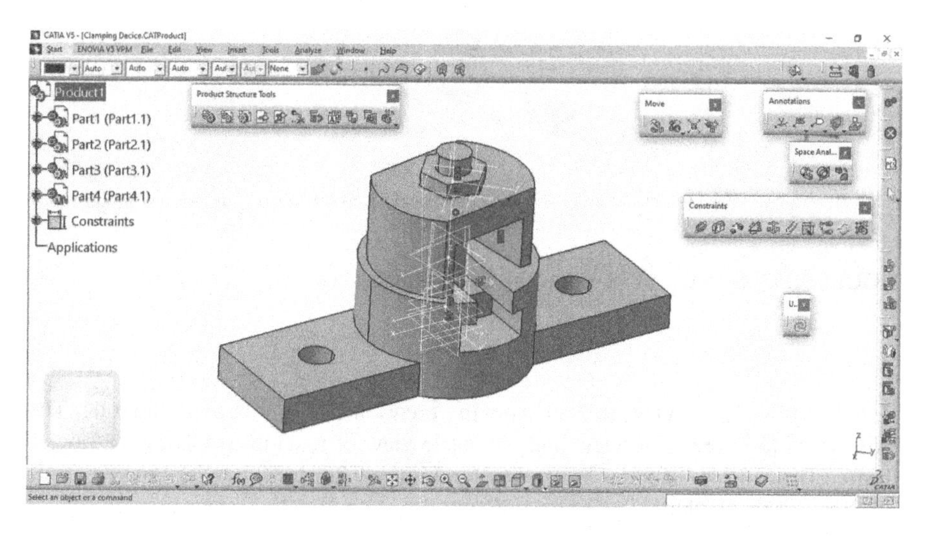

FIGURE 9.17 Selection of assembled clamping device.

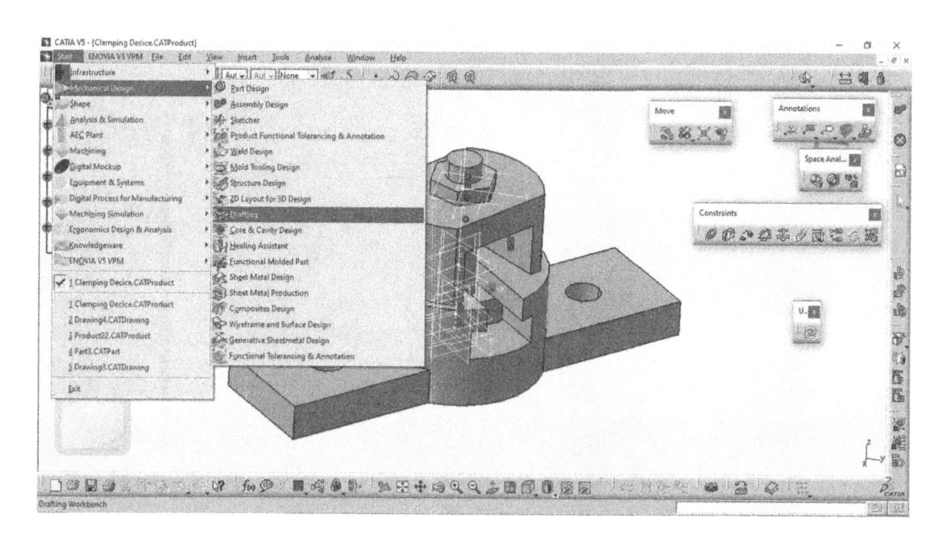

FIGURE 9.18 Selection of drafting.

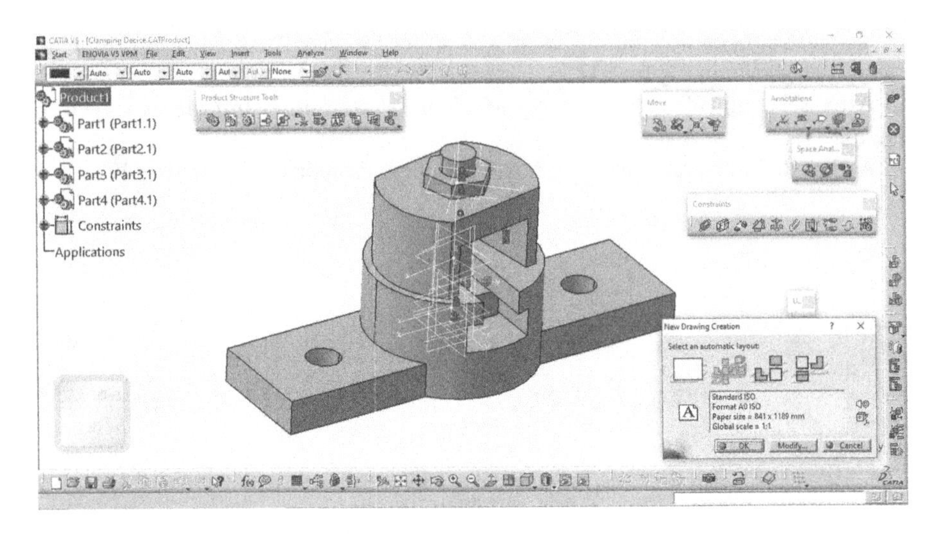

FIGURE 9.19 Selection of New drawing creation dialog box.

Select the second view in New drawing creation dialog box and click OK. The front view, top view, side view and isometric view of assembled Clamping Device drafted Drawing will appear as provided in Figure 9.20.

Now select the Generate Dimension option to generate the dimensions in Figure 9.20. Now Figure 9.20 will have all the dimensions as shown in Figure 9.21.

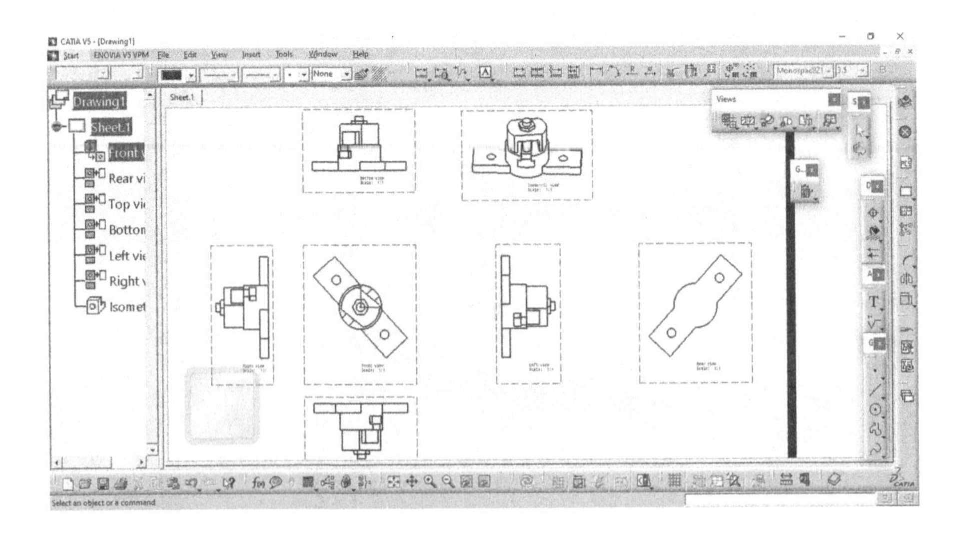

FIGURE 9.20 Drafted drawing of assembled clamping device.

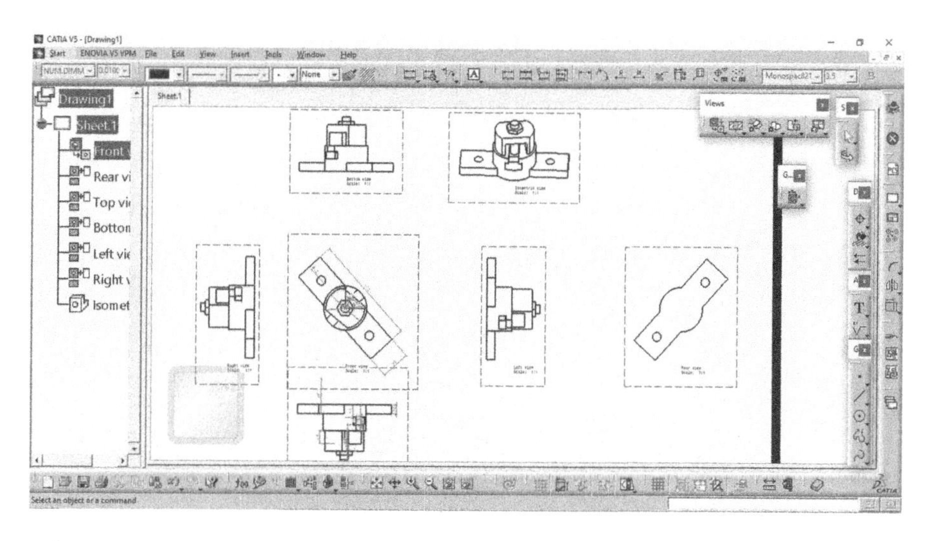

FIGURE 9.21 Dimension generation of assembled clamping device.

Index

M

O

P

Q

R

S

T

U

V

W

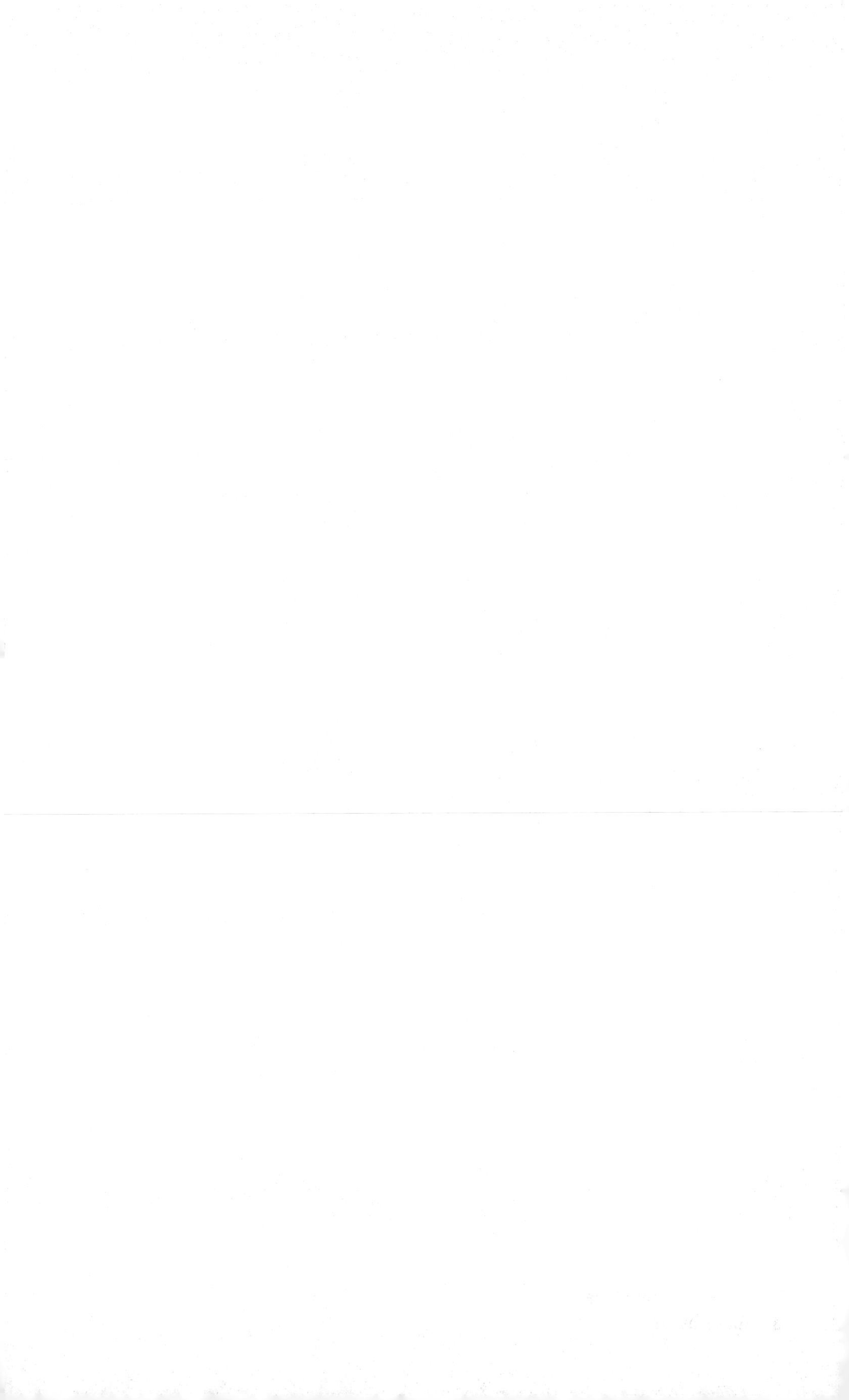

Printed in the United States
By Bookmasters